Microbial Biofilms

Microbial Biofilms

Applications and Control

Edited by
Naga Raju Maddela and Aransiola Sesan Abiodun

CRC Press
Taylor & Francis Group
Boca Raton   London

CRC Press is an imprint of the
Taylor & Francis Group, an **informa** business

First edition published 2022
by CRC Press
6000 Broken Sound Parkway NW, Suite 300, Boca Raton, FL 33487–2742

and by CRC Press
4 Park Square, Milton Park, Abingdon, Oxon, OX14 4RN

CRC Press is an imprint of Taylor & Francis Group, LLC

Library of Congress Cataloging-in-Publication Data
A catalog record for this book has been requested

ISBN: 978-1-032-02632-9 (hbk)
ISBN: 978-1-032-02735-7 (pbk)
ISBN: 978-1-003-18494-2 (ebk)

DOI: 10.1201/9781003184942

Typeset in Times LT Std
by Apex CoVantage, LLC

Contents

PART I Microbiology Biofilms and Environmental Occurrence

PART II   Applications of Microbial Biofilms

PART III   Control of Microbial Biofilms

Foreword

Microbial biofilms is a sore subject, and microbial biofilms exhibit both positive and negative effects in several fields such as medicine, food industries, water and wastewater treatment sectors, agriculture, aquaculture, etc. Biofilms are an important lifestyle of microorganisms, but unfortunately, the form of growth and survival of many bacteria are incompletely understood. Thus, it is becoming harder to design an effective mitigation strategy to control unwanted biofilms such as pathogenic biofilms causing infectious diseases in humans and aquaculture, biofouling in membrane bioreactors used in the water and wastewater treatment facilities, etc. Another important issue is 'antibiotic resistance' in microorganisms; therefore, many infectious diseases are more difficult, as antibiotics used to treat the infections are becoming less effective. Global health, food security, and development today are vastly threatened by antibiotic resistance in microorganisms. This scenario greatly warrants searching for a sustainable biofilm mitigation strategy.

In view of these issues, the present edited book titled *Microbial Biofilms: Applications and Control* has been well structured with the following three areas: Part I, 'Microbiology Biofilms and Environmental Occurrence', Part II, 'Applications of Microbial Biofilms', and Part III, Control of Microbial Biofilms: Emerging Methods'. To justify these areas, the editors have done a splendid job by inviting 51 subject experts from six different countries to contribute to this volume. This collection will provide a great deal of assistance in understanding microbial biofilms from different angles, such as biofilm formation in drug-resistant pathogenic staphylococcus, the role of quorum sensing in microbial biofilm formation, microbial mats ecosystems, succession of bacterial communities in environmental biofilm structures, oral biofilm of hospitalized patients, roles of biofilms in corrosion, medical applications of microbial cellulose biofilms in the field of urology, algal biofilms in bioremediation, microbial biofilms in water sanitation and treatment of effluents, application of marine biofilms, use of nanotechnology for the biofilm mitigation, postbiotics of lactic acid bacteria in biofilm mitigation, and control of microbial biofilms in the oil and gas industry. I truly believe that this volume will have a wider readership and will serve the researchers, industrialists, technicians, and students for a considerable length of time.

Vicente Véliz Briones, Ph.D.
Rector, Universidad Técnica de Manabí
Portoviejo, Ecuador
20 August 2021

Preface

This book titled *Microbial Biofilms: Applications and Control* was aimed to provide detail scientific updates on the implications and control of microbial biofilm. Microbial biofilms are considered to have both positive and negative effects, and hence, require the new scientific innovations to control such effect(s). In such a manner, this book in its present form has been designed so that the enthusiast can be well aware about the implications of microbial biofilms and recent ways to control such biofilm implications.

This book takes into consideration new ways of controlling microbial biofilm from the environment. These new ways include using a phage, nanotechnology, and – most especially – using newly discovered microbial enzymes to mitigate the effects of the said biofilm. Microbial biofilm communities, which are composed of living, reproducing microorganisms, are ubiquitous. They can be found in all habitats where surface, water, and nutrients are present. From the frozen deserts of the Antarctic, to the depths of the ocean, and to the interstices of rock buried thousands of feet below the earth's surface, biofilms have been found in abundance. Microorganisms that form biofilms consist of mixtures of many species of bacteria, as well as fungi, algae, yeasts, protozoa, and other microorganisms, along with non-living debris and corrosion products. One common example of a biofilm is dental plaque, a slimy build-up of bacteria that forms on the surfaces of teeth. Pond scum is another example. Biofilms form when bacteria adhere to surfaces in aqueous environments and begin to excrete a slimy, glue-like substance that can anchor them to a variety of materials including metals, plastics, soil particles, medical implant materials, and – most significantly – human or animal tissue.

Taking into consideration the wide scope and effects of microbial biofilm, the book reflects relevant new areas in the study and recent advances made by researchers across the world on the efficacy, danger, and control of microbial biofilm in all habitats. This book also has been designed to serve as a source of information hub about modern sciences of surface microorganisms and their general implication to students and researchers of this field. To unfold the importance of this book title, three parts have been designed with 16 chapters: Part I, 'Microbiology Biofilms and Environmental Occurrence'; Part II, 'Applications of Microbial Biofilms'; and Part III, 'Control of Microbial Biofilms: Emerging Methods'. Part I consists of six chapters. Chapter 1 is a general overview of microbial biofilms. Chapter 2 reveals the biofilm formation in drug resistant pathogens *Staphylococcus*. Chapter 3 discusses the role of quorum sensing in microbial biofilm formation. Chapter 4 and Chapter 5 inform about the microbial mats ecosystem and succession of bacterial communities in environmental biofilm structures, respectively. Oral biofilm of hospitalized patients is the topic of Chapter 6, the last chapter of this part. Part II consists of six chapters on the roles of biofilms in corrosion (Chapter 7), microbial cellulose biofilms: medical applications in the field of urology (Chapter 8), algal biofilms and role in bioremediation (Chapter 9), impact and application of microbial biofilms in water sanitation (Chapter 10), versatility of bacterial cellulose and its promising application as support for catalyst used in the treatment of effluents (Chapter 11), and application of marine biofilms: an emerging thought to explore (Chapter 12). Part III consists of four chapters, which focus on use of nanotechnology for biofilm mitigation (Chapter 13), postbiotics produced by lactic acid bacteria: current and recent advanced strategies for combating pathogenic biofilms (Chapter 14), impact and control of microbial biofilms in oil and gas industry (Chapter 15), and the futurity of microbial biofilm research (Chapter 16).

Prof. Naga Raju Maddela, Ph.D.
Universidad Técnica de Manabí
Portoviejo, Ecuador

Aransiola Sesan Abiodun, M.Tech., Ph.D.
National Biotechnology Development Agency (NABDA)
Ogbomoso, Nigeria

Acknowledgments

We greatly acknowledge the support of the chapter contributors for their valuable contributions and timely responses during this project. Without their enthusiasm and support, this volume would not be ready in the scheduled time, and therefore, we really appreciate their cooperation and collaboration. We also thank anonymous reviewers for their constructive criticism, which helped us improve the quality of this book by inviting experts to contribute the additional chapters. We greatly acknowledge the Taylor and Francis Group editorial and production teams for their valuable support; without their guidelines, this project would not have been finished in such a very short time. It is our honor to work with them, honestly. Finally, yet importantly, we are very much thankful to the colleagues at Universidad Técnica de Manabí (Ecuador) and National Biotechnology Development Agency (Nigeria) for their unconditional support and for the provision of valuable suggestions at the time of book proposal and final book preparation.

Naga Raju Maddela, Ph.D.
Universidad Técnica de Manabí
Portoviejo, Ecuador

Aransiola Sesan Abiodun, M.Tech., Ph.D.
National Biotechnology Development Agency (NABDA)
Ogbomoso, Nigeria

About the Editors

 **Naga Raju Maddela** received his M.Sc. (1996–1998) and Ph.D. (2012) in microbiology from Sri Krishnadevaraya University, Anantapuramu, India. During his doctoral program in the area of environmental microbiology, he investigated the effects of industrial effluents/insecticides on soil microorganisms and their biological activities and worked as a faculty in microbiology for 17 years, teaching undergraduate and postgraduate students. He received 'Prometeo Investigator Fellowship' (2013–2015) from Secretaría de Educación Superior, Ciencia, Tecnología e Innovación (SENESCYT), Ecuador, and 'Postdoctoral Fellowship' (2016–2018) from Sun Yat-sen University, China. He also received external funding from 'China Postdoctoral Science Foundation' in 2017, and internal funding from 'Universidad Técnica de Manabí' in 2020, and worked in the area of environmental biotechnology, participated in 20 national/international conferences, and presented research data in China, Cuba, Ecuador, India, and Singapore. Currently, he is working as a full professor at the Facultad de Ciencias de la Salud, Universidad Técnica de Manabí, Portoviejo, Ecuador. He has published six books, ten chapters and 50 research papers.

 **Aransiola, Sesan Abiodun** obtained his first degree (B.Tech) and Master's degree (M.Tech) from the Federal University of Technology, Minna, Niger State, Nigeria in microbiology in 2009 and environmental microbiology in 2014, respectively. Currently, Abiodun is concluding his Ph.D. studentship at Federal University of Technology, Minna, working on 'Microbial and Vermicompost-Assisted Phytoremediation of Heavy Metals Polluted Soil'. He has published more than 40 referred articles in professional journals, including five book chapters. Also, he is a senior scientific officer at Bioresources Development Centre, National Biotechnology Development Agency, Nigeria. His area of interest is environmental microbiology, with research area in phytoremediation, biosorption and bioremediation of soil-contaminated environments. Abiodun is a member of the Nigerian Society for Microbiology and the American Society for Microbiology.

List of Contributors

Abioye O.P.
Department of Microbiology
Federal University of Technology
Minna, Nigeria

Helenise Almeida do Nascimento
Departamento de Engenharia Química
Universidade Federal de Pernambuco
Recife, Brasil

Alex Leandro Andrade de Lucena
Departamento de Engenharia Química
Universidade Federal de Pernambuco
Recife, Brasil

Aransiola S.A.
Bioresources Development Centre
National Biotechnology Development
    Agency
P.M.B. 3524 Onipanu
Ogbomoso, Nigeria

Auta S.H.
Department of Microbiology
Federal University of Technology
Minna, Nigeria

Babaniyi R.B.
Bioresources Development Centre
National Biotechnology Development
    Agency
P.M.B. 3524 Onipanu, Ogbomoso, Nigeria

Pablo Bogino
Instituto de Biotecnología Ambiental y
    Salud (INBIAS), CONICET
Departamento de Biología Molecular
Facultad de Ciencias Exactas
Físico-Químicas y Naturales, Universidad
    Nacional de Río Cuarto (UNRC)
Río Cuarto, Córdoba, Argentina

Karina Carvalho de Souza
Programa de Pós-Graduação em
    Ciências dos Materiais
Universidade Federal de Pernambuco
Recife, Brasil

Patricia Cerrutti
Grupo de Biotecnología y Materiales
    Biobasados
Instituto de Tecnología en
    Polímeros y Nanotecnología
    (ITPN-UBA-CONICET)
Facultad de Ingeniería, Universidad de
    Buenos Aires
Las Heras 2214 (CP 1127AAR), Buenos
    Aires, Argentina

Sai Pavan Chilumoju
Centre for Biotechnology
Institute of Science and Technology
Jawaharlal Nehru Technological
    University
Hyderabad, India

Ricardo Sergio Couto de Almeida
Department of Microbiology
State University of Londrina, Brasil

Karolinny Cristiny de Oliveira Vieira
Faculty of Health Science
University of Western São Paulo
Presidente Prudente, SP/Brazil

Maria Vitoria Minzoni De Souza Iacia
University of Western São Paulo
Presidente Prudente, SP/Brazil

Chinedu Enemalu
Department of Microbiology
Federal University of Technology
Minna, Nigeria

Lucimeire Fernandes Correia
University of Western São Paulo
Presidente Prudente, SP/Brazil

María Laura Foresti
Instituto de Tecnología en
    Polímeros y Nanotecnología
    (ITPN-UBA-CONICET)
Facultad de Ingeniería, Universidad de
    Buenos Aires
Las Heras 2214 Buenos Aires,
    Argentina

Letícia Franco Gervasoni
Faculty of Health Science
University of Western São Paulo
Presidente Prudente, SP/Brazil

Daniel Gana
Department of Microbiology
Federal University of Technology
Minna, Nigeria

Walter Giordano
Instituto de Biotecnologia
    Ambiental y Salud (INBIAS),
    CONICET Departamento de Biologia
    Molecular
Facultad de Ciencias Exactas Fisico-
    Quimicas y Naturales
Universidad Nacional de Rio Cuarto
    (UNRC) Rio Cuarto, Cordoba,
    Argentina

Archana Giri
Centre for Biotechnology
Institute of Science and Technology
Jawaharlal Nehru Technological
    University
Hyderabad, India

Marina Gomes Silva
Departamento de Engenharia
    Química
Universidade Federal de Pernambuco
Recife, Brasil

Jakir Hossain
Department of Marine Fisheries and
    Oceanography
Sher-e-Bangla Agricultural
    University
Dhaka, Bangladesh

Md. Foysul Hossain
Department of Aquatic Environment
    and Resource Management
Sher-e-Bangla Agricultural
    University
Dhaka, Bangladesh

Ijah U.J.J.
Department of Microbiology
Federal University of
    Technology
Minna, Nigeria

Roksana Jahan
Department of Marine Fisheries and
    Oceanography
Sher-e-Bangla Agricultural
    University
Dhaka Bangladesh

Lizziane Kretli Winkelströter
Faculty of Health Science
University of Western São
    Paulo
Presidente Prudente,
    SP/Brazil

Palakeerti Srinivas Kumar
Center for Biotechnology
Institute of science and Technology,
    JNTUH
Kukatpally, Hyderabad 500085,
    Telangana, India

Erika Kushikawa Saeki
Regional Laboratory of Presidente
    Prudente
Adolfo Lutz Institute
Presidente Prudente, SP/Brazil

Anjaneyulu Musini
Centre for Biotechnology, Institute of
   Science and Technology
Jawaharlal Nehru Technological
   University
Hyderabad, India

Sai Manisha
Center for Biotechnology
Institute of science and Technology,
   JNTUH
Kukatpally, Hyderabad Telangana, India

Musa O.I.
Department of Microbiology
Federal University of Technology
P.M.B. 65, Niger State, Nigeria

Umar Faruk J. Meeranayak
Department of Studies in Microbiology
   and Biotechnology

Karnatak University
Dharwad, India

Naga Raju Maddela
Departamento de Ciencias Biológicas
Facultad de Ciencias de la Salud
Universidad Técnica de Manabí
Portoviejo-130105, Manabí, Ecuador

Daniella Carla Napoleão
Departamento de Engenharia
   Química
Universidade Federal de Pernambuco
Recife, Brasil

Fiorela Nievas
Instituto de Biotecnología Ambiental y
   Salud (INBIAS), CONICET
Departamento de Biología Molecular
Facultad de Ciencias Exactas
Físico-Químicas y Naturales
Universidad Nacional de Río Cuarto
   (UNRC)
Río Cuarto, Córdoba, Argentina

Oluwafemi Adebayo Oyewole
Department of Microbiology
Federal University of Technology
Minna, Nigeria

Olusegun Julius Oyedele
Centre Co-Ordinator, Bioresources
   Development Centre
National Biotechnology Development
   Agency
P.M.B. 3524 Onipanu, Ogbomoso, Nigeria

Rahul R. Patil
P.G. Department of Botany
Basavaprabhu Kore College
Chikodi, Karnataka, India

Ramat Onyeneoyiza Raji
Department of Microbiology Federal
University of Technology Minna,
Nigeria

Pabbati Ranjit
Center for Biotechnology
Institute of science and Technology,
   JNTUH
Kukatpally, Hyderabad 500085,
   Telangana, India

Kondakindi Venkateswar Reddy
Center for Biotechnology
Institute of Science and Technology,
   JNTUH
Kukatpally, Hyderabad Telangana, India

Joan Manuel Rodriguez-Diaz
Departamento de Procesos Químicos
Facultad de Ciencias Matemáticas
Físicas y Químicas, Universidad
   Técnica de Manabí
Portoviejo, Ecuador

Rayany Magali da Rocha Santana
Departamento de Engenharia Química
Universidade Federal de Pernambuco
Recife, Brasil

Thainah Bruna Santos Zambrano
Pathology Department
Universidad San Gregorio de Portoviejo
Ecuador

Andrea Sordelli
Instituto de Medicina Traslacional
    e Ingeniería Biomédica (IMTIB)
Hospital Italiano de Buenos Aires (HIBA)
CONICET, Instituto Universitario HIBA
Potosí 4240 (CP 1199)
Buenos Aires, Argentina

Rubén Jaime Szwom
Department of Postgraduate in
    Biomedicine
Instituto Universitario Italiano de
    Rosario (IUNIR)
Argentina

Maribel Tupa
Grupo de Biotecnología y Materiales
    Biobasados, Instituto de Tecnología
    en Polímeros y Nanotecnología
    (ITPN-UBA-CONICET)
Facultad de Ingeniería, Universidad de
    Buenos Aires
Las Heras 2214 (CP 1127AAR), Buenos
    Aires, Argentina

Glória Maria Vinhas
Departamento de Engenharia
    Química
Universidade Federal de
    Pernambuco
Recife, Brasil

Victor-Ekwebelem M. O.
Department of
    Microbiology
Alex Ekwueme Federal
    University
Ndufu-Alike, Ikwo, Ebonyi State,
    Nigeria

Japhet Gaius Yakubu
Department of Microbiology
Federal University of
    Technology
Minna, Nigeria

Part I

Microbiology Biofilms and Environmental Occurrence

1 Microbial Biofilm, Applications, and Control
Editorial Overview

*Aransiola S.A., Musa O.I., Babaniyi R.B.,
and Naga Raju Maddela*

CONTENTS

DOI: 10.1201/9781003184942-2

1.1 MICROBIAL BIOFILM

Biofilms are microbial totals that comprise bacterial cells and a self-delivered lattice principally made out of proteins, lipids, polysaccharides, and nucleic (Yin *et al.*, 2019). A biofilm is an efficient, collaborating local area of microorganisms immobilized in a self-blended grid (extracellular polymeric substances [EPS]), which is a combination of polymers discharged by microorganisms and assumes the essential part in the progression of supplements inside a biofilm framework. Biofilms can be a relationship of a solitary or various type of microorganisms, parasites, green growth, and archaea. Biofilm arrangement and EPS creation can change both in design and substance, contingent upon the ecological conditions (Yin *et al.*, 2019).

Biofilms are utilized as a biomarker (biofilm catalysts) for assessing the nature of stream water defiled with heavy metals (Pool *et al.*, 2013). EPS created by microbial biofilms has been utilized as a coagulant in the treatment of wastewater for the expulsion of natural and inorganic pollutants (Flemming *et al.*, 2016). A few microbial biofilms have accounted for the sterilization of contaminations from the climate (Guezennec *et al.*, 2012; Turki *et al.*, 2017). It is accepted that when polluted water goes through the biofilm, microorganisms in the biofilm will eat (and subsequently eliminate) the unsafe natural materials from the polluted water. Biofilms are utilized in water and wastewater treatment innovations, and are effective for bioremediation as they ingest, immobilize, and debase different natural toxins.

1.1.1 An Outline of Biofilm

The connection of microorganisms to the human tooth surface was found by Antony van Leuwenhoek (He and Shi, 2009), and interestingly, the expression "biofilm" was coined by Bill Costerton in 1978 (Costerton *et al.*, 1999). Biofilm is pervasive and typically requires sodden surfaces and nonsterile conditions for development. Both blended and single bacterial species may exist in a biofilm development. In a multispecies biofilm, numerous kinds of positive/synergistic (coaggregation, formation, and insurance to annihilation by antimicrobial specialists) and negative/hostile (bacteriotoxin creation, bringing down of pH) cooperation happens (Butler and Boltz, 2014).

1.1.2 Structure of Biofilm

Biofilms are efficient, collaborating local areas of microorganisms at biotic or abiotic surfaces which are immobilized in a self-incorporated grid known as EPS that secure against pollutants, natural burdens, and savage protozoa. EPS involves various parts like proteins, polysaccharides, nucleic acids or lipids, surfactants, and humic substances. Water is a significant segment of biofilm (B97%), which has a fundamental influence in the progression of supplements inside the biofilm framework (Flemming *et al.*, 2016). Bacterial strain, ecological development conditions, and supplement accessibility are the main considerations that decide the synthesis of EPS.

1.1.3   POLYSACCHARIDES

A large portion of the EPS involves polysaccharides, from which unbiased polysaccharides like hexose and pentose and uronic acids are in the lion's share. EPS attributes are controlled by uronic acids or basic utilitarian gatherings, either natural or inorganic. A few strains of *Xanthomonas campasteris* have been accounted for the creation of xanthan, a polysaccharide (Vu *et al.*, 2009). Overproduction of the Psl polysaccharide implies biofilm design, which prompts the upgraded cell surface and intercellular attachment of *Pseudomonas aeruginosa* (Ma *et al.*, 2012). Alginate is an all-around contemplated polyanion polysaccharide delivered by *P. aeruginosa* biofilms (Orgad *et al.*, 2011).

1.1.4   PROTEIN MATRIX

Related proteins are another pivotal piece of the EPS part in light of the fact that the creation and debasement of proteins are identified with changing conditions in biofilms. The EPS of *P. aeruginosa* biofilms was found to comprise cell flotsam and jetsam determined discharged proteins, and most proteins are related with extracellular film vesicles (Toyofuku *et al.*, 2012).

Pellicles are a kind of biofilm shaped by *Bacillus subtilis*, which glide at the air/fluid interface (Romero *et al.*, 2018). *B. subtilis* biofilm protection from fluid wetting and gas entrance is potentially because of the surface-dynamic protein BslA in the biofilm network, which is significant for proper biofilm improvement. BslA protein is needed to keep up the surface microstructure of the biofilm and restrain the deficiency of surface repellency. A few examinations of *P. aeruginosa* biofilms recommend that the EPS protein contributes not exclusively to supplement procurement, yet in addition gives a safeguard against ecological pressure and shields from microorganisms and hunters (Zhang *et al.*, 2015). During *P. aeruginosa* contamination, the extracellular elastase and lipase compounds go about as a destructiveness factor, finishing the supplement prerequisite of biofilm microorganisms by the corruption of host tissue. Biofilm is discovered wherever a different assortment of extracellular compounds are present in common biological systems (Flemming *et al.*, 2016), which structure stable buildings that give protection from warm denaturation, drying out, and proteolysis measures.

1.1.5   EXTRACELLULAR DNA

Extracellular DNA (eDNA) is a piece of EPS produced during lysis of the phones inside the biofilm. Beforehand, it was accepted that eDNA assumes no part in biofilm, yet these days, it has been demonstrated that it is an essential piece of the biofilm. The length of eDNA pieces in biofilm are around 100–10,000 bp (Romero *et al.*, 2018). Genomic DNA and eDNA are primarily not indistinguishable, but rather have a few similitudes that have been demonstrated by succession examination of eDNA and genomic DNA. The spatial course of action of eDNA in sea-going bacterium biofilm structures a filamentous organization that firmly supports and separates such contemplations.

It appears to be the case that a filamentous organization of eDNA contributes not exclusively to cell correspondence, yet in addition to cell development inside the biofilm and could utilize fibers as nanowires for electron move (Martins *et al.*, 2010).

Film vesicles are significant segments of the biofilm, which essentially comprise lytic compounds. Layer vesicles can tie outer segments in a biofilm, and with this, its catalysts can separate the polymeric compound, give supplements, and help to build opposition by killing the impact of some unfriendly specialists.

Layer vesicles can go about as transporters of hereditary materials in relationship with comparable estimated phages and infections and furthermore upgrade quality trade properties (Nasarabadi *et al.*, 2019). Because of the arrival of film vesicles from the biofilm, the lattice actuation turns out to be more powerful and flexible.

1.2   DEVELOPMENT/FORMATION – MECHANISM

Biofilm is a huge instrument of microbial development in the climate and is a method of advancement particular from creating planktonic life forms (Butler and Boltz, 2014). The marvel of biofilm development happens under different conditions that require generally damp surfaces that can be either living/normal or non-living/human-made.

Valuable biofilm networks have been recorded in numerous spots – for example, streaming channels of wastewater treatment plants, the nutritious waterways of well evolved creatures, and streams. In nature, biofilms are found at all three kinds of interfaces: strong/fluid, strong/air, or fluid/fluid interfaces. For instance, biofilms framed at fluid/fluid interfaces have been applied in hydrocarbon debasement like fills, oils, and mechanical coolants (Percival *et al.*, 2011).

EPS segments are tied by means of physicochemical holding powers for the duration of the existence pattern of biofilm, for dealing with the biofilm design. In the first place, the planktonic microorganisms join to foundation with electrostatic and van der Waals forces toward the start of biofilm arrangement. At that point, ionic association powers or hydrogen powers are included to build the attachment of microscopic organisms. For the most part, four stages are engaged with the engineering of the biofilm, what start with planktonic bacterial connection with unsterilized wet foundation and end with separation. Between these means, microcolony arrangement and three-dimensional construction advancement of biofilm are likewise included (Jamal *et al.*, 2018). Microcolonies contain single or consortia of bacterial cell networks, and primarily these are smooth or unpleasant, level, soft, filamentous, or mushroom-molded encased by watermade up of shortcomings (Ghanbari *et al.*, 2016).

In a microcolony, the EPS content relies on the investment of bacterial species. It has been discovered that microcolonies comprise 75–90% EPS content, while the rest consists of cells. Each progression is shown in Figure 1.1.

1.2.1   MICROBIAL ATTACHMENT TO THE SURFACE

At first, the bacterial connection to the surface is supported by a fluid medium (e.g. water, blood) and is covered by a molding layer framed by polymers of that medium.

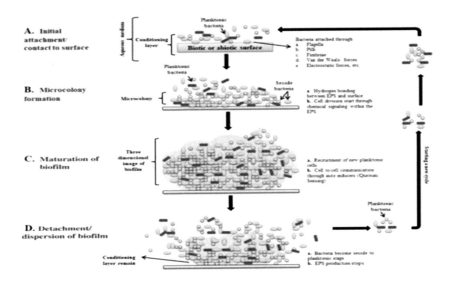

FIGURE 1.1 Biofilm life cycle: attachment, microcolony formation, maturation of biofilm (formation of three-dimensional structures), and detachment in clumps.

This molding layer is natural and structures form inside a couple moments of connection and keep on developing for a few hours. Cell surface extremities like pili and flagella are stringy constructions that help in bacterial connection to the surface at this phase of biofilm (Figure 1.2). Other actual powers like van der Waal's powers, electrostatic collaboration, different variables, and so on, are additionally included for bacterial bond to the surface.

1.2.2 MICROCOLONY DEVELOPMENT

After bacteria bond to the unsterilized surface, the connection becomes stable, and afterward, bacterial cells separate and duplicate through explicit synthetic motioning inside EPS to shape a microcolony (Giorno and D'Agostino, 2015). At the end of the day, microcolonies are considered as bunches of cells, and the spatial course of action of these microcolonies comparative with one another decides the mind-boggling design and capacity of the biofilm. Ordinarily, extraordinary micro communities of bacterial settlements take part in biofilm design. This administration helps in sharing supplements, conveyance of the metabolic items, and clearing of the finished results (Boelee *et al.*, 2011). For instance, in an anaerobic biofilm, a perplexing natural compound is changed over at long last into two finished results, methane (CH_4) and carbon dioxide (CO_2), by the activity of three bacterial species: first, fermentative microorganisms assault complex natural mixtures and produce corrosive and liquor; at that point, acetogenic microscopic organisms utilize these items as their substrates; and in the last phase of the response, interaction methanogens produce methane gas by using the acetic acid derivation, carbon dioxide, and hydrogen particles.

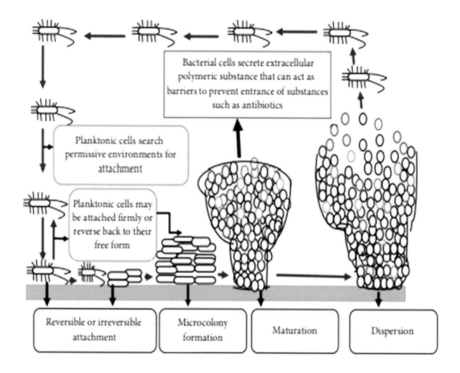

FIGURE 1.2   Biofilm formation and structure.

1.2.3 MATURATION AND DESIGN

Biofilm development is the following phase of the biofilm engineering measure in which bacterial cells inside a biofilm are imparting to one another through auto inducer flags and will achieve the necessary cell thickness level (Vasudevan, 2014). These auto inducer signals favor the majority detecting (QS) marvel in biofilm, encouraging the statement of explicit quality items that assume a fundamental part in EPS development at this development phase of the biofilm. Since the huge piece of a three-dimensional biofilm is its EPS framework, aquous environment for interstitial shortcomings are likewise shaped in the EPS. These go about as a circulatory framework in biofilm and furthermore help in supplements dissemination, and the discharge of metabolic finished results from the bacterial networks of microcolonies (Karatan and Watnick, 2009).

1.2.4 DETACHMENT/SCATTERING OF BIOFILM

At the point when the separation happens in common examples, the augmentation and cell division measure in the biofilm's bacterial cells starts quickly toward the beginning of separation, so sessile cells can be changed over into their motile structure (Flemming and Wingender, 2010). Some bacterial species don't deliver EPS and either scatter straightforwardly into the climate or mechanical pressure might be associated with this interaction (Boudarel *et al.*, 2018).

Biofilm dispersal isn't generally deliberate; uninvolved scattering, like disintegration and sloughing, can happen because of hydrodynamic or shear powers in the fluid climate.

Layers of biofilm might be severed because of a common scraped area, supplement hardship, and hydrodynamic powers brought about by the speed of the fluid (Flemming *et al.*, 2016).

1.2.5 MECHANISM IN BIOFILM FORMATION

Other than ecological conditions, another interaction directs biofilm arrangement and conduct, known as QS. QS is a cycle that assists with beginning cell to cell flagging and eventually alters the cell conduct in a planned manner (Raina *et al.*, 2010). During QS, microbes inside the biofilm are incorporating explicit signals and are found to gauge cell populace thickness. The signs likewise control the bacterial quality articulation when the cell thickness arrives at a limit. QS signal atoms fundamentally come into a general classification of compound classes, including oligopeptides, N-acylhomoserine lactones (AHL), and autoinducer-2 (AI-2). Auto-inducer synthetic substances inside biofilm end up being beneficial for the entire populace as a result of managing the water supply, controlling the supplement fare to singular cells, and waste expulsion. The motile and sessile ways of life of microbes are dictated by their metabolic, physiological, and phenotypic attributes that react to natural signs, which are conveyed by flagging pathways at the phase of biofilm arrangement. Cyclic di-GMP (c-di-GMP), which is a subsequent courier, goes about as a transducing signal for a lot of this intracellular flagging (Valentini and Filloux, 2016). The convergence of c-di-GMP inside cells relies on the overall capacity of two proteins, diguanylate cyclase and phosphodiesterase, which have a place with two chemical families.

The higher cell thickness of the biofilm triggers the QS interaction, which manages the biofilm's contamination, arrangement, and scattering by controlling the creation of exoenzymes and exotoxins. Toward the start of QS flagging, an auto-inducing peptide initiates the histidine kinase, which phosphorylates a reaction controller protein that is needed to tie the four advertisers on the chromosome. One of these four advertisers is for the operon of the flagging pathway, so this is auto-inducible, and the second orchestrates the RNA-III atom needed for the creation of the two exotoxins and exoenzymes by protein RNA associations (Kavanaugh and Horswill, 2016). Likewise, *P. aeruginosa* contains four QS flagging frameworks called Las, Rhl, PQS, and IQS frameworks, and each has its own sign and administrative protein. In the Las framework, N-(3-oxododecanoyl)- homoserine lactone (3-oxo-C12-HSL) is the sign, and LasR is the protein receptor, while the sign in the Rhl framework is N-butyrylhomoserine lactone (C4-HSL) and its receptor is RhlR protein. PQS flagging framework orchestrates the 2-heptyl-3-hydroxy-4-quinolone signal particle and the PqsR protein fills in as a receptor. The Iqs framework has its own 2-(2-hydroxyphenyl)-thiazole-4-carbaldehyde signal atom for the QS interaction (Lee and Zhang, 2015). The qualities – liable for protein and polysaccharide transportation, iron and sulfur digestion, auxiliary metabolites combination, and stress reaction – are upregulated during biofilm arrangement. Biofilm-scattered cells are transcriptionally not quite the same as planktonic cells, however, nearer to their parent cells and expressing better versatile phenotypic characters for colonization at the new ecological site (Guilhen *et al.*, 2016).

1.3   APPLICATIONS OF MICROBIAL BIOFILMS

Despite the negative impacts, bacterial biofilms have also been applied in several areas (Rosche *et al.*, 2009). The formation of bacterial biofilms is often important in agricultural and other industrial settings (Bogino *et al.*, 2013; Berlanga and Guerrero, 2016). These beneficial biofilms are currently used as biological control agents against phytopathogens and biofertilizers to enhance crop production (Timmusk *et al.*, 2017), for bioremediation treatment of hazardous pollutants (Irankhah *et al.*, 2019), for wastewater treatment (Ali *et al.*, 2018), for protection of marine ecosystem (Naidoo and Olaniran, 2013), and for prevention of corrosion (Jayaraman *et al.*, 1997; Martinez *et al.*, 2015). Although biofilms can be beneficial to agriculture and industry, people's understanding of the harmful side of biofilms has been far better than the benefits for decades. Therefore, the beneficial aspects of biofilms will have great development prospects in the future.

1.3.1   AGRICULTURAL APPLICATIONS

Plant growth and biocontrol agents are enhanced by bacteria forming biofilms, thereby leading to improved crop production by renewal of rhizobacteria, either by direct or indirect mechanisms (Table 1.1). Biofilm promotes healthy growth of

TABLE 1.1
Beneficial Biofilm-Forming Bacteria on Plant Growth and Soil Health

Microbes in Biofilms	Crop Inoculated	Effect	References
Alcaligenes faecalis	Rice disease	Suppression	Kakar *et al.* (2018)
Anabaena Azotobacter chroococcum	Mung bean	Fresh weight, dry weight, root length	Prasanna *et al.* (2014)
Trichoderma viride and Anabaena torulosa	Cotton	Enhanced plant growth	Kakar *et al.* (2018)
Trichoderma viride Mesorhizobium ciceri	Chickpea	Enhanced enzymatic activity	Das *et al.* (2017)
Brevibacterium halotolerans	Wheat	Amelioration of salinity stress	Ansari and Ahmad (2018)
Bacillus salmalaya	Oil palm	Improved growth quality, adaptation of the seedlings	Azri *et al.* (2018)
Bacillus megaterium	Barley	Removal of salinity stress	Kasim *et al.* (2016)

plants by fervent double secretion of growth hormones such as indole carboxylic acid (Suman *et al.*, 2016). Plant growth-promoting microbes (PGPMs) such as bio-fertilizers are a life-supporting microbiome enabling solubilization of phosphorus, potassium and zinc (Verma *et al.*, 2017). Biological nitrogen fixation, production of siderophores, hydrolytic enzymes, hydrogen cyanide (HCN), and ammonia (Yadav *et al.*, 2018) microbes have potential and varied functions like solubilizing differ-ent macronutrients, emitting atmospheric chemical elements, and suppressing varied phytopathogens (Yadav *et al.*, 2018). Biofilms have been studied for oxidizing herbi-cides and pesticides (Divjot *et al.*, 2020).

1.3.2   WASTEWATER TREATMENT

Human health and water quality is improved by biological treatment; the microbial communities in the biofilm break down organic materials, carbonaceous materials, and pathogens in the wastewater. Floc and solid materials are separated from the liquid and the treated water could be used for irrigation, recreation, and drinking. Biofilm systems of wastewater treatment has several benefits, such as small space requirements, low hydraulic retention time, easily adaptability to environmental changes, ease of operation, increased biomass residence time, degradation of recal-citrant wastes, reduction of microbial growth rate, and reduced sludge generation (Annapurna and Abhay, 2020).

1.3.3   BIOREMEDIATION AND BIODEGRADATION

Soil and water contaminated with xenobiotic organic compounds, chlorinated phenols, polychlorinated biphenyls, polycyclic aromatic hydrocarbons, along-side ethene and perchloroethylene, and polychlorinated dibenzodioxins, has been mediated. *Cyanophyte* species are capable of heavy metal removal by intracellu-lar uptake through surface assimilation of phosphates mediation. Biofilm fashioned by *Stenotrophomonas acidaminiphila* is capable of degrading polycyclic aromatic hydrocarbons, phenanthrene, and pyrene. (Divjot *et al.*, 2020).

1.3.4   MEDICAL APPLICATION OF BIOFILMS

Extracellular activities of microbial biofilm are mainly responsible for antimicrobial drug treatment with a high-level of tolerance or restriction by acting as a physical barrier or tolerance for the permeation of the antibiotic. The mechanism of biofilm-associated antimicrobial resistance seems to be multifactorial and may different from organism to organism. Nutrient and oxygen depletion, along with waste product accumulation, within the biofilm might cause some bacteria to enter a stationary state in which they are less susceptible to growth but rely on antimicrobials for kill-ing biofilms and protecting microorganism pH from being altered, its osmolarity, prevention of nutrients scarcity, and mechanical and shear forces, but also blocks the access of bacterial biofilm communities from antibiotics and a host's immune cells (Brij *et al.*, 2021).

1.3.5 MICROBIAL BIOFILM IN CHRONIC INFECTION TREATMENT

In the case of acute infection, organisms are short lived and are cleared by the host immune system with or without antimicrobial drug treatment. In contrast, chronic infections may be prolonged due to provision of an excellent niche for lengthy inter-action of the infective microbes for establishing synergistic interactions leading to biofilm formation and attainable cistron transfer. For instance, microbial biofilm is thought to play a vital role in prolonging the chronicity of diabetic chronic wounds. Varied Gram-positive and Gram-negative bacterium square measures are celebrated to be inhabitants of chronic wound beds and often manufacture mixed microbial infections because of mixed microbial biofilm (Brij *et al.*, 2021).

1.3.6 MODULATION OF IMMUNE RESPONSE

Microbial biofilm is capable of modulating the host immunologic response toward incursive pathogens. Meanwhile, inflammatory response pressed by the host is targeted against infectious microorganisms and is meant to safeguard the host cells and destroy the incursive pathogen(s). Distinctively, its square measures many clinical things involving chronic infections, while the unhealthy immunologic response against the pathogen(s) posed lot of damages than helpful to the host cells. Often, the presence of microbial bio-film is the underlying cause for the misdirected attack (McInnes *et al.*, 2014).

1.3.7 INDUSTRIES

Microbial biofilm have been used industrially to supply a spread of valuable com-pounds, like carboxylic acid, table oil, polyose, and electricity generation, whereby the energy is made throughout the degradation of organic and inorganic compounds in industrial waste material; that square measure used by microbial fuel cells to sup-ply electricity (Divjot *et al.*, 2020).

1.4 PROBLEMS ASSOCIATED WITH MICROBIAL BIOFILMS

Based on study, it has been recorded that close to 40–80% of bacterial cells on earth can form biofilms (Flemming and Wuertz, 2019). Biofilm formation is detrimental in several situations (Donlan and Costerton, 2002; Dobretsov *et al.*, 2006; Coughlan *et al.*, 2016). For example, in food industries, pathogenic bacteria are able to form biofilms inside of processing facilities, leading to food spoilage and endangering consumer's health (Galie *et al.*, 2018). In hospital settings, biofilms have also been shown to persist on medical device surfaces and on patient tissues, causing persistent infections (Dongari-Bagtzoglou, 2008; Percival *et al.*, 2015). In view of the serious impact of biofilms on human health and other aspects, researchers and the public have long focused on prevention and control of harmful biofilms.

1.4.1 BIOFILM THICKNESS

Thickness and cell density are factors primarily considered for the analysis or usage of microbial biofilms Mathur *et al.* (2017). Porosity of the film is an essential parameter that determines the structure and strength of biofilm. It has conjointly been ascertained that thinner biofilm has fewer pores with higher cell density than the structure of thick biofilms. These pores give a platform to biofilm for solid liquid partitioning, so the degradation potential of phenol by biofilm is equal to their thickness, and their thickness is usually recommended for use as a resource for economical tools in bioremediation (Crampon *et al.*, 2018).

1.4.2 POLLUTANTS NATURE

Micropollutants that are positively charged (90%) and at pH 7.5 have shown vital sorption potential compared with micropollutants that are negative or neutral charged compounds for activated sludge. It's been ascertained that pH acts as a deciding factor for the charge and sorption potential of micropollutants. As an example, the robust ionic interaction between positively charged macrolides and negatively charged cell surface happens because of the protonation of the radical of the macrolides (Torresi *et al.*, 2017).

1.4.3 BIOCORROSION

Biofilms can instigate biocorrosion of metal pipes (Wang *et al.*, 2012), which then modify the standard of water due to the mobilization of the hooked-up biofilms into water. Oil-producing plants face several issues, such as pipe corrosion, blockage of filtration instruments, and oil spoilage. These issues are as a result of biofilms formation by sulfate-reducing microorganism (Romero *et al.*, 2018).

1.4.4 BIOFOULING

Different industrial issues arise due to biofouling – for instance, lagging of filters membrane and birth infections. Biofouling has negatively affected the oil industry. It causes damping or infection of filter membranes; expansion of biofilm-clogged membrane bioreactors that are accessible instruments for the treatment of effluent. The evolution of biofilm on a membrane filter leads to biofouling (Nasuno *et al.*, 2017).

1.4.5 INFECTION

Many microbial biofilms are highly infectious and therefore cause many diseases. Antimicrobial agents like ciprofloxacin are often used for the treatment of infections caused by *P. aeruginosa* biofilm. The development of biofilm on contact lenses is related to corneal infections. Several surgical instruments like scalps, drips, and catheters are the first supply for the event of biofilm and related infections such as nosocomial infection. The basic concern is the ability of biofilm to suppress numerous antibiotics. Biofilm-forming methicillin-resistant *S. aureus* (MRSA) has generated resistance against all antibiotic programs (Schmieden *et al.*, 2018). The sources for MRSA development were the patients themselves. Biofilm-forming microbes have the potential to contain and neutralize antimicrobial agents.

TABLE 1.2

Different Substrates for Biofilm Formation

Biofilm Formers	Substrate	References
Staphylococcus aureus ATCC 25923, P. aeruginosa ATCC 27853, and E. coli ATCC 25922	Contact lenses	Torresi *et al.* (2017)
Sulfolobus metallicus DSM 6482	Sulfur prills and cubes	Crampon *et al.* 2018
Algal biofilm	Cotton rope	Romero *et al.* (2018)
B. subtilis N4-pHT01-nit	Polyethylene	Kragh *et al.* (2016)
Bacillus sp. Mcn4	Polyurethane foam, cotton fibers, cellophane film, brick particles, ceramic particles, glass beads	Romero *et al.* (2018)

Various microbes can create biofilm on food merchandise and raw food materials. The EPS matrix of biofilm is accountable for the usual persistence of microbes within the food business (Annapurna and Abhay, 2020). Due to EPS enhancement, biofilms in several ways metabolized complex food materials via extracellular enzymes, guarding the cells from the impact of ototoxic compounds and transferring the signals cell to cell. Also, biofilm edges because of the contribution of EPS against desiccation, disinfectants, and different ototoxic chemicals (Flemming *et al.*, 2016). Formation of biofilms on food materials or instruments is related to major foodborne diseases; biofilm-forming microorganisms secrete toxins that cause single or multiple intoxications. Soft drinks manufacturing industries have suffered contamination effects of biofilm formed by *Asaia* species, a contaminant persistent even with the use of preservatives (Galie *et al.*, 2018).

Biofilms formation depends on availability and nature of substrates (Table 1.2). Depending on the suitability of the substrate, a bacterium attaches itself to the substrate with the aid of flagella and pili while feeding on the substrate. Biofilm formation depends on the knowledge of known microorganism's potential to grow on a particular substrate. If one of these is missing, biofilm formation will not be possible (Romero *et al.*, 2018).

1.5 ANTI-BIOFILM STRATEGIES

There are many strategies for the control of biofilms, as they do cause severe complications in many areas such as food industries, water and wastewater treatment plants, medical devices, disease development, etc. These strategies are mainly based on the following principles: (i) targeting to inhibit the initial attachment biofilm-forming bacteria on the surface; (ii) disrupting the biofilms during the process of maturation; (iii) interfering the quorum sensing (QS), which is a bacterial communication system vital in the biofilm formation. Details of anti-biofilm strategies (Subhadra *et al.*, 2018) are shown in Figure 1.3. There is a significant advancement in understanding the role of QS (Maddela *et al.*, 2019; Maddela and Meng, 2020) and extracellular polymeric substances (Maddela *et al.*, 2018) in the formation of biofilm in the recent past. However, additional investigations are needed to understand the basic mechanisms by which the QS system regulates the biofilm formation (Subhadra *et al.*, 2018).

Several naturally available substances are excellent sources of anti-biofilm molecules, which target the bacteria (Roy *et al.*, 2018), implying that there is a

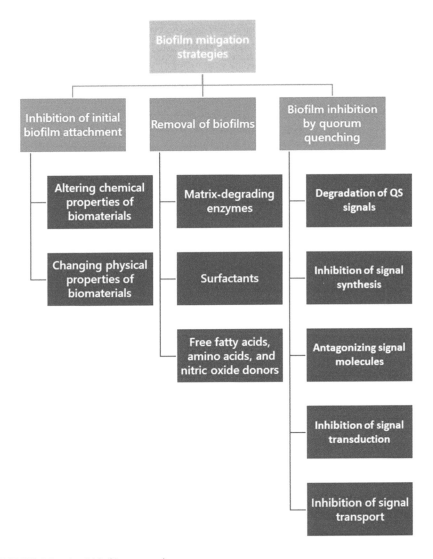

FIGURE 1.3 Anti-biofilm strategies.

possibility to control the biofilms in a sustainable way. For example, epigallocat-echin gallate (EGCG) obtained from *Camellia sinesis* (green tea) can effectively mitigate the biofilm formation by different bacteria such as *Aacinetobacter baumannnii, Pseudomonas aeruginosa, Staphylococcus aureus, Escherichia coli,* and *Stenotrophomonas maltophilia* (Roy *et al.,* 2018). Likewise, many natural molecules have proven anti-biofilm propensities, such as ellagic acid, esculetin, fisetin, reserpine, quercetin, linoleic acid, berberine, chitosan, eugenol, curcumin, synthetic halogenated furanone (F-56) derived from natural furanone, peptide 1018, CFT073 group-II capsular polysaccharide (Serotype K2), Pel polysaccharide, Psl polysaccharide, sophorolipid (biosurfactant), colistin (polymyxin E), polymyxin B, etc.

There are many other strategies to prevent biofilm formation that are under investigation. Active surface topography designed by micrometer-scale pillars and programmable electromagnetic fields keep the antifouling activity (so-called anti-biofilm propensity) for up to one month (Gu *et al.,* 2020). Bacterial poly(lysine)-polyoxometalate(POM) hybrid derivatives have a propensity to prevent the biofilms; such derivatives can be produced by using POMs redox-active molecular metal-oxide anions and α-amino acid N-carboxyanhydrides (Soria-Carrera *et al.,* 2021). Another emerging approach is the use of a precision-structured diblock copolymer brush which has two blocks such as anti-bifouling block and bactericidal block. Covalently grafting block polymer derived functional materials at the surfaces of polyurethane catheters were found to show significant anti-biofilm properties (Hou *et al.,* 2020). There are other emerging biofilm strategies that are under consideration, such as drug carrier biointerfaces (Finbloom *et al.,* 2020), nanomaterial-based therapeutics for the control of antibiotic-resistant bacterial infections (Makabenta *et al.,* 2021), nanostructured surfaces (Linklater *et al.,* 2021), physical approaches for the control of bacterial colonization on polymer surfaces (Echeverria *et al.,* 2020), etc.

Nonetheless, there is much research into the mitigation of biofilms, which is one of the active research areas at present. It is worth noting that approximately 1 million cases of catheter-associated urinary tract infections (CAUTI) are reported each year, which are due to bacterial biofilms (https://en.wikipedia.org/wiki/Biofilm_prevention), which implies the importance of investigating new and sustainable biofilm mitigation strategies.

1.6 ABOUT THIS VOLUME

Keeping in view the occurrence, implications, and problems associated with biofilms, this volume has been designed to give priority to a wide range of topics with the latest insights. How biofilms are formed by drug-resistant *Staphylococcus,* significance of QS in biofilm formation, microbial mats ecosystem, how bacterial communities succeed in environmental biofilms, and detailed information about the oral biofilms of hospitalized patients are the topics of Part I of this volume. Such insights will aid in the deeper understanding of biofilm formation mechanisms and provide sufficient clues as to how to mitigate unwanted biofilms. Under the part of applications of biofilms, chapters that have been included are biofilms in corrosion, urological applications of microbial cellulose biofilms, algal biofilms for the restoration of contaminated sites, microbial biofilms in water sanitation, promising applications of bacterial cellulose in the treatment of effluents, and applications of marine biofilms. Thus, Part II is devoted to providing the details of emerging applications of microbial biofilms for environmental sustainability. At the same time, we would like to provide the latest insights over the mitigation of unwanted biofilms, this has been covered in part three of this volume. Chapters of Part III include different biofilm mitigation strategies based on nanotechnology and lactic acid bacteria, and biofilm control in the oil and gas industry. In the final chapter of this volume, the futurity of microbial biofilm research has been presented. In total, 51 subject experts from six different countries (Argentina, Bangladesh, Brazil, Ecuador, India, and Nigeria) have contributed to this volume.

1.7 CONCLUSION

In this chapter, outlines of microbial biofilms structure and components, mechanisms of biofilm formation, environmental applications of microbial biofilms, biofilm-associated problems in different environmental settings, and biofilm mitigation strategies have been discussed; thus, it could be an introductory chapter of this volume. To justify its title, this volume consists of three important areas – occurrence, applications, and mitigation – of microbial biofilms. At the end, a detailed structure of this volume and the contributors has been presented.

REFERENCES

Ali J, Sohail A, Wang L, Rizwan HM, Mulk S, Pan G (2018) Electro-microbiology as a promising approach towards renewable energy and environmental sustainability. *Energies.* 11:18–22.

Annapurna M, Abhay R (2020) Recent advances in the application of biofilm in bioremediation of industrial wastewater and organic pollutants. *Microorganisms for Sustainable Environment and Health.* 5:81–181. doi: 10.1016/B978-0-12-819001-2.00005-X

Ansari FA, Ahmad I (2018) Plant growth promoting attributes and alleviation of salt stress to wheat by biofilm forming *Brevibacterium sp.* FAB3 isolated from *rhizospheric* soil. *Saudi J Biol Sci.* 8:003. https://doi.org/10.1016/j.sjbs.

Azri MH, Ismail S, Abdullah R (2018) An endophytic *bacillus* strain promotes growth of oil palm seedling by fine root biofilm formation. *Rhizosphere.* 5:1–7.

Berlanga M, Guerrero R (2016) Living together in biofilms: the microbial cell factory and its biotechnological implications. *Microb Cell Fact.* 15:165. doi: 10.2210/pdb4bhu/pdb.

Boelee N, Temmink H, Janssen M, Buisman C, Wijffels RJ (2011) Nitrogen and phosphorus removal from municipal wastewater effluent using microalgal biofilms. *Water Res.* 45:5925–5933.

Bogino PC, Oliva Mde L, Sorroche FG, Giordano W (2013) The role of bacterial biofilms and surface components in plant-bacterial associations. *Int J Mol Sci* 14:15838–15859. doi:10.3390/ijms140815838

Boudarel H, Mathias JD, Blaysat B, Grédiac MJN (2018) Toward standardized mechanical characterization of microbial biofilms: analysis and critical review. *NPJ Biofilms Microbiom.* 4:115.

Brij PS, Sougate G, Ashwini C (2021) Development, dynamics and control of antimicrobial-resistant bacterial biofilms: a review. *Environ Chem Lett.* doi:10.1007/s10311

Butler C, Boltz J (2014) *Biofilm processes and control in water and wastewater treatment.* https://link.springer.com/referenceworkentry/10.1007%2F978-3-319-73645-7_137

Costerton JW, Stewart PS, Greenberg EP (1999) Bacterial biofilms: a common cause of persistent infections. *Sci.* 284:1318–1322.

Coughlan LM, Cotter PD, Hill C, Alvarez-Ordonez A (2016) New weapons to fight old enemies: novel strategies for the (Bio)control of bacterial biofilms in the food industry. *Front Microbiol.* 7:1641. doi: 10.3389/fmicb.2016.01641

Crampon M, Hellal J, Mouvet C, Wille G, Michel C, Wiener A, *et al.* (2018) Do natural biofilm impact nZVI mobility and interactions with porous media? A column studies. *Sci Total Environ.* 610:709–719.

Das K, Rajawat MVS, Saxena AK, Prasanna R (2017) Development of *Mesorhizobium ciceri*-based biofilms and analyses of their antifungal and plant growth promoting activity in chickpea challenged by Fusarium wilt. *Indian J Microbiol.* 57:48–59.

Divjot K, Ranaa KL, Neelam TK, Yadava AN, Ali AR, Saxena AK (2020) Microbial biofilms: functional annotation and potential applications in agriculture and allied sectors. *Microbial Biofilms.* 283–301. doi:10.1016/B978-0-444-64279-0.00018–9

Dobretsov S, Dahms HU, Qian PY (2006) Inhibition of biofouling by marine microorganisms and their metabolites. *Biofouling.* 22:43–54.

Dongari-Bagtzoglou A (2008) Pathogenesis of mucosal biofilm infections: challenges and progress. *Expert Rev Anti Infect Ther.* 6:201–208. doi: 10.1586/ 14787210.6.2.201

Donlan RM, Costerton JW (2002) Biofilms: survival mechanisms of clinically relevant microorganisms. *Clin Microbiol Rev.* 15:167–193.

Echeverria C, Torres MDT, Fernández-García M, de la Fuente-Nunez C, Muñoz-Bonilla A (2020) Physical methods for controlling bacterial colonization on polymer surfaces. *Biotechnol Adv.* 107586.

Finbloom JA, Sousa F, Stevens MM, Desai TA (2020) Engineering the drug carrier biointerface to overcome biological barriers to drug delivery. *Adv Drug Delivery Rev.* 167:89–108.

Flemming HC, Wingender J (2010) The biofilm matrix. *Nat Rev Microbiol.* 8:623.

Flemming HC, Wingender J, Szewzyk U, Steinberg P, Rice SA, Kjelleberg S (2016) Biofilms: an emergent form of bacterial life. *Nat Rev Microbiol.* 14:563.

Flemming HC, Wuertz S (2019) Bacteria and archaea on Earth and their abundance in biofilms. *Nat Rev Microbiol.* 17:247–260.

Galie S, García-Gutiérrez C, Miguélez EM, Villar CJ, Lombó F (2018) Biofilms in the food industry: health aspects and control methods. *Front Microbiol.* 9:898.

Ghanbari A, Dehghany J, Schwebs T, Müsken M, Häussler S, Meyer-Hermann M (2016) Inoculation density and nutrient level determine the formation of mushroom-shaped structures in pseudomonas aeruginosa biofilms. *Sci Rep.* 6:32097.

Giorno L, D'Agostino N (2015) Bacterial biofilm formation. In *Encyclopedia of membranes.* Springer, 15.

Gu H, Lee SW, Carnicelli J, Zhang T, Ren D (2020) Magnetically driven active topography for long-term biofilm control. *Nature Comm.* 11(1):1–11.

Guezennec AG, Michel C, Joulian C, Dictor MC, Battaglia-Brunet F (2012) *Treatment of arsenic contaminated mining water using biofilms.* Interfaces Against Pollution.

Guilhen C, Charbonnel N, Parisot N, Gueguen N, Iltis A, Forestier C *et al.* (2016) Transcriptional profiling of Klebsiella pneumoniae defines signatures for planktonic, sessile and biofilm-dispersed cells. *BMC Genom.* 17:237.

He XS and Shi WY (2009) Oral microbiology: past, present and future. *Inter J Oral Sci.* 1(2):47–58.

Hou Z, Wu Y, Xu C, Reghu S, Shang Z, Chen J, Pranantyo D, Marimuth K, De PP, Ng OT, Pethe K (2020) Precisely structured nitric-oxide-releasing copolymer brush defeats broad-spectrum catheter-associated biofilm infections in vivo. *ACS Central Sci.* 6(11):2031–2045.

Irankhah S, Ali A, Mallavarapu M, Soudi MR, Subashchandrabose S, Gharavi S *et al.* (2019) Ecological role of Acinetobacter calcoaceticus GSN3 in natural biofilm formation and its advantages in bioremediation. *Biofouling.* 35:377–391.

Jamal M, Ahmad W, Andleeb S, Jalil F, Imran M, Nawaz MA (2018) Bacterial biofilm and associated infections. *J Chin Med Assoc.* 81:711.

Jayaraman A, Cheng ET, Earthman JC, Wood TK (1997) Importance of biofilm formation for corrosion inhibition of SAE 1018 steel by axenic aerobic biofilms. *J Ind Microbiol Biotechnol.* 18:396–401.

Kakar K, Nawaz Z, Cui Z, Almoneafy A, Ullah R, Shu QY (2018) Rhizosphere-associated *Alcaligenes* and *Bacillus* strains that induce resistance against blast and sheath blight diseases, enhance plant growth and improve mineral content in rice. *J Appl Microbiol.* 124:779–796.

Karatan E, Watnick P (2009) Signals, regulatory networks, and materials that build and break bacterial biofilms. *Microbiol Mol Biol Rev.* 73:310–347.

Kasim WA, Gaafar RM, Abou-Ali RM, Omar MN, Hewait HM (2016). efect of biofilm forming plant growth promoting rhizobacteria on salinity tolerance in barley. *Annals of Agricultural Sciences.* 61(2):217–227.

Kavanaugh JS, Horswill AR (2016) Impact of environmental cues on staphylococcal quorum sensing and biofilm development. *J Biol Chem.* 291:1255612564.

Kragh KN, Hutchison JB, Melaugh G, Rodesney C, Roberts AE, Irie Y (2016) Role of multicellular aggregates in biofilm formation. *mBio.* 7:237–00216.

Lee J, Zhang L (2015) The hierarchy quorum sensing network in *pseudomonas aeruginosa. Protein Cell.* 6:26–41.

Linklater DP, Baulin VA, Juodkazis S, Crawford RJ, Stoodley P, Ivanova EP (2021) Mechano-bactericidal actions of nanostructured surfaces. *Nat Rev Microbiol.* 19(1):8–22.

Ma L, Wang S, Wang D, Parsek MR, Wozniak DJ (2012) The roles of biofilm matrix polysaccharide Psl in mucoid pseudomonas aeruginosa biofilms. *FEMS Immunol Med Microbiol.* 65:377–380.

Maddela NR, Meng F (2020) Discrepant roles of a quorum quenching bacterium (Rhodococcus sp. BH4) in growing dual-species biofilms. *Sci Total Environ.* 713:136402.

Maddela NR, Sheng B, Yuan S, Zhou Z, Villamar-Torres R, Meng F (2019) Roles of quorum sensing in biological wastewater treatment: a critical review. *Chemosphere.* 221:616–629.

Maddela NR, Zhou Z, Yu Z, Zhao S, Meng F (2018) Functional determinants of extracellular polymeric substances in membrane biofouling: experimental evidence from pure-cultured sludge bacteria. *Appl Environ Microb.* 84(15):e00756.

Makabenta JMV, Nabawy A, Li CH, Schmidt-Malan S, Patel R, Rotello VM (2021) Nanomaterial-based therapeutics for antibiotic-resistant bacterial infections. *Nat Rev Microbiol.* 19(1):23–36.

Martinez P, Vera M, Bobadilla-Fazzini RA (2015) Omics on bioleaching: current and future impacts. *Appl Microbiol Biotechnol.* 99:8337–8350. doi: 10.1007/s00253-015-6903-8

Martins M, Uppuluri, P, Thomas DP, Cleary IA, Henriques M, Lopez-Ribot JL (2010) Presence of extracellular DNA in the candida albicans biofilm matrix and its contribution to biofilms. *Mycopathologia.* 169:323–331.

Mathur A, Parashar A, Chandrasekaran N, Mukherjee A (2017) Nano-TiO2 enhances biofilm formation in a bacterial isolate from activated sludge of a waste water treatment plant. *Int Biodeter Biodegrad.* 116:17–25.

McInnes RL, Cullen BM, Hill KE (2014) Contrasting host immuno-inflammatory responses to bacterial challenge within venous and diabetic ulcers. *Wound Repair Regen.* 22(1):58–69. doi: 10.1111/wrr.12133.

Naidoo S, Olaniran AO (2013) Treated wastewater effluent as a source of microbial pollution of surface water resources. *Int J Environ Res Public Health.* 11:249–270. doi: 10.3390/ijerph110100249

Nasarabadi A, Berleman JE, Auer M (2019) Outer membrane vesicles of bacteria: structure, biogenesis, and function. In *Biogenesis of fatty acids, lipids and membranes.* Springer, 593–607.

Nasuno E, Abe Y, Iimura KI, Ohno M, Okuda T, Nishijima W *et al.* (2017) *Isolation of biofilmforming bacteria from the secondary effluent of the wastewater treatment plant and its ability to produce N-acylhomoserine lactone as quorum sensing signal.* Paper presented at Applied Mechanics and Materials, Trans Tech Publications.

Orgad O, Oren Y, Walker SL, Herzberg M (2011) The role of alginate in pseudomonas aeruginosa EPS adherence, viscoelastic properties and cell attachment. *Biofouling.* 27:787798.

Percival SL, Malic S, Cruz H, Williams DW (2011) Introduction to biofilms. In *Biofilms and veterinary medicine.* Springer, 4168.

Percival SL, Suleman L, Vuotto C, Donelli G (2015) Healthcareassociated infections, medical devices and biofilms: risk, tolerance and control. *J Med Microbiol.* 64:323–334.

Pool JR, Kruse NA, Vis ML (2013) Assessment of mine drainage remediated streams using diatom assemblages and biofilm enzyme activities. *Hydrobiologia.* 709:01116.

Prasanna R, Triveni S, Bidyarani N, Babu S, Yadav K, Adak A, Khetarpal S, Pal M, Shivay YS, Saxena AK (2014) Evaluating the efficacy of *cyanobacterial* formulations and biofilmed inoculants for leguminous crops. *Arch Agron Soil Sci.* 60:349–366.

Raina S, Odell M, Keshavarz TJ (2010) Quorum sensing as a method for improving sclerotiorin production in *penicillium sclerotiorum. J Biotechnol.* 148:91–98.

Romero CM, Martorell P, López AG, Peñalver CN, Chaves S, Mechetti M (2018) Architecture and physicochemical characterization of Bacillus biofilm as a potential enzyme immobilization factory. *Colloids Surf. B Biointerfaces.* 162:246–255.

Rosche B, Li XZ, Hauer B, Schmid A, Buehler K (2009) Microbial biofilms: a concept for industrial catalysis? *Trends Biotechnol.* 27:636–643.

Roy R, Tiwari M, Donelli G, Tiwari V (2018) Strategies for combating bacterial biofilms: A focus on anti-biofilm agents and their mechanisms of action. *Virulence.* 9(1):522–554.

Schmieden DT, Vázquez SJ, Sangüesa HC, van der Does M, Idema T, Meyer AS (2018) Printing of patterned, engineered E. coli biofilms with a low-cost 3D printer. *ACS Synth Biol.* 7:328–11337.

Soria-Carrera H, Franco-Castillo I, Romero P, Martín S, de la Fuente JM, Mitchell SG, Martín-Rapún R (2021) On-POM ring-opening polymerisation of n-carboxyanhydrides. *Angewandte Chemie International Edition.* 60(7):3449–3453.

Subhadra B, Kim DH, Woo K, Surendran S, Choi CH (2018) Control of biofilm formation in healthcare: recent advances exploiting quorum-sensing interference strategies and multidrug efflux pump inhibitors. *Materials.* 11(9):1676.

Suman A, Yadav AN, Verma P (2016) Endophytic microbes in crops: diversity and beneficial impact for sustainable agriculture. In Singh D, Abhilash P, Prabha R (Eds.), *Microbial inoculants in sustainable agricultural productivity, research perspectives.* Springer-Verlag, 117–143.

Timmusk S, Behers L, Muthoni J, Muraya A, Aronsson AC (2017) Perspectives and challenges of microbial application for crop improvement. *Front. Plant Sci.* 8:49. doi: 10.3389/fpls.2017.00049

Torresi E, Polesel F, Bester K, Christensson M, Smets BF, Trapp S, *et al.* (2017) Diffusion and sorption of organic micropollutants in biofilms with varying thicknesses. *Water Res.* 123:388–400.

Toyofuku M, Roschitzki B, Riedel K, Eberl L (2012) Identification of proteins associated with the pseudomonas aeruginosa biofilm extracellular matrix. *J Proteome Res.* 11:4906–4915.

Turki Y, Mehri I, Lajnef R, Rejab AB, Khessairi A, Cherif H (2017) Biofilms in bioremediation and wastewater treatment: characterization of bacterial community structure and diversity during seasons in municipal wastewater treatment process. *Environ Sci Pollut Res Int.* 24:3519–3530.

Valentini M, Filloux A (2016) Biofilms and cyclic di-GMP (c-di-GMP) signaling: lessons from Pseudomonas aeruginosa and other bacteria. *J Biol Chem.* 291:12547–12555.

Vasudevan R (2014) Biofilms: microbial cities of scientific significance. *J Microbiol Exp.* 1:00014.

Verma P, Yadav AN, Khannam KS, Saxena AK, Suman A (2017) Potassium-solubilizing microbes: diversity, distribution, and role in plant growth promotion. In Panpatte DG, Jhala YK, Vyas RV, Shelat HN (Eds.), *Microorganisms for green revolution in: microbes for sustainable crop production.* Springer, vol. 1, 125–149. https://doi.org/10.1007/978-981-10-6241-4-7.

Vu B, Chen M, Crawford R, Ivanova E (2009) Bacterial extracellular polysaccharides involved in biofilm formation. *Molecules.* 14:2535–2554.

Wang H, Hu C, Hu X, Yang M, Qu J, (2012) Effects of disinfectant and biofilm on the corrosion of cast iron pipes in a reclaimed water distribution system. *Water Res.* 46:1070–1078.

Yadav AN, Kumar V, Prasad R, Saxena AK, Dhaliwal HS (2018) *Microbiome in crops: diversity, distribution and potential role in crop improvement through microbial biotechnology.* Elsevier, 305–332.

Yin W, Wang Y, Liu L, He J (2019) Biofilms: the microbial "protective clothing" in extreme environments. *Int J Mol Sci.* 20:3423.

Zhang W, Sun J, Ding W, Lin J, Tian R, Lu L (2015) Extracellular matrix-associated proteins form an integral and dynamic system during *pseudomonas aeruginosa* biofilm development. *Front Cell Infect Microbiol.* 5:40.

2 Biofilm Formation in Drug-Resistant Pathogen *Staphylococcus aureus*

Anjaneyulu Musini, Sai Pavan Chilumoju, and Archana Giri

CONTENTS

2.1 INTRODUCTION

Bacteria are found all around the planet. *Staphylococcus aureus* is the most common among them. Sir Alexander Ogston, a Scottish surgeon, originally named these germs *Staphylococcus* (from the Greek staphylos "grape" and kokkos "berry or seed") in

DOI: 10.1201/9781003184942-3

1882 (Wilson, 1987). It is almost universally known *S. aureus*, so named because of the color of the pigmented colonies (aureus means gold-colored in Latin) (Rosenbach, 1884). *Staphylococcus aureus* is a gram-positive bacterium that colonizes healthy nasal mucosa and skin of healthy individuals (Wertheim *et al.*, 2005). Humans appear to be the ancestral host of *S. aureus*, according to molecular studies and pathogenesis (Lowder *et al.*, 2009). Owing to its resistance to numerous antibiotics and tendency to build biofilms, methicillin-resistant *S. aureus* (MRSA) is a particularly dangerous pathogen, in addition to the conventional hospital-acquired MRSA (HA-MRSA), and the emergence of community-acquired MRSA (CA-MRSA) represents a significant risk to immune-compromised patients and healthy people (Ayami Sato *et al.*,2019). Biofilm-forming bacteria are known to be able to persist in the presence of strong antimicrobial concentrations (Lebeaux *et al.*, 2014). Biofilm-associated infections account for more than 80% of nosocomial infections, with *Staphylococcus aureus* being the most common pathogen (Römling *et al.*, 2012).

Methicillin and vancomycin resistance are two of the most notable antibiotic resistances developed by *Staphylococcus aureus*. MRSA is developed when methicillin-susceptible *S. aureus* (MSSA) acquires the methicillin-resistance gene mecA by horizontal gene transfer mediated by the staphylococcal cassette chromosome (SCC) of a mobile genetic element (Ito *et al.*,1999). MecA was previously the only known genetic marker for MRSA. However, we now have to be concerned about MRSA carrying mecB or mecC in the clinical laboratory. According to recent research, the frequency of mecC-mediated methicillin resistance in human MRSA isolates ranges from 0–2.8% (Deplano *et al.*, 2014; Cuny *et al.*, 2011; Petersen *et al.*, 2013). MRSA acquires certain key antibiotic resistance characteristics through spontaneous mutations. The most well-known examples are rifampin and fluoroquinolone resistance. Furthermore, vancomycin resistance – which has cast a pall on anti-MRSA treatment in recent years – has been acquired by mutation. Clinical isolates of MRSA strains with reduced susceptibility to vancomycin (vancomycin intermediate-resistant *S. aureus* [VISA]) and, more recently, with high-level vancomycin resistance (vancomycin-resistant *S. aureus* [VRSA]) have been described in the clinical literature (Gardete and Tomasz, 2014).

2.2   MICROBIAL BIOFILM FORMATION

A biofilm is a sessile community of microorganisms made up of cells adhering to a biotic or abiotic surface, immersed in a self-produced extracellular matrix, and with changed phenotypic and genotypic traits that allow adaptation to harsh environmental circumstances. The formation of a biofilm may also give protection from hazardous substances found in the environment, such as antibiotics (Anwar *et al.*, 1992). Biofilms are formed in numerous phases, according to genetic analyses of single-species biofilms. Depending on the environment in which the biofilm has evolved, noncellular elements such as mineral crystals, corrosion particles, clay or silt particles, or blood components may be detected in the biofilm matrix. The substratum (i.e. texture or roughness, hydrophobicity, surface chemistry, charge, and preconditioning film), the medium (i.e. nutrient levels, ionic strength, temperature, pH, flow rate, and presence of antimicrobial agents), and intrinsic cell properties (i.e. cell surface hydrophobicity, extracellular appendages, extracellular polymeric substances (EPS), signaling molecules) all influence biofilm development (Donlan, 2002; Renner, 2011).

Biofilm production is a multi-step (typically cyclic) process that involves numerous bacterial species (Al-Ahmad *et al.*, 2010). Biofilms have different functions and appearances in different contexts, yet they always come from the same series of events (Escher and Characklis, 1990; Van Loosdrecht *et al.*, 1990). As indicated in Figure 2.1, the biofilm production process involves the following steps involved in bacteria including *S. aureus*, conditioning film components adsorption, attachment of microorganisms, growth, maturation, biofilm detachment, and expansion colonization of new niches.

2.2.1 CONDITIONING FILM COMPONENTS ADSORPTION

In the solid-liquid interface between a surface and an aqueous environment in which organic matter (such as blood, milk, urine, seawater, tear fluid, and saliva) is optimal for microbe adhesion and development, substratum surfaces are initially covered with a layer of adsorbed, organic molecules, which is referred to as a "conditioning film" (Gristina *et al.*, 1987; Schneider *et al.*, 1994; Donlan, 2002). When compared to microbes, molecules adsorb to a substratum comparatively quickly (Escher and Characklis, 1990; Gristina *et al.*, 1987). The production of these conditioning coatings on surfaces exposed to saltwater was initially reported by Loeb and Neihof (1975). The rate and degree of adhesion may also be influenced by the surface's physicochemical characteristics.

2.2.2 ATTACHMENT OF MICROORGANISMS

As the second phase in biofilm development, bacteria are transported toward a substratum surface.

1. A reversible cell attachment to a preconditioning layer is generated on an abiotic or biotic surface by a weak interactions (i.e. Van der Waals forces) (Bos *et al.*, 1999; Donlan, 2002); microbial adhesion – either of single organisms or of (co)aggregates – can occur, which is typically reversible at first but becomes permanent over time, for example, due to the adhering microbes' excretion of exopolymer compounds (Dufrêne *et al.*, 1996; Neu *et al.*, 1990).
2. Microbial pair co-adhesion: only a few adhering, sessile microorganisms can induce the attachment of other planktonic bacteria that are still floating. This can happen when sessile microorganisms slow down an approaching planktonic microorganism, increasing its chances of adhering to the substratum surface, as seen commonly under flow, or when strong attractive interactions between sessile and planktonic bacteria, known as co-adhesion (Bos *et al.*, 1999).
3. An irreversible adsorption to the surface is mediated by various attachment structures (e.g. flagella, fimbriae, lipopolysaccharides, and adhesive proteins),

hydrophilic/hydrophobic interactions, electrostatic forces, and Lewis's acid–base interactions (Bos *et al.*, 1999; Donlan, 2002); microorganism attachment to surfaces is a complicated process with numerous factors that influence the outcome. Surfaces that are rougher, more hydrophobic, and covered with surface "conditioning" coatings are the most vulnerable for attachment (Donlan *et al.*, 2002).

2.2.3 GROWTH

2.2.3.1 Cell Proliferation and ECM Production

Adsorbed cell growth is resulted from the development of a self-produced EPS matrix, which is mostly constituted of proteins, polysaccharides, and extracellular DNA (eDNA) (Branda *et al.*, 2005; Flemming *et al.*, 2007). The growth and division of the initially connected cells promotes the formation of micro-colonies (primary colonizers) (Dufour *et al.*, 2012).

2.2.3.2 Biofilm Maturation

Matured biofilms (characterized by the presence of macro-colonies) with water channels that effectively transfer nutrients and signaling molecules inside the biofilm, which can be flat or mushroom-shaped (Dufour *et al.*, 2012:24; Hall-Stoodley *et al.*, 2004).

2.2.4 BIOFILM DETACHMENT

Some biofilm cells can detach singly or in clumps to survive when nutrients become scarce, or simply to expand and colonize new niches; dispersion of biofilms happens as a result of environmental changes and is influenced by growth circumstances. (O'Toole *et al.*, 2000).

2.2.5 EXPANSION AND COLONIZATION OF NEW NICHES

Detachment appears to be a dynamic process that permits new niches to be spread and colonized (Srey *et al.*, 2013; Sauer *et al.*, 2002). Furthermore, malnutrition is seen as a cause of separation, allowing bacteria to seek out a nutrient-rich environment (O'Toole *et al.*, 2000).

2.3 BIOFILM FORMATION IN *STAPHYLOCOCCUS AUREUS*

Staphylococcus aureus is one of the most common foodborne pathogens, and it may be detected in food processing plants on rare occasions (Pastoriza *et al.*, 2002). It has the ability to stick to surfaces and create biofilms in this state: initial attachment to the biotic and the abiotic surfaces.

2.3.1 BIOTIC SURFACES

S. aureus attachment to biotic surfaces necessitates considerably more precise interactions mediated by a wide range of cell wall–anchored proteins. Foster *et al.*

(2014) considered classifying CWA (cell wall–anchored) proteins into four categories based on the existence of motifs identified through structure–function research: MSCRAMM (microbial surface components recognizing adhesive matrix molecule family), NEAT motif family (near iron transporter), Protein A (tandemly repeated three-helical bundle), and the G5–E repeat family.

The best-understood biofilm formation processes re SasG- and Aap-dependent biofilm development. The iron-regulated surface protein A (IsdA), an NEAT motif protein capable of capturing heme from hemoglobin under iron-restricted circumstances in the host, as well as the *S. aureus* surface protein G (SasG), a G5–E repeat protein, are all implicated in attachment to desquamated epithelial cells (Clarke *et al.*, 2006, 2009; Roche *et al.*, 2003).

Other MSCRAMMs, such as the fibronectin-binding proteins A (FnBPA) and B (FnBPB) and the bone sialoprotein binding protein, are also implicated in extracellular matrix adherence (Burke *et al.*, 2011:40; Geoghegan *et al.*, 2013; McCourt *et al.*, 2014; Vazquez *et al.*, 2011).

2.3.2 ABIOTIC SURFACES

S. aureus attachment to abiotic surfaces, on the other hand, mostly depends on the physicochemical characteristics of the cell and the contact surface. Hydrophobic and electrostatic interactions have a role in the initial attachment of staphylococci, although certain bacterial surface components, such as autolysins or teichoic acids, have also been implicated (Gross *et al.*, 2001; Heilmann *et al.*, 1997). Surprisingly, high ionic strength environments, such as seawater in fisheries or saline serum in hospitals, can enhance *S. aureus* adherence to some abiotic surfaces.

2.4 PRODUCTION OF EXTRACELLULAR POLYMERIC MATRICES

Extracellular polymeric matrices are polysaccharides, proteins, and extracellular DNA.

2.4.1 POLYSACCHARIDES

The biofilm matrix contains a significant amount of polysaccharide intracellular adhesion (PIA) (Götz, 2002), a PNAG(Poly-N-acetylglucosamine) with partially deacetylated residues. For the staphylococci, PNAG/PIA is considered to be the most essential biofilm matrix component (Caiazza and O'Toole, 2003; Foster and Höök, 1998; Gross *et al.*, 2001; Götz, 2002; Vuong *et al.*, 2000). An operon containing the four genes *icaA*, *icaD*, *icaB*, and *icaC* contains the genetic components coding for poly-N-acetylglucosamine production (intercellular adhesin). The products of the chromosomal ica operon (also known as icaADBC) in most *S. aureus* strains synthesize PNAGs (Cramton *et al.*, 1999; Fitzpatrick *et al.*, 2005; Maira-Litrán *et al.*, 2002). The involvement of the ica operon in *S. aureus* biofilm formation is complicated, and it is supposed to be strain- and environment-dependent (O'Gara, 2007). In both *S. aureus* and coagulase negative staphylococci, PIAs play a crucial role in biofilm formation and cohesiveness, and they are engaged in vivo in an immune evasion

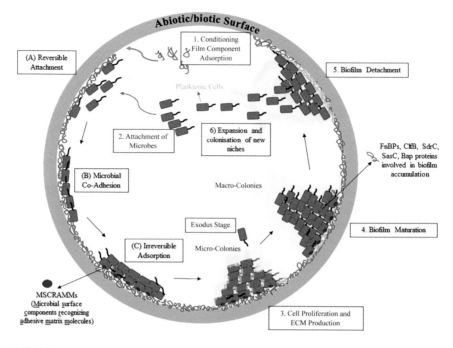

FIGURE 2.1   Biofilm formation in *S. aureus.*

mechanism. Different strains exhibit distinct phenotypes, regardless of whether or not this molecule is present; the purpose of PIAs in *S. aureus* is yet unknown. Teichoic acids are now recognized to be part of the biofilm matrix, although not being frequently linked with it (Archer *et al.*, 2011; Payne and Boles, 2016). Teichoic acids are composed of a phosphate chain alternating with ribitol or glycerol, which are generally replaced by d-alanine and N-acetylglucosamine residues to give the cell surface a positive net charge (Gross *et al.*, 2001).

2.4.2   PROTEINS

So far, many proteins have been found in the biofilm matrix of *S. aureus.* MSCRAMMs may be expressed by *S. aureus* in a number of ways, such as biofilm-associated protein (bap) (Carme Cucarella *et al.*, 2001). Fibrinogen-binding protein clumping factors A and B (clfA/clfB) (Louise O'Brien, 2002a, 2002b), fibronectin-binding proteins A and B (fnbpA/fnbpB) (Philippe *et al.*, 2015), elastin-binding protein (Ebps) (Campoccia *et al.*, 2009) serine-aspartate repeat (Sdr) family proteins, fibrinogen-binding protein (fib) (Shannon & Flock, 2004) collagen-binding protein (cna), laminin-binding protein (eno) (Azara *et al.*, 2017), each factor's proportionate influence appears to be strain or condition specific.

In *S. aureus,* a distinct surface-wall associated protein, SasG, has been linked to cell accumulation and biofilm formation; however, this process is dependent on Zn^{2+}

(Corrigan *et al.*, 2007; Geoghegan *et al.*, 2010). Similarly, Schroeder *et al.* (2009) discovered that the surface protein SasC plays a crucial role in the development of clusters and intercellular adhesion (ica) in *S. aureus* cells during host infection.

Secreted proteins, notably phenol-soluble modulins (PSMs), regulate the development of *S. aureus* biofilms. PSMs have a role in the development of channels in mature biofilms, which allow nutrients to reach deeper levels (Periasamy *et al.*, 2012). The quorum sensing system accessory gene regulator (agr), which is activated by auto-inducing peptides, induces the production of these surfactant-like molecules. (Boles and Horswill, 2008; Lauderdale *et al.*, 2009; Yarwood *et al.*, 2004). Furthermore, Merino *et al.* (2009) discovered that when produced at high levels, the multifunctional virulence factor SpA plays a key role in the development of biofilm-associated illnesses by promoting bacterial aggregation and biofilm formation without being covalently attached to the cell wall. (Jeffrey *et al.*, 2019) reported that membrane-attached lipoproteins may interact with the eDNA in the biofilm matrix and enhance biofilm development using *Staphylococcus aureus* biofilms, suggesting that lipoproteins could be a target for new treatments aiming at disrupting bacterial biofilms, and they also added secreted eDNA-binding proteins and membrane-attached lipoproteins to the electrostatic net model, which can act as anchor points between eDNA in the biofilm matrix and the bacterial cell surface.

2.4.3 Extracellular DNA

In *S. aureus* and other staphylococci species, eDNA has been postulated as an important structural component of the biofilm matrix (Izano *et al.*, 2008). Because of its negative electric charge, eDNA can serve as a binding agent in cell–cell, cell–surface, and cell–host interactions (Lister and Horswill, 2014). Recent research suggests that when *S. aureus* develops biofilms on abiotic surfaces, extracellular DNA (eDNA) may have a bigger role than MSCRAMM adhesin proteins (Moormeier *et al.*, 2014). eDNA is involved in the development and maturation of biofilms, as well as the preservation of mature formations. This dual role of eDNA was discovered in research by Izano *et al.* (2008), who found that adding deoxyribonuclease I to an *S. aureus* culture reduced biofilm development while also dispersing pre-existing biofilm on polystyrene plates. Furthermore, it has been shown that eDNA can play a protective function since its depolymerization makes *S. aureus* biofilms more susceptible to antimicrobials, owing to a profound change in the matrix structure. It has been observed that eDNA may interact with other molecules in biofilm matrices, functioning as a cohesion molecule that acts as a scaffold connecting various components, enhancing the stability of final structure (Huseby *et al.*, 2010; Schwartz *et al.*, 2016). Because eDNA is abundant in the biofilm matrix, it offers a significant genetic pool that can facilitate horizontal gene transfer, particularly in biofilms containing several strains with diverse features (Flemming *et al.*, 2007; Schwartz *et al.*, 2016). As a result, it may facilitate the transmission and spread of antimicrobial resistances, posing a major concern in clinically related biofilms.

S. aureus biofilm formation has been demonstrated to follow a five-stage developmental process based on micro fluidic flow-cell devices and time-lapse microscopy, viz attachment, multiplication, exodus, maturation, and dispersal (Moormeier *et al.*, 2014). Exodus stage of biofilm development is an early dispersal event that coincides with micro colony formation and results in the restructuring of the biofilm (Figure 2.1). Approximately six hours after the start of the multiplication stage, the cells are released in a coordinated manner. Exodus occurs concurrently with the creation of micro-colonies and leads in biofilm reorganization in *S. aureus* (Moormeier *et al.*, 2014).

2.5   BIOFILM DISPERSION

The final phase in a biofilm's life cycle is cell detachment, which allows *S. aureus* to colonize new niches (O'Toole *et al.*, 2000). *S. aureus*, on the other hand, is known to employ self-controlling processes, such as broad-spectrum and specialized mechanisms, to enable the release of biofilm cells into the environment. Surprisingly, the efficiency of each of these processes is strongly reliant on the polymeric matrix's composition (Chaignon *et al.*, 2007; Kiedrowski *et al.*, 2011). Protease secretion is a more selective dispersion strategy of *S. aureus* that results in the breakdown of proteinaceous matrix components and, as a result, leads to biofilm breakup. *S. aureus* produces ten distinct secreted proteases (Shaw *et al.*, 2004), including the V8 serine protease (SspA), the aureolysin (Aur), and the staphopains (SspB and ScpA). Both FnBPs and Bap proteins can be degraded by SspA (McGavin *et al.*, 1997; O'Neill *et al.*, 2008; Martí *et al.*, 2009). *Saccharomyces cerevisiae* secretes thermostable nucleases. *S. aureus* appears to be involved in biofilm dispersion during infection (Berends *et al.*, 2010; Olson *et al.*, 2013), as well as the development of channels within the biofilm (Kiedrowski *et al.*, 2014). The potential of these dispersion enzymes as a method for controlling the formation of biofilms by bacterial pathogens such as *S. aureus*, for the search of other dispersal molecules, must be explored further.

2.6   GENES RESPONSIBLE FOR BIOFILM FORMATION

Fahimeh *et al.* (2016) carried out a study using 110 *S. aureus* strains obtained from three main hospitals. CRA and multiplex PCR were used to investigate antibiotic resistance patterns, phenotypes, and biofilm-forming genes (M-PCR). Fahimeh *et al.* (2016) discovered that MRSA was present in 103 out of 110 samples (93.6%). The 12 genes involved in biofilm formation were found to have a high frequency: icaA (34.2%), icaB (29.7%), icaC (69.3%), icaD (54.8%), fnbA (38.1%), fnbB (46.6%), fib (39.9%), clfA (41.4%), clfB (44.1%), ebps (26.5%), cna (18.3%), and eno (29.6%) (Fahimeh *et al.*, 2016), including the icaABCD and other related genes (Mirzaee *et al.*, 2015; Vancraeynest *et al.*, 2004). Because the icaA and icaD genes are required for intercellular adhesion, it is easy to infer that these genes play a crucial role in biofilm development and also for the synthesis of the bacterial multilayer. These genes, on the other hand, are linked to the production of slime and biofilm (Ammendolia *et al.*, 1999; Arciola *et al.*, 2001). The results of recent studies comparing biofilm cells to planktonic cells revealed that the ica gene is required for the start of biofilm formation (Resch *et al.*, 2005).

2.7 ANTIBIOTIC-RESISTANT AND ASSOCIATED GENES

S. aureus isolates were tested by Qi *et al.* (2019) for antibiotic resistance and associated genes, as well as biofilm-forming potential. Antibiotic resistance in *S. aureus* is commonly thought to be linked to particular resistance genes. *S. aureus* resistance to erythromycin, clindamycin, and inducible clindamycin was shown to be mostly attributable to the *ermA* or *ermC* genes, according to (Ng *et al.*, 2001; Jarajreh *et al.*, 2017). Ng *et al.* (2001) discovered that the tetracycline resistance of *S. aureus* was linked to the tetM and tetK genes. Several antibiotics are often used in veterinary medicine, including lactams (penicillin), macrolides (erythromycin), and lincosamide (clindamycin) (Cháfer-Pericás *et al.*, 2010). *S. aureus* isolates with phenotypic resistance to the five most common antibiotics – namely penicillin, erythromycin, clindamycin, tetracycline, and inducible clindamycin – were found to be positive for the resistance gene associated, implying that targeting the resistance gene can be used as a useful reference for microbial resistance evaluation (Qi *et al.*, 2019).

The earliest evidence of antibiotic resistance in *S. aureus* was discovered in the early 1940s when PRSA (penicillin-resistant *S. aureus*) was discovered. The blaZ gene in PRSA encodes lactamase (Bla), which hydrolyzes the lactam ring of penicillin and causes resistance to it. In the late 1950s, methicillin was given as a therapy for PRSA when it became common in hospitals and the community. Unfortunately, within two years of clinical usage, the first MRSA strains were discovered. MRSA is to blame for a number of hospital-acquired *S. aureus* infections, as well as morbidity and mortality. In MRSA, the mecA gene produces a low-affinity penicillin-binding protein (PBP2a or PBP2') (McGuiness *et al.*, 2017). In addition to the factors that increase antibiotic resistance in microorganisms, such as antimicrobial selectivity, microbial characteristics, increased antibiotic use, susceptible hosts, and infection control program errors, biofilm formation in *S. aureus* causes an increase in antibiotic resistance when compared to the planktonic state (Bhattacharya *et al.*, 2015:141; Asadi *et al.*, 2014). Resistance to new antibiotics is rising in bacteria, such as quinolone resistance in *Streptococcus pneumoniae*, ESBL resistance in gram-negative bacteria, clarithromycin resistance in *Helicobacter pylori*, and so on (Kargar *et al.*, 2011, 2014; Ghorbani-Dalinia *et al.*, 2015). Vancomycin is the most often prescribed antibiotic for *S. aureus* biofilm infections (Bhattacharya *et al.*, 2015). The first VRSA (vancomycin-resistant *S. aureus*) was discovered in the United States in 2002 (McGuiness *et al.*, 2017). The vanA operon, which was inserted into a staphylococcal resident plasmid by transposon Tn1546 from the VRE (vancomycin-resistant Enterococci) plasmid, is responsible for 'a' here. complete vancomycin resistance in *S. aureus* (MIC 16 g/ml). New antibiotics or combinations of other medicines, such as rifampin plus linezolid or daptomycin, are required due to increasing resistance of biofilms to vancomycin in *S. aureus* (Bhattacharya *et al.*, 2015).

2.8 QUORUM SENSING IN *S. AURUES*

In *S. aureus*, quorum sensing has been reported. In the coccus, the agr quorum–sensing system and the LuxS quorum–sensing system are both present. These mechanisms appear to be critical to the survival of a *S. aureus* biofilm.

   The colonization of bacteria on surfaces promotes cell growth stability and allows favorable cell–cell interactions such as quorum sensing and genetic exchange (Daniels *et al.*, 2004; Davey and O'Toole, 2000; Elias and Banin, 2012; Hall-Stoodley *et al.*, 2004; Watnick and Kolter, 2000a, 2000b). Low-molecular-mass signaling molecules known as auto-inducers (AI) influence gene expression, metabolic cooperativity and competition, physical contact, and bacteriocin synthesis in biofilm cells, offering a mechanism for self-organization and regulation (Daniels *et al.*, 2004; Donlan, 2002; Elias and Banin, 2012; Parsek and Greenberg, 2005; van-Houdt and Michiels, 2010). The effects of quorum sensing are reliant on AI concentration, which rises in a cell density-dependent way (Parsek and Greenberg, 2005). Meanwhile, the exchange of mobile genetic elements between adjacent biofilm cells allows for the acquisition of additional antibiotic resistance or virulence characteristics, as well as environmental survival capacities (Madsen *et al.*, 2012). Surface-attached quorum sensing is activated by AIP-I, whereas it is blocked by surface-attached TrAIP-II (Minyoung *et al.*, 2017). The use of quorum sensing – also known as the accessory gene regulator (agr) system, which is widespread among staphylococci – is one possible colonization resistance strategy (Morgan *et al.*, 2020). Biofilm-associated *S. aureus* causes a number of severe illnesses in which the accessory gene regulator (agr) quorum sensing mechanism is considered to play a key role. (Yarwood *et al.*, 2004) investigated agr's role in biofilm formation, as well as agr-dependent transcription. Disruption of agr expression had no discernable effect on biofilm development in certain circumstances, while it prevented or increased biofilm formation in others. There is some data on the link between agr expression and *S. aureus* biofilms. Pratten *et al.* (2001) reported minimal difference between wild-type *S. aureus* and an agr mutant in adhesion to uncoated or fibronectin-coated glass. The P2 and P3 promoters control two divergent operons in the agr locus. The agrBDCA gene, which codes for the RNAII transcript, is found in the P2 operon. P3 is responsible for the transcription of RNAIII, the agr locus' effector molecule. In addition, RNAIII is used to translate hemolysin, a secreted virulence factor encoded by hld. AgrA and AgrC make up a two-part regulatory mechanism that reacts to autoinducing cyclic octapeptide. AgrB converts the agrD product into the autoinducing cyclic octapeptide. This system, in conjunction with other regulatory elements like SarA, promotes transcription from both the P2 and P3 promoters, resulting in increased intracellular RNAIII concentrations. RNAIII increases the expression of secreted virulence factors while decreasing the expression of many surface adhesins, such as protein A and the fibronectin-binding protein, in batch culture. Vuong *et al.* (2000) investigated the relationship between a functioning agr system and *S. aureus* clinical isolates' capacity to adhere to polystyrene under static circumstances. Under these circumstances, just 6% of the isolates with a functional agr system developed a biofilm, compared to 78% of the agr-defective isolates. Spinning-disk reactor biofilm cells of agr D mutant were particularly susceptible to rifampin but not to oxacillin (Pratten *et al.*, 2001). Previous research has linked acyl-HSL quorum sensing to the biofilm formation of various proteobacteria species under specific circumstances (Davies *et al.*, 1998; Huber *et al.*, 2001, 2002; Merritt *et al.*, 2003; Steidle *et al.*, 2002). Quorum sensing may play a part in biofilm formation under some situations, but it may not play an evident role in biofilm development under other conditions. Agr expression likely affects the attachment of inoculum cells to a surface through its regulation of cell surface-associated adhesins, as the agr system is thought to exert regulatory control over an agr expression and likely affects the attachment of inoculum cells to a surface through its regulation of cell surface-associated adhesins (Pratten *et al.*, 2001). Findings of oscillations in GFP fluorescence in flow-cell biofilms, along with the fact that the oscillations were not limited to the expression of an agr reporter, point to a wider model of gene expression in the biofilm that is linked to cell survival and detachment.

2.8.1 AGR QUORUM SENSING IN *S. AUREUS*

In the detachment phase of the biofilm formation cycle, the agr quorum–sensing system is critical. Detachment permits cells to build biofilms on additional locations, allowing a biofilm-associated infection to spread throughout the host organism. The agr quorum–sensing system in *S. aureus* controls the synthesis of PSMs, which facilitate cell detachment from biofilms due to their amphipathic character. One study found that a *Staphylococcus* mutant lacking a functional agr system developed a thicker biofilm, which was most likely due to the cells' inability to detach from the biofilm (Kong *et al.*, 2006). The agr quorum–sensing system permits cells to break from a biofilm and disseminate infection by producing PSMs. The agr quorum–sensing mechanism has also been demonstrated to improve a staphylococcal biofilm's ability to defend itself against the immune system of humans. The capacity of a *Staphylococcus* mutant without a functional agr system to properly respond to host immunological responses was shown to be significantly hampered. In comparison to a *Staphylococcus* with a functional agr system, the mutant was less capable of inducing chemotaxis (the ejection) of human neutrophils and was less resistant to human antimicrobial peptides. The mutant was also thought to be less efficient in dealing with the respiratory burst – or the production of reactive oxygen species, such as hydrogen peroxide – by human neutrophils, because the agr system regulates the overall oxidative stress responses of staphylococcal bacteria. As a result, the agr system is critical for a *Staphylococcus* biofilm's capacity to defend itself against the immune system (Kong *et al.*, 2006).

2.8.2 LuxS QUORUM SENSING IN *S. AUREUS*

The density of a *S. aureus* biofilm is affected by the LuxS quorum sensing system. The LuxS protein controls the synthesis of PIA, a chemical that makes it easier for cells in a biofilm to stick together. A *Staphylococcus* mutant without a working LuxS system developed a thicker biofilm than a wild-type *Staphylococcus* with a working LuxS system. The LuxS quorum sensing system, like the agr system, encourages cells to separate from biofilms (Kong *et al.*, 2006).

2.9 BIOFILM FORMATION AND DRUG RESISTANCE

Biofilms may influence drug resistance in pathogenic bacteria (Frieri *et al.*, 2017). Antibiotics can't get through biofilms, which improves bacterial resistance

(Jolivet-Gougeon *et al.*, 2014). For both MSSA and MRSA isolates, biofilm formation has been categorized as a significant defensive mechanism and pathogenic characteristic, both as a way of surviving in the environment (e.g. on a hospital ward) and in the host during infections (Amorena *et al.*, 1999). The high bacterial cell density inside the biofilm increases the number of antibiotic-resistant bacteria that may be chosen, improves horizontal genetic exchange, and increases the frequency of mutation (Lazăr and Chifiriuc2010). Antimicrobial drugs are unable to reach their targets due to the biofilm matrix acting as a barrier. In addition, the matrix has a short-term protective impact. Positively charged antibiotics, such as aminoglycosides and polypeptides, may have delayed in their penetration into biofilms when coupled to negatively charged biofilm matrix polymers. Astha and Jain (2012) demonstrate that staphylococcal isolates with a proclivity for biofilms are more resistant to antibiotics and hence are more difficult to treat. The development of biofilm, the expression of biofilm-related genes, and antibiotic resistance have all been linked in research (Motaharesadat *et al.*, 2020). Another research included in the current review (Yousefi *et al.*, 2018), found a link between biofilm development and antibiotic resistance in *S. aureus* isolates (nitrofurantoin, tetracycline, erythromycin, and ciprofloxacin). They also discovered that bacteria that produce biofilms had a greater rate of antibiotic resistance. In addition, (Mahmoudi *et al.* (2019) found that MRSA strains had a considerably greater incidence of biofilm development. Ahmadrajabi *et al.* (2017) found no difference in biofilm production capabilities between MSSA and MRSA isolates, but they did identify a link between biofilm development and antibiotic resistance. Similarly, Goudarzi *et al.* (2019) and Jahanshahi *et al.* (2018) found that MRSA isolates had double the capacity of MSSA isolates to produce biofilms.

Antibiotics clump together in the biofilm matrix, accounting for up to 25% of its weight. This matrix acts as a barrier to antibiotics reaching the bacteria in the biofilm's deeper layers. The mobility of antibiotics in the matrix is decreased, and the inner layer does not get enough antibiotic. The concentration of nutrients and metabolic substrates varies between the biofilm's outer and interior layers (Stewart, 2003). The concentration of oxygen in the core of biofilms is low. Oxygen has a crucial role in metabolite activity, and reduced oxygen availability increases resistance to antibiotics such amino glycosides, according to previous reports (HΦiby *et al.*, 2010; Parastan *et al.*, 2020). Reduced nutrient levels in biofilms cause an increase in guanine nucleotide-guanosine 3,5-bis-pyrophosphate (ppGpp) accumulation and a decrease in RNA (tRNA and rRNA) production. Cell growth rates in biofilms are heterogeneous, and this is due to cellular enzyme production inside the biofilm (Poulsen *et al.*, 1993). Many antibiotics, such as penicillin, are growth dependent, meaning they only affect bacteria that are growing (Tuomanen *et al.*, 1986). Resistant persister cells make up the bacterial cells in the biofilm, which are multidrug resistant (MDR) (Parastan *et al.*, 2020). In *Pseudomonas aeruginosa*, the Agr system plays a key role in MRSA heterogeneous resistance, as well as resistance to hydrogen peroxide and kanamycin (Hassett *et al.*, 2000). Antibiotic-degrading enzymes such as beta-lactamases, efflux pumps, and some gene products whose expressions are altered by quorum sensing as a stress response might help biofilms to become more antibiotic-tolerant (Antunes *et al.*, 2010). Beta-lactamases help biofilms withstand beta-lactam antibiotics. Bacterial beta-lactamases are a crucial component in the biofilm that causes resistance to beta-lactam antibiotics (Bjarnsholt *et al.*, 2011).

2.10   PREVENTING AND INHIBITING BIOFILMS

There are a number of approaches for removing adhesions and inhibiting bacterial formation in biofilm structures. In reality, the goal of these techniques is to restore cells from biofilms to planktonic states, lowering bacterial virulence and antibiotic resistance (Rosenthal *et al.*, 2014). Antiadhesion chemicals, innovative medical devices, destructive enzymes, bacteriophage, and vaccinations are the most important treatment approaches. *S. aureus* may adhere to abiotic surfaces such as glass, metals, and plastic (used in medical devices or implants), as well as biotic surfaces such as host tissue. The attachment of *S. aureus* to surfaces is reliant on bacterial MSCRAMMs for host protein. Surfaces should be coated with anti-adhesions such as arylrhodamines, calcium chelators, silver nanoparticles, chitosan, and others to inhibit *S. aureus* adherence to surfaces through MSCRAMMs (Bhattacharya *et al.*, 2015). Proteases, nucleases, hyaloronate lyase, dispersin B, and lysostaphin are examples of enzymes that may degrade and suppress biofilms by various methods (Rosenthal *et al.*, 2014). Proteases released by *S. aureus* enzymes are positively regulated by the agr system. Nuclease1 (Nuc1) and nuclease2 (Nuc2) are two nucleases produced by *S. aureus* (Nuc2). Exogenous treatment with the Nuc enzyme inhibits the development of *S. aureus* biofilms, while *S. aureus* mutants lacking Nuc create bigger biofilms (Tong *et al.*, 2015). Nuc2 has lesser action than Nuc1, destroys localized biofilms, and may generate a channel inside biofilms. Another enzyme that cleaves the 1–4 glycosidic link of hyaluronic acid in *S. aureus* and disseminates cutaneous infection in a mouse model is hyaluronate lyases, which is encoded by the gene hysA. Actino bacillus actinomycetem comitans, a dental pathogen, produces dispersin B and acts as a N-acetyl-glucoseaminidase. When dispersin B is introduced to staphylococcal biofilms, it can cleave PIA in the biofilm matrix, causing the biofilm to disintegrate quickly and increasing antimicrobial susceptibility (Kaplan *et al.*, 2004). With or with out antimicrobials, the fibrinolytic agents plasmin, streptokinase, nattokinase, and TrypLE, a recombinant trypsin-like protease, induce efficiently dispersed existing *S. aureus* biofilms and kill bacterial cells being released from the biofilm, respectively, and have no cytotoxicity. MRSA and MSSA strains can be prevented from forming biofilms by using DNase I (Hogan *et al.*, 2008).

Biofilm development in MRSA and MSSA is inhibited by nanoparticles (NPs) such as chitosan. Chitosan is a non-toxic polymer that has long been utilized as a powerful antibacterial agent and now helps to inhibit MRSA adherence. The antibiofilm activity of the layer-by-layer coated NPs is 94% in *S. aureus* and 40% in *E. coli*, respectively. These nanoparticles have no impact on human cells and can be utilized in conjunction with a little amount of antibacterial agent (Ivanova *et al.*, 2018).

Medicinal herbs with action against MDR and biofilm development are a novel method for controlling and inhibiting biofilm formation in *S. aureus*. *Cinnamomum glaucescens* (MIC50 = 369 mg/ml and BIC50 = 246 mg/ml), *Acacia pennata* (MIC50 = 369 mg/ml and BIC50 = 246 mg/ml), *Ficus hispida* (MIC50 = 369 mg/ml and BIC50 = 246 mg/ml), and *Holigarna caustica* (MIC50 = 369 mg/ml and

BIC50 = 246 mg/ml) were among the Indian medicinal plants studied by Panda *et al.* (2020).

2.11 VACCINES BASED ON BIOFILMS AND ADHESIONS

The QS system is a promising target for the development of new anti-infective drugs, as well as immunotherapeutic techniques. The use of virus-like particles to find immunogenic mimics of quorum sensing peptides is becoming more common. The findings showed that a VLP-based epitope identification strategy to vaccine development targeting agr signaling disruption might be effective against *S. aureus* (O'Rourke *et al.*, 2014). Antibody activity to each vaccination component is higher in conjugative vaccines that contain two antigens. ClfA and PNAG are used in one of these vaccinations. Currently, vaccination candidates for SA3Ag (*S. aureus* three-antigen) and SA4Ag (*S. aureus* four-antigen) exist. Capsular CP5, CP8, and ClfA are antigens found in SA3Ag; CP5, CP8, ClfA, and MntC make up the SA4Ag vaccine. In healthy individuals, these two vaccinations have a high immunogenicity (Xu *et al.*, 2017; Creech *et al.*, 2020). ClfA has been proven preclinically as one of the best adhesion antigens against *S. aureus*. Several firms have licensed ClfA antigens for use in clinical trials of multi-antigen vaccines (Anderson *et al.*, 2016; Levy *et al.*, 2015). Bacteria attach to surfaces via adhesions in the biofilm state, therefore preventing *S. aureus* from adhering to biotic and abiotic surfaces, one way to avoid pathogenesis. Antibiofilm vaccinations are being developed to inhibit the production of *S. aureus* biofilms. In the biofilm matrix, extracellular polysaccharides or cell wall–related proteins may generate a protective immunological response against *S. aureus* infections (Giersing *et al.*, 2016). In a kidney infection model, active or passive vaccination using poly–Nsuccinyl-1–6-glucoseamin (PNSG) as a surface polysaccharide–induced resistance against S. aureus (McKenney *et al.*, 1999). Only PNAG-1 is immunogenic, and immune response to this antigen is varied. PNAG is an immunogenic molecule that may be split into three fractions: PNAG-I, PNAG-II, and PNAG-III. Furthermore, not all *S. aureus* strains generate PNAG/PIA; therefore, additional study is needed in this area (Maira-Litrán *et al.*, 2002).

2.12 CONCLUSION

Biofilm development is important for *S. aureus* survival and dispersal in environments that may pose a threat to human health. *S. aureus* is one of the most prevalent bacteria found on the human epidermis; it can enter and attach to tissue or medical equipment if the physical barrier, such as the skin, is disrupted. Many *S. aureus* strains are drug resistant (MRSA), and *S. aureus* is the most common pathogen that causes persistent biofilm-associated illness. In most situations, treatment for this type of persistent infection, particularly MRSA, is ineffective. Understanding *S. aureus* biofilm is essential for developing new approaches to prevent biofilm development and/or eliminate biofilm for all of these reasons. To identify efficient control measures against *S. aureus*, it is necessary to fully understand all of the underlying mechanisms involved in the creation of biofilms, cell-to-cell communication, resistance to external stressors, and cell dispersion from mature structures, among others. By regulating bacterial physiology and virulence processes, QS systems in *Staphylococcus* have a huge influence on pathogen success during infection. As anti-adhesions and surgery to remove biofilms are costly, researchers are working on developing DNA vaccines and recombinant proteins based on biofilms. After successful human trials, appropriate molecular identification of biofilm genes may stimulate the development of vaccines against *S. aureus* infections in the future. The medical community believed that the war against microbes had been won after the groundbreaking discovery of antibiotics. However, the battle had only just begun, as bacteria may produce biofilms, which are resistant structures. Biofilm is the true foe, and it completely alters the situation.

REFERENCES

Ahmadrajabi, R., S. Layegh-Khavidaki, D. KalantarNeyestanaki, and Y. Fasihi (2017) Molecular analysis of immune evasion cluster (IEC) genes and intercellular adhesion gene cluster (ICA) among methicillin-resistant and methicillin-sensitive isolates of Staphylococcus aureus. *J Prevent Med Hyg.* 58:E308. doi: https://doi.org/10.15167/2421-4248/jpmh2017.58.4.711

Al-Ahmad, M. Wiedmann-Al-Ahmad, J. Faust, M. Bachle, M. Follo, M. Wolkewitz, C. Hannig, E. Hellwig, C. Carvalho, R. Kohal (2010) Biofilm formation and composition on different implant materials in vivo. *Journal of Biomedical Material research.* doi: https://doi.org/10.1002/jbm.b.31688

Ammendolia MG, Di Rosa R, Montanaro L, Arciola CR, Baldassarri L (1999) Slime production and expression of the slime-associated antigen by staphylococcal clinical isolates. *J Clin Microbiol.* 37(10):3235–8. doi: 10.1128/JCM.37.10.3235-3238.1999

Amorena B, Gracia E, Monzón M, *et al.* (1999) Antibiotic susceptibility assay for Staphylococcus aureus in biofilms developed in vitro. *J Antimicrob Chemother.* 44:43–55. doi: 10.1093/jac/44.1.43

Anderson, A.S., Scully, I.L., Buurman, E.T., Eiden, J., Jansen, K.U (2016) Staphylococcusaureus clumping factor A remains a viable vaccine target for prevention of S. aureus infection. *mBio.* 7:e02232. doi: https://doi.org/10.1128/mBio.00225-16.

Antunes LC, Ferreira RB, Buckner MM and Finlay BB. Quorum sensing in bacterial virulence. Microbiol (2010) 156:2271–2282. doi: 10.1099/mic.0.038794-0

Anwar, H., Strap, J.L., Costerton, J.W (1992) Establishment of aging biofilms: possible mechanism of bacterial resistance to antimicrobial therapy. *Antimicrob Agents Chemother.* 36: 1347–1351. doi: 10.1128/aac.36.7.1347

Archer, N.K., Mazaitis, M.J., Costerton, J.W., Leid, J.G., Powers, M.E., Shirtliff, M.E (2011) Staphylococcus aureus biofilms: properties, regulation, and roles in human disease. *Virulence.* 2:445–459. doi: 10.4161/viru.2.5.17724

Arciola CR, Baldassarri L, Montanaro L (2001) Presence of icaA and icaD genes and slime production in a collection of staphylococcal strains from catheter-associated infections. *J Clin Microbiol.* 39(6):2151–2156. doi: 10.1128/JCM.39.6.2151-2156.2001

Asadi, S., Kargar, M., Solhjoo, K., Najafi, A., Ghorbani-Dalini, S (2014) The association ofvirulence determinants of uropathogenic Escherichia coli with antibiotic resistance. Jundishapur J. *Microbiol.* 7. doi: 10.5812/jjm.9936

Astha, A, Jain A (2012) Association between drug resistance & production of biofilm in staphylococci. *Ind J Med Res.* 135(4):562–564. PMCID: PMC3385245.

Ayami Sato, Tetsuo Yamaguchi, Masakaze Hamada, Daisuke Ono, Shiro Sonoda, Takashi Oshiro, Makoto Nagashima, Keisuke Kato, Shinichi Okazumi, RyojiKatoh, Yoshikazu Ishii, and Kazuhiro Tateda (2019) Morphological and Biological Characteristics of *Staphylococcus aureus* Biofilm Formed in the Presence of Plasma. *Microbial Drug Resistance*. doi: 10.1089/mdr.2019.0068

Azara, Elisa, Longheu, Carla, Sanna, Giovanna, Tola, Sebastiana (2017) Biofilm formation and virulence factor analysis of Staphylococcus aureus isolates collected from ovine mastitis. *Journal of applied Microbiol*. 123. doi: 10.1111/jam.13502

Berends, E.T., Horswill, A.R., Haste, N.M., Monestier, M., Nizet, V., Kockritz-Blickwede, M (2010) Nuclease expression by Staphylococcus aureus facilitates escape from neutrophil extracellular traps. J. *Innate Immun*. 2:576–586. doi: 10.1159/000319909

Bhattacharya, M., Wozniak, D.J., Stoodley, P., Hall-Stoodley, L (2015) Prevention and treatment of Staphylococcus aureus biofilms. Expert Rev. *Anti-Infect Ther*. 13,1499–1516. doi: https://doi.org/10.1586/14787210.2015.1100533

Boles, Blaise, Horswill, Alexander (2008) Agr-Mediated Dispersal of Staphylococcus aureus Biofilms. *PLoS pathogens*. 4. e1000052. doi: https://doi.org/10.1371/journal.ppat.1000052

Bos, Rolf, van der Mei, Henny, Busscher, Henk (1999) Physico-chemistry of initial microbial adhesive interactions – Its mechanisms and methods for study. *FEMS Microbiol reviews*. 23.179–230. doi: https://doi.org/10.1111/j.1574-6976.1999.tb00396.x

Branda, Steven, Vik, Shild, Friedman, Lisa, Kolter, Roberto (2005) Biofilms: The Matrix Revisited. Trends in microbiology. 13.20–6. doi: 10.1016/j.tim.2004.11.006

Burke, Fiona, Dipoto, Antonella, Speziale, Pietro, Foster, Timothy (2011) The A domain of fibronectin-binding protein B of Staphylococcus aureus contains a novel fibronectin binding site. *The FEBS journal*. 278.2359–71. doi: 10.1111/j.1742-4658.2011.08159.x

Caiazza, Nicky, O'Toole, G (2003) Alpha-Toxin Is Required for Biofilm Formation by Staphylococcus aureus. *Journal of bacteriology*. 185.3214–7. doi: 10.1128/JB.185.10.3214-3217.2003

Campoccia, Davide, Speziale, Pietro, Ravaioli, Stefano, Cangini, Ilaria, Rindi, Simonetta, Pirini, Valter, Montanaro, Lucio, Arciola, Carla (2009) The presence of both bone sialoprotein-binding protein gene and collagen adhesin gene as a typical virulence trait of the major epidemic cluster in isolates from orthopedic implant infections. *Biomaterials*. 30.6621–8. doi: 10.1016/j.biomaterials.2009.08.032

Chafer-Pericas, Consuelo, Maquieira, Angel, Puchades, Rosa (2010) Fast Screening Methods to Detect Antibiotic Residues in Food Samples. *TrAC Trends in Analytical Chemistry*. 29.1038–1049. doi: 10.1016/J.TRAC.2010.06.004

Chaignon, P, Sadovskaya, Irina, Ragunah, Ch, Kaplan, Jeffrey, Jabbouri, Said (2007) Susceptibility of staphylococcal biofilms to enzymatic treatments depends on their chemical composition. *Applied Microbiol and biotechnology*. 75.125–32: doi: 10.1007/s00253-006-0790-y

Clarke, Simon, Andre, Guillaume, Walsh, Evelyn, Dufrêne, Yves, Foster, Timothy, Foster, Simon (2009) Iron-Regulated Surface Determinant Protein A Mediates Adhesion of Staphylococcus aureus to Human Corneocyte Envelope Proteins. *Infection and immunity*. 77.2408–16. doi: 10.1128/IAI.01304-08

Clarke, Simon, Brummell, Kirsten, Horsburgh, Malcolm, McDowell, Philip, Mohamad, Sharifah, Stapleton, Melanie, Acevedo, Jorge, Read, Robert, Day, Nicholas, Peacock, Sharon, Mond, James, Kokai-Kun, John, Foster, Simon (2006) Identification of In Vivo – Expressed Antigens of Staphylococcus aureus and Their Use in Vaccinations for Protection against Nasal Carriage. *The Journal of infectious diseases*. 193.1098–108. doi: https://doi.org/10.1086/501471

Corrigan, Rebecca, Rigby, David, Handley, Phoebe, Foster, Timothy (2007) The role of Staphylococcus aureus surface protein SasG in adherence and biofilm formation. *Microbiol (Reading, England)* 153.2435–46. doi: 10.1099/mic.0.2007/006676-0

Cramton, Sarah, Gerke, Christiane, Schnell, Nikki, Nichols, Wright, Götz, Friedrich (1999) The Intercellular Adhesion (ica) Locus Is Present in Staphylococcus aureus and Is Required for Biofilm Formation. *Infection and immunity.* 67.5427–33. doi: 10.1128/ IAI.67.10.5427-5433.1999

Creech, C.B., Frenck, R.W., Fiquet, A., Feldman, R., Kankam, M.K., Pathirana, S., *et al.* (2020) Persistence of Immune Responses Through 36 Months in Healthy Adults After Vaccination With a Novel Staphylococcus aureus 4-Antigen Vaccine (SA4Ag) 7OFID. doi: 10.1093/ofid/ofz532

Cucarella, Carme, Solano, Cristina, Valle, Jaione, Amorena, Beatriz, Lasa, Iñigo, Penadés, José (2001) Bap, a Staphylococcus aureus Surface Protein Involved in Biofilm Formation. *Journal of bacteriology.* 183.2888–96. doi: https://doi.org/10.1128/ JB.183.9.2888-2896.2001

Cuny, Christiane, Layer, Franziska, Strommenger, Birgit, Witte, Wolfgang (2011) Rare Occurrence of Methicillin-Resistant Staphylococcus aureus CC130 with a Novel mecA Homologue in Humans in Germany. *PloS one.* 6. e24360. doi: 10.1371/journal. pone.0024360

Daniels, Ruth, Vanderleyden, Jos, Michiels, Jan (2004) Quorum sensing and swarming in bacteria. FEMS Microbiol Rev. *FEMS Microbiol reviews.* 28.261–89. doi: 10.1016/j. femsre.2003.09.004

Davey, Mary, O'toole, George (2001) Microbial Biofilms: from Ecology to Molecular Genetics. *Microbiol and molecular biology reviews*: MMBR. 64.847–67. doi: 10.1128/ MMBR.64.4.847-867.2000

Davies, D.D.G., Parsek, Matthew, Pearson, James, Iglewski, Barbara, Costerton, J.W., Greenberg, E.P. (1998) The Involvement of Cell-to-Cell Signals in the Development of a Bacterial Biofilm. Science (New York, N.Y.) 280.295–8. doi: 10.1126/ science.280.5361.295

Deplano, Ariane, Vandendriessche, Stien, Nonhoff, Claire, Denis, Olivier (2014) Genetic diversity among methicillin-resistant Staphylococcus aureus isolates carrying the mecC gene in Belgium. *The Journal of antimicrobial chemotherapy.* 69. doi: 10.1093/ jac/dku020

Donlan, Rodney (2002) Biofilms: Microbial Life on Surfaces. *Emerging infectious diseases.* 8. 881–90. doi: 10.3201/eid0809.020063

Dufour, Delphine, Leung, Vincent, Levesque, Celine (2010) Bacterial biofilm: structure, function, and antimicrobial resistance. *Endodontic Topics.* 22. doi: 10.1111/j.1601-1546.2012.00277.x

Dufrêne, Yves, Boonaert, Christophe, Rouxhet, Paul (1996) Adhesion of Azospirillum brasilense: Role of proteins at the cell-support interface. Colloids and Surfaces B: *Biointerfaces.* 7. 113–128. doi: 10.1016/0927-7765(96)01288-X

Elias, Peretz, Sivan, Banin, Ehud (2012) Multi-species biofilms: Living with friendly neighbors. *FEMS Microbiol reviews.* 36. doi: 10.1111/j.1574-6976.2012.00325.x

Escher A, Characklis WG (1990) Modeling the initial events in biofilm accumulation. In Characklis WG, Marshall KC (Eds.), *Biofilms.* John Wiley and Sons, 445–486.

Fahimeh Nourbakhsh & Namvar, Amirmorteza (2016) Detection of genes involved in biofilm formation in Staphylococcus aureus isolates. *GMS hygiene and infection control.* 11. Doc07. doi: 10.3205/dgkh000267

Fitzpatrick, Fidelma, Humphreys, H, O'Gara, James (2005) The genetics of staphylococcal biofilm formation-will a greater understanding of pathogenesis lead to better management of device-related infection? Clin Microbiol Infect. *Clinical Microbiol and infection: the official publication of the European Society of Clinical Microbiol and Infectious Diseases.* 11.967–73. doi: 10.1111/j.1469-0691.2005.01274.x

Flemming, Hans-Curt, Neu, Thomas, Wozniak, Daniel (2007) The EPS matrix: The "House of Biofilm cells". *Journal of bacteriology.* 189.7945–7. doi: 10.1128/JB.00858-07

Foster, Timothy, Geoghegan, Joan, Ganesh, Vannakambadi, Höök, Magnus (2014) Adhesion, invasion and evasion: The many functions of the surface proteins of Staphylococcus aureus. *Nat Rev Microbiol.* 12.49–62. doi: 10.1038/nrmicro3161

Foster, Timothy, Höök, Magnus (1998) Foster, T.J., Hook, M. Surface protein adhesins of Staphylococcus aureus. *Trends Microbiol.* 6:484–488. doi: 10.1016/S0966-842X(98)01400-0

Frieri, M, Kumar, K, Boutin, A (2017) Antibiotic resistance. *J Infect Public Health.* 10:369–378. doi:10.1016/j.jiph.2016.08.007

Gardete, Susana, Tomasz, Alexander (2014) Mechanisms of vancomycin resistance in *Staphylococcus aureus. Clin Invest.* 124(7):2836–2840. doi:10.1172/JCI68834

Geoghegan, Joan, Corrigan, Rebecca, Gruszka, Dominika, Speziale, Pietro, O'Gara, James, Potts, Jennifer, Foster, Timothy (2010) Role of Surface Protein SasG in Biofilm Formation by Staphylococcus aureus. *Journal of bacteriology.* 192. 5663–73. doi: 10.1128/JB.00628-10

Geoghegan, Joan, Monk, Ian, O'Gara, James, Foster, Timothy (2013) Subdomains N2N3 of Fibronectin Binding Protein A Mediate Staphylococcus aureus Biofilm Formation and Adherence to Fibrinogen Using Distinct Mechanisms. *Journal of bacteriology.* 195. doi: 10.1128/JB.02128-12

Ghorbani-Dalini, Sadegh, Kargar, Mohammad, Doosti, Abbas, Abbasi, Pejman, Sarshar, Meysam (2015) Molecular Epidemiology of ESBL Genes and Multi-Drug Resistance in Diarrheagenic Escherichia Coli Strains Isolated from Adults in Iran. *Iranian journal of pharmaceutical research: IJPR.* 14.

Giersing, B.K., Dastgheyb, S.S., Modjarrad, K., Moorthy, V (2016) Status of vaccine research and development of vaccines for Staphylococcus aureus. *Vaccine.* 34,2962–2966. doi: 10.1016/j.vaccine.2016.03.110

Götz, Friedrich (2002) Staphylococcus and biofilms. *Molecular Microbiol.* 43. 1367–78. doi: 10.1046/j.1365-2958.2002.02827.x

Goudarzi, Mehdi, Mohammadi, Anis, Amirpour, Anahita, Fazeli, Maryam, Nasiri, Mohammad Javad, Hashemi, Ali, Goudarzi, Hossein (2019) Genetic diversity and biofilm formation analysis of Staphylococcus aureus causing urinary tract infections in Tehran, Iran. *The Journal of Infection in Developing Countries.* 13. 777–785. doi: 10.3855/jidc.11329

Gristina, A, Sherk, HH (2004) Biomaterial-centered infection – microbial adhesion versus tissue integration – (Reprinted from Science, vol. 237, pg. 1588–1595, 1987) *Clinical Orthopaedics and Related Research.* 4–12. doi: 10.1097/01.blo.0000145156.89115.12

Gross, Matthias, Cramton, Sarah, Götz, Friedrich, Peschel, Andreas (2001) Key Role of Teichoic Acid Net Charge in Staphylococcus aureus Colonization of Artificial Surfaces. *Infection and immunity.* 69. 3423–6. doi: 10.1128/IAI.69.5.3423-3426.2001

Hall-Stoodley, Luanne, Costerton, J, Stoodley, Paul (2004) Bacterial Biofilms: From the Natural Environment to Infectious Diseases. *Nat Rev Microbiol.* 2. 95–108. doi: 10.1038/nrmicro821

Hassett D, Ma J, Elkins J, Mcdermott T, Ochsner, U *et al.* (2000) Quorum sensing in pseudomonas aeruginosa controls expression of catalase and superoxide dismutase genes and mediates biofilm susceptibility to hydrogen peroxide. *Molecular Microbiol.* 34:1082–1093.

Heilmann, C., Hussain, M., Peters, G., Götz, F.(1997) Evidence for autolysin-mediated primary attachment of Staphylococcus epidermidis to a polystyrene surface. *Mol Microbiol.* 24:1013–1024. doi: 10.1046/j.1365-2958.1997.4101774.x

Hogan, S., O'Gara, J.P., O'Neill, E. (2008) Novel treatment of Staphylococcus aureus device related infections using fibrinolytic agents. *Antimicrob Agents Chemother.* 62:e2008–17. Doi: https://doi.org/10.1128/AAC.02008-17.

HΦiby, N., Bjarnsholt, T., Givskov, M., Molin, S.R., Ciofu, O. (2010) Antibiotic resistance of bacterial biofilms. *Int J Antimicrob Agents*. 35:322–332. Doi: https://doi.org/10.1016/j. ijantimicag.2009.12.011.

Huber, Birgit, Riedel, Kathrin, Hentzer, Morten, Heydorn, Arne, Gotschlich, Astrid, Givskov, Michael, Molin, Soeren, Eberl, Leo. (2001) The cep quorum-sensing system of Burkholderia cepacia H111 controls biofilm formation and swarming motility. *Microbiol (Reading, England)* 147.2517–28. Doi: 10.1099/00221287-147-9-2517

Huber, Birgit, Riedel, Kathrin, Köthe, Manuela, Givskov, Michael, Molin, Soeren, Eberl, Leo. (2002) Genetic analysis of functions involved in the late stages of biofilm development in Burkholderia cepacia H111. *Molecular Microbiol*. 46.411–26. Doi: 10.1046/j.1365–2958.2002.03182.x

Huseby, Medora, Kruse, Andrew, Digre, Jeff, Kohler, Petra, Vocke, Jillian, Mann, Ethan, Bayles, Kenneth, Bohach, Gregory, Schlievert, Patrick, Ohlendorf, Doug, Earhart, Cathleen. (2010) Beta toxin catalyzes formation of nucleoprotein matrix in staphylococcal biofilms. *Proceedings of the National Academy of Sciences of the United States of America*. 107.14407–12. Doi: 10.1073/pnas.0911032107

Ito, T, Katayama, Yuki, Hiramatsu, K. (1999) Cloning and Nucleotide Sequence Determination of the Entire mec DNA of Pre-Methicillin-Resistant Staphylococcus aureus N315. *Antimicrobial agents and chemotherapy*. 43.1449–58. Doi: 10.1128/AAC.43.6.1449

Ivanova, A., Ivanova, K., Hoyo, J., Heinze, T., Sanchez-Gomez, S., Tzanov, T. (2018) Layer-by-layer decorated nanoparticles with tunable antibacterial and antibiofilmproperties against both Gram-positive and Gramnegative bacteria. ACS Appl. *Mater. Interfaces*. 10:3314–3323. Doi: https://doi.org/10.1021/acsami.7b16508

Izano, Era, Amarante, Matthew, Kher, William, Kaplan, Jeffrey. (2008) Differential Roles of Poly-N-Acetylglucosamine Surface Polysaccharide and Extracellular DNA in Staphylococcus aureus and Staphylococcus epidermidis Biofilms. *Applied and environmental Microbiol*. 74.470–6. Doi: 10.1128/AEM.02073–07

Jahanshahi, Aidin, Zeighami, Habib, Haghi, Fakhri. (2018) Molecular Characterization of Methicillin and Vancomycin Resistant Staphylococcus aureus Strains Isolated from Hospitalized Patients. *Microbial Drug Resistance*. 24. Doi: 10.1089/mdr.2018.0069

Jarajreh, Dua, Aqel, Amin, Alzoubi, Hamed, Al-Zereini, Wael. (2017) Prevalence of inducible clindamycin resistance in methicillin-resistant Staphylococcus aureus: *The first study in Jordan*. JIDC. 11.350. Doi: 10.3855/jidc.8316

Jeffrey Kavanaugh, Flack, Caralyn, Lister, Jessica, Ricker, Erica, Ibberson, Carolyn, Jenul, Christian, Moormeier, Derek, Delmain, Elizabeth, Bayles, Kenneth, Horswill, Alexander. (2019) Identification of Extracellular DNA-Binding Proteins in the Biofilm Matrix. *mBio*. 10. Doi: 10.1128/mBio.01137–19

Jolivet-Gougeon, Anne, Bonnaure-Mallet, Martine. (2014) Biofilms as a mechanism of bacterial resistance. *Drug Discovery Today: Technologies*. 11. Doi: 10.1016/j. ddtec.2014.02.003

Kaplan, J.B., Velliyagounder, K., Ragunath, C., Rohde, H., Mack, D., Knobloch, J.K., *et al*., 2004. Genes involved in the synthesis and degradation of matrix polysaccharide inActinobacillus actinomycetemcomitans andActinobacillus pleuropneumoniae biofilms. *J. Bacteriol*. 186:8213–8220. Doi: https://doi.org/10.1128/JB.186.24.8213-8220.2004.

Kargar, M, Ghorbani-Dalini, S, Doosti, Abbas, Souod, Negar. (2011) Real-time PCR for Helicobacter pylori quantification and detection of clarithromycin resistance in gastric tissue from patients with gastrointestinal disorders. *Research in Microbiol*. 163.109–13. Doi: 10.1016/j.resmic.2011.11.005

Kargar, M., Moein Jahromi, F., Doosti, A., Mohammadalipour, Z., Lorzadeh, Sh. (2014) Resistance to different generations of quinolones in Streptococcus pneumoniae strainsisolated from hospitals in Shiraz. Comp. *Clin Pathol*. 24. Doi: https://doi.org/10.1007/s00580-014-1937-3.

Kiedrowski, Megan, Crosby, Heidi, Hernandez, Frank, Malone, Cheryl, McNamara II, James, Horswill, Alexander. (2014) Staphylococcus aureus Nuc2 Is a Functional, Surface-Attached Extracellular Nuclease. *PloS one*. 9. e95574. Doi: 10.1371/journal. pone.0095574

Kiedrowski, Megan, Kavanaugh, Jeffrey, Malone, Cheryl, Mootz, Joe, Voyich, Jovanka, Smeltzer, Mark, Bayles, Kenneth, Horswill, Alexander. (2011) Nuclease Modulates Biofilm Formation in Community-Associated Methicillin-Resistant Staphylococcus aureus. *PloS one*. 6. e26714. Doi: 10.1371/journal.pone.0026714

Kong, Kok-Fai, Vuong, Cuong, Otto, Michael. (2006) Staphylococcus quorum sensing in biofilm formation and infection. *International journal of medical Microbiol*: IJMM. 296.133–9. Doi: 10.1016/j.ijmm.2006.01.042

Lauderdale, Katherine, Boles, Blaise, Cheung, Ambrose, Horswill, Alexander. (2009) Interconnections between Sigma B, agr, and Proteolytic Activity in Staphylococcus aureus Biofilm Maturation. *Infection and immunity*. 77.1623–35. Doi: 10.1128/IAI.01036–08

Lazăr, Veronica, Chifiriuc, Mariana (2010) Architecture and physiology of microbial biofilms. *Roum Arch Microbiol Immunol*. 69:95–107.

Lebeaux, David, Ghigo, Jean-Marc, Beloin, Christophe. (2014) Biofilm-Related Infections: Bridging the Gap between Clinical Management and Fundamental Aspects of Recalcitrance toward Antibiotics. *Microbiol and molecular biology reviews*: MMBR. 78.510–543. Doi: 10.1128/MMBR.00013–14

Levy, Jack, Licini, Laurent, Haelterman, Edwige, Morris, Philip, Lestrate, Pascal, Damaso, Silvia, Belle, Pascale, Boutriau, Dominique. (2015) Safety and immunogenicity of an investigational 4-component Staphylococcus aureus vaccine with or without AS03 B adjuvant: Results of a randomized phase I trial. *Human vaccines, immunotherapeutics*. 11. Doi: 10.1080/21645515.2015.1011021

Lister, Jessica, Horswill, Alexander. (2014) Staphylococcus aureus biofilms: Recent developments in biofilm dispersal. *Frontiers in cellular and infection Microbiol*. 4.178. Doi: 10.3389/fcimb.2014.00178

Loeb, George, Neihof, Rex. (1975) Marine Conditioning Films. Doi: 10.1021/ba-1975–0145. ch016

Lowder, Bethan, Guinane, Caitriona, Ben Zakour, Nouri, Weinert, Lucy, Conway Morris, Andrew, Cartwright, Robyn, Simpson, A, Rambaut, Andrew, Nübel, Ulrich, Fitzgerald, J. (2009) Recent human-to-poultry host jump, adaptation, and pandemic spread of Staphylococcus aureus. *Proceedings of the National Academy of Sciences of the United States of America*. 106.19545–50. Doi: 10.1073/pnas.0909285106

Madsen, Jonas, Burmølle, Mette, Hansen, Lars, Sørensen, Søren. (2012) The interconnection between biofilm formation and horizontal gene transfer. *FEMS immunology and medical Microbiol*. 65.183–95. Doi: 10.1111/j.1574–695X.2012.00960.x

Mahmoudi, Hassan, Chiniforush, Nasim, Soltanian, Ali-Reza, Alikhani, Mohammad, Bahador, Abbas. (2019) Biofilm formation and antibiotic resistance in meticillin-resistant and meticillin-sensitive Staphylococcus aureus isolated from burns. *Journal of Wound Care*. 28.66–73. Doi: 10.12968/jowc.2019.28.2.66

Maira-Litrán, Tomás, Kropec, Andrea, Abeygunawardana, Chitrananda, Joyce, Joseph, Mark, George, Goldmann, Donald, Pier, Gerald. (2002) Immunochemical Properties of the Staphylococcal Poly-N-Acetylglucosamine Surface Polysaccharide. *Infection and immunity*. 70.4433–40. Doi: 10.1128/IAI.70.8.4433–4440.2002

Martí, Miguel, Trotonda, María, Tormo-Mas, M Angeles, Vergara-Irigaray, Marta, Cheung, Ambrose, Lasa, Iñigo, Penadés, José. (2009) Extracellular proteases inhibit protein-dependent biofilm formation in Staphylococcus aureus. *Microbes and infection/Institut Pasteur*. 12.55–64. Doi: 10.1016/j.micinf.2009.10.005

McCourt, Jennifer, O'Halloran, Dara, McCarthy, Hannah, O'Gara, James, Geoghegan, Joan. (2014) Fibronectin binding proteins are required for biofilm formation by community-associated methicillin resistant Staphylococcus aureus strain LAC. *FEMS Microbiol Letters*. 353. Doi: 10.1111/1574–6968.12424

McGavin, Martin, Zahradka, C, Rice, K., Scott, J. (1997) Modification of the Staphylococcus aureus fibronectin binding phenotype by V8 protease. Infection and immunity. 65.2621–8. Doi: 10.1128/IAI.65.7.2621-2628.1997

McGuinness, Will, Malachowa, Natalia, DeLeo, Frank. (2017) Vancomycin Resistance in Staphylococcus aureus. *The Yale Journal of Biology and Medicine*. 90.269–281.

McKenney, D., Pouliot, K.L., Wang, Y., Murthy, V., Döring, Ulrich M., *et al.* (1999) Broadly protective vaccine for Staphylococcus aureus based on an in vivo-expressed antigen. *Science*. 284:1523–1527. Doi: https://doi.org/10.1126/science.284.5419.1523.

Merino, Nekane, Toledo-Arana, Alejandro, Vergara-Irigaray, Marta, Valle, Jaione, Solano, Cristina, Calvo, Enrique, Lopez, Juan, Foster, Timothy, Penadés, José, Lasa, Iñigo. (2009) Protein A-Mediated Multicellular Behavior in Staphylococcus aureus. *Journal of bacteriology*. 191.832–43. Doi: 10.1128/JB.01222–08

Merritt, Justin, Qi, Fengxia, Goodman, Steven, Anderson, Maxwell, Shi, Wenyuan. (2003) Mutation of luxS Affects Biofilm Formation in Streptococcus mutans. *Infection and immunity*. 71.1972–9. Doi: 10.1128/IAI.71.4.1972–1979.2003.

Minyoung, Kim, Zhao, Aishan, Wang, Ashley, Brown, Zechariah, Muir, Tom, Stone, Howard, Bassler, Bonnie. (2017) Surface-Attached Molecules Control Staphylococcus aureus Quorum Sensing and Biofilm Development. *Nature Microbiol*. 2. Doi: 10.1038/nmicrobiol.2017.80

Mirzaee, Mohsen, Peerayeh, Shahin, Behmanesh, Mehrdad, Moghadam, Mahdi. (2015) Relationship Between Adhesin Genes and Biofilm Formation in Vancomycin-Intermediate Staphylococcus aureus Clinical Isolates. *Current Microbiol*. 70. Doi: 10.1007/s00284-014-0771-9

Moormeier, Derek, Bose, Jeffrey, Horswill, Alexander, Bayles, Kenneth. (2014) Temporal and Stochastic Control of Staphylococcus aureus Biofilm Development. *mBio*. 5. Doi: 10.1128/mBio.01341–14

Morgan, Brown, Kwiecinski, Jakub, Cruz, Luis, Shahbandi, Ali, Todd, Daniel, Cech, Nadja, Horswill, Alexander. (2020) Novel peptide from commensal Staphylococcus simulans blocks MRSA quorum sensing and protects host skin from damage. *Antimicrobial Agents and Chemotherapy*. 64. Doi: 10.1128/AAC.00172–20

Motaharesadat, Hosseini, Shapouri-Moghaddam, Abbas, Derakhshan, Solmaz, Hashemipour, Seyed, haddadi fishani, Mehdi, Pirouzi, Aliyar. (2020) Correlation Between Biofilm Formation and Antibiotic Resistance in MRSA and MSSA Isolated from Clinical Samples in Iran: A Systematic Review and Meta-Analysis. *Microbial Drug Resistance*. 26. Doi: 10.1089/mdr.2020.0001

Neu, Thomas, Marshall, Kevin. (1990) Bacterial Polymers: Physicochemical Aspects of Their Interactions at Interfaces. *Journal of biomaterials applications*. 5.107–33. Doi: 10.1177/088532829000500203

Ng, L.K., Martin, Irene, Alfa, Michelle, Mulvey, Michael. (2001) Multiplex PCR for the detection of tetracycline resistant genes. *Molecular and cellular probes*. 15.209–15. Doi: 10.1006/mcpr.2001.0363

O'Brien, Louise, Kerrigan, Steve, Kaw, Gideon, Hogan, Michael, Penadés, José, Litt, David, Fitzgerald, Des, Foster, Timothy, Cox, Dermot. (2002a) Multiple mechanisms for the activation of human platelet aggregation by Staphylococcus aureus: Roles for the clumping factors CLFA and CLFB, the serine-aspartate repeat protein sdre and protein A. *Molecular Microbiol*. 44.1033–44. Doi: 10.1046/j.1365–2958.2002.02935.x

O'Brien, Louise, Walsh, Evelyn, Massey, Ruth, Peacock, Sharon, Foster, Timothy. (2002b) Staphylococcus aureus clumping factor B (ClfB) promotes adherence to human type I cytokeratin 10: Implications for nasal colonization. *Cellular Microbiol.* 4.759–70. Doi: 10.1046/j.1462–5822.2002.00231.x

O'Gara, James. (2007) Ica and beyond: Biofilm mechanisms and regulation in Staphylococcus epidermidis and Staphylococcus aureus. *FEMS Microbiol letters.* 270.179–88. Doi: 10.1111/j.1574–6968.2007.00688.x

Olson, Michael, Nygaard, Tyler, Ackermann, Laynez, Watkins, Robert, Zurek, Oliwia, Pallister, Kyler, Griffith, Shannon, Kiedrowski, Megan, Flack, Caralyn, Kavanaugh, Jeffrey, Kreiswirth, Barry, Horswill, Alexander, Voyich, Jovanka. (2013) Staphylococcus aureus Nuclease Is an SaeRS-Dependent Virulence Factor. *Infection and immunity.* 81. Doi: 10.1128/IAI.01242–12

O'Neill, Eoghan, Pozzi, Clarissa, Houston, Patrick, Humphreys, Hilary, Robinson, D, Loughman, Anthony, Foster, Timothy, O'Gara, James. (2008) A Novel Staphylococcus aureus Biofilm Phenotype Mediated by the Fibronectin-Binding Proteins, FnBPA and FnBPB. *Journal of bacteriology.* 190.3835–50. Doi: 10.1128/JB.00167–08.

O'Rourke, John, Daly, Seth, Triplett, Kathleen, Peabody, David, Chackerian, Bryce, Hall, Pamela. (2014) Development of a Mimotope Vaccine Targeting the Staphylococcus aureus Quorum Sensing Pathway. *PloS one.* 9. e111198. Doi: 10.1371/journal.pone.0111198

O'Toole, George, Kaplan, Heidi, Kolter, Roberto. (2000) Biofilm Formation as Microbial Development. *Annual review of Microbiol.* 54.49–79. Doi: 10.1146/annurev.micro.54.1.49

Panda, S.K., Das, R., Lavigne, R., Luyten, W. (2020) Indian medicinalplant extracts tocontrol multidrug-resistant S. aureus, including in biofilms. S. Afr. J. Bot. 128,283–291. https://doi.org/10.1016/j.sajb.2019.11.019.

Parastan, R., Kargar, M., Solhjoo, k, Kafilzadeh, F. (2020) A synergistic association betweenadhesion-related genes and multidrug resistance patterns of Staphylococcusaureus isolates from different patients and healthy individuals. JGAR. https://doi.org/10.1016/j.jgar.2020.02.025. In Press.

Parsek, M.R., Greenberg, E.P., (2005) Sociomicrobiology: the connections between quorum sensing and biofilms. *Trends Microbiol.* 13:27–33. https://doi.org/10.1016/j.tim.2004.11.007

Pastoriza, L., Cabo, M.L., Berna´ rdez, M., Sampedro, G., Herrera, J.R (2002) Combined effects of modified atmosphere packaging and lauricacid on the stability of pre-cooked fish products during refrigerated storage. Eur. Food Res. Technol. 215:189–193. https://dx.doi.org/10.1007%2Fs00294-015-0527-5

Payne, D.E., Boles, B.R., (2016) Emerging interactions between matrix components during biofilm development. *Curr Genet.* 62:137–141. https://dx.doi.org/10.1007%2Fs00294-015-0527-5

Periasamy, S., Chatterjee, S.S., Cheung, G.Y., Otto, M., (2012) Phenol-soluble modulins in staphylococci: what arethey originally for? *Commun Integr Biol.* 5:275–277. https://doi.org/10.4161/cib.19420

Petersen A, Stegger M, Heltberg O, Christensen J, Zeuthen A, Knudsen LK, (2013) Epidemiology of methicillin-resistant *Staphylococcus aureus* carrying the novel mecC gene in Denmark corroborates a zoonotic reservoir with transmissionto humans. Clin Microbiol Infect Official Publ Eur Soc *Clin Microbiol Infect Dis*.19:E16e22. https://doi.org/10.1111/1469-0691.12036

Philippe Herman-Bausier, Sofiane El-Kirat-Chatel, Timothy J Foster, Joan A Geoghegan, Yves F Dufrêne (2015) *Staphylococcus aureus* Fibronectin-Binding Protein A Mediates Cell-Cell Adhesion through Low-Affinity Homophilic Bonds.6(3):e00413–15. doi: 10.1128/mBio.00413-15.

Poulsen, L.K., Ballard, G., Stahl, D.A., (1993) Use of rRNA fluorescence in situ hybridizationfor measuring the activity of single cells in young and established biofilms. Appl. Environ. Microbiol. 59:1354–1360.

Pratten, J., S. J. Foster, P. F. Chan, M. Wilson, and S. P. Nair. (2001) Staphylococcus aureus accessory regulators: expression within biofilms and effect on adhesion. Microbes Infect. 3:633–637. DOI: 10.1016/s1286–4579(01)01418–6

Qi Lin, Honghu Sun,*, Kai Yao, Jiong Cai, Yao Ren and Yuanlong Chi (2019) The Prevalence, Antibiotic Resistance and BiofilmFormation of Staphylococcus aureus in BulkReady-To-Eat Foods. DOI: 10.3390/biom9100524

Renner, L.D., Weibel, D.B. (2011) Physicochemical regulation of biofilm formation. *MRS Bulletin*. 36:347–355.

Resch A, Rosenstein R, Nerz C, Götz F. (2005) Differential geneexpression profiling of Staphylococcus aureus cultivated underbiofilm and planktonic conditions. *Appl Environ Microbiol*. 71(5):2663–76. DOI: 10.1128/AEM.71.5.2663–2676.2005

Roche, F.M., Meehan, M., Foster, T.J. (2003) The Staphylococcus aureus surface protein SasG and its homologues promote bacterial adherence to human desquamated nasal epithelial cells. *Microbiol*. 149:2759–2767. DOI: 10.1099/mic.0.26412–0

Römling U, Balsalobre C. (2012) Biofilm infections, their resilience to therapy and innovative treatment strategies. *J Intern Med*. 2012. 272(6):541–61. DOI: 10.1111/joim.12004

Rosenbach FJ (1884) *Microorganisms in the wound infections diseases of man*. J.F. Bergmann, 18.

Rosenthal CB, Mootz JM, Horswill AR (2014) Staphylococcus aureus biofilm formation and inhibition. In Rumbagh KP, Ahmad I (Eds.), *Antibiofilm agents from diagnosis to treatment*. E-Publishing Inc, 233–260.

Sauer, K., Camper, A. K., Ehrlich, G. D., Costerton, J. W., Davies, D. G. (2002) *Pseudomonas aeruginosa* displays multiple phenotypes during development asa biofilm. *Bacteriology*. 184(4):1140e1154. DOI: 10.1128/jb.184.4.1140–1154.2002

Schneider, R.P. and Marshall, K.C. (1994) Retention of theGram-negative marine bacterium SW8 on surfaces ^ e¡ects ofmicrobial physiology, substratum nature and conditioning films. *Colloids Surfaces B Biointerfaces*. 2(4):387–396. https://doi.org/10.1016/0927-7765(94)80002-2

Schroeder, K., Jularic, M., Horsburgh, S.M., Hirschhausen, N., Neumann, C., Bertling, A., Schulte, A., Foster, S., Kehrel, B.E., Peters, G., Heilmann, C., (2009) Molecular characterization of a novel Staphylococcus aureus surface protein (SasC) involved in cell aggregation and biofilm accumulation. *PLoS One*. 4:e7567. https://doi.org/10.1371/journal.pone.0007567

Schwartz, K., Ganesan, M., Payne, D.E., Solomon, M.J., Boles, B.R., (2016) Extracellular DNA facilitates the for¬mation of functional amyloids in Staphylococcus aureus biofilms. Mol. Microbiol. 99:123–134. DOI: 10.1111/mmi.13219

Shannon, Oonagh, Flock, Jan-Ingmar. (2004) Extracellular fibrinogen binding protein, Efb, from Staphylococcus aureus binds to platelets and inhibits platelet aggregation. Thrombosis and haemostasis. 91.779–89. DOI:10.1160/TH03–05–0287

Shaw, L., Golenka, E., Potempa, J., Foster, S.J., (2004) The role and regulation of the extracellular proteases of Staphylococcus aureus. *Microbiology*. 150:217–228.

Srey, S., Jahid, I.K., Ha, S. (2013) Biofilm formation in food industries: a food safety concern. *Food Control*. 31:572–585.

Steidle, A., M. Allesen-Holm, K. Riedel, G. Berg, M. Givskov, S. Molin, andL. Eberl. (2002) Identification and characterization of an N-acylhomoserine lactone-dependent quorum-sensing system in Pseudomonas putida strain IsoF. Appl. Environ. Microbiol. 68:6371–6382.

Stewart, P.S. (2003) Diffusion in biofilms. J. Bacteriol. 185:1485–1491. https://doi.org/10.1128/JB.185.5.1485-1491.2003.

Tong, S.Y.C., Davis, J.S., Eichenberger, E., Holland, T.L., Fowler Jr., V.G., (2015) Staphylococcus aureus infections: epidemiology, pathophysiology, clinical manifestations, and management. *Clin Microbiol Rev.* 28:303 361. https://doi.org/10.1128/CMR.00134-14.

Tuomanen, E., Cozens, R., Tosch, W., Zak, O., Tomasz, A., (1986) The rate of killing of Escherichia coli by beta lactam antibiotics is strictly proportional to the rate of bacterial-growth. *J Gen Microbiol.* 132:1297–1304. https://doi.org/10.1099/00221287-132-5-1297

Vancraeynest D, Hermans K, Haesebrouck F. (2004) Genotypic andphenotypic screening of high and low virulence Staphylococcusaureus isolates from rabbits for biofilm formation andMSCRAMMs. Vet Microbiol.103(3–4):241–7. DOI:10.1016/j.vetmic.2004.09.002

van-Houdt, R., Michiels, C.W., (2010) Biofilm formation and the food industry, a focus on the bacterial outer sur¬face. J. Appl. Microbiol. 109:1117–1131. https://doi.org/10.1111/j.1365-2672.2010.04756.x

Van Loosdrecht, M.C.M., Lyklema, J., Norde, W. and Zehnder, A.J.B. (1990) Influences of interfaces on microbial activity. *Microbiol Rev.* 54:75–87. https://doi.org/10.1128/mr.54.1.75-87.1990

Vazquez, V., Liang, X., Horndahl, J.K., Ganesh, V.K., Smeds, E., Foster, T.J., Hook, M., (2011) Fibrinogen is a ligand for the Staphylococcus aureus microbial surface components recognizing adhesive matrix molecules (MSCRAMM) bone sialoprotein-binding protein (Bbp). *J Biol Chem.* 286:29797–29805. https://doi.org/10.1074/jbc.m110.214981

Vuong, C., Saenz, H.L., Götz, F., and Otto, M. (2000) Impact of the agr quorum-sensing system on adherence to polystyrene in Staphylococcus aureus. J. Infect. Dis. 182:1688–1693. https://doi.org/10.1086/317606

Watnick P.I., and Kolter R. (2000a) Steps in the development of a Vibrio cholerae biofilm. Mol. Microbiol.341999586–595. https://doi.org/10.1128/JB.182.10.2675-2679.2000

Watnick, P., and Kolter, R. (2000b) Biofilm, city of microbes. J. Bacteriol. 182:2675–2679. https://doi.org/10.1128/JB.182.10.2675-2679.2000

Wertheim H.F., Melles D.C, Vos M.C, van Leeuwen W., van Belkum A., Verbrugh H.A, and Nouwen J.L. 2005. The role of nasal carriage in *Staphylococcus aureus* infections. *Lancet Infect Dis.* 5:751–762. https://doi.org/10.1016/s1473-3099(05)70295-4

Wilson, LG. (1987) The early recognition of streptococci as causes of disease. *Medical History.* 31:403–414. https://doi.org/10.1017/S0025727300047268

Xu, X., Zhu, H., Lv, H. (2017) Safety of Staphylococcus aureus four-antigen and threeantigenvaccines in healthy adults: a meta-analysis of randomized controlled trials(?). Hum. Vaccin. Immunother. 14:314–321. https://doi.org/10.1080/21645515.2017.1395540.

Yarwood, J.M., Bartels, D.J., Volper, E.M., Greenberg, E.P., (2004) Quorum sensing in Staphylococcus aureus biofilms. *J. Bacteriol.* 186:1838–1850. https://doi.org/10.1128/jb.186.6.1838-1850.2004

Yousefi, N., S. Yazdansetad, A. Ardebili, M. Saki, and E. Najjari. (2018) Detection of intercellular adhesion (ica) genes involved in biofilm and slime formation in clinical isolates of Staphylococcus aureus harboringmecA gene. J BabolUniver Med Sci. 20:27–35. http://dx.doi.org/10.18869/acadpub.jbums.20.6.27

3 The Role of Quorum Sensing in Microbial Biofilm Formation

Oluwafemi Adebayo Oyewole, Ramat Onyeneoyiza Raji, and Japhet Gaius Yakubu

CONTENTS

DOI: 10.1201/9781003184942-4

3.1 INTRODUCTION

The study of microbial development revealed that microorganisms are able to interact and differentiate in a complex way. Biofilm, simply explained as the formation of microbial communities that attach to surfaces and are embedded in a self-produced extracellular matrix has been studied to be an excellent model system for the study of microbial development (Solano *et al.*, 2014). Biofilm occurs on both natural and artificial environments, and adheres strongly to surfaces with the help of exopolymeric substances, which ensures a complex interaction among microbial cells and protects them against environmental stress and antimicrobials. In addition, many bacteria are able to interact among one another and carry out different complex social behaviors (Carradori *et al.*, 2020). Biofilms are formed through a multistep process that includes cell attachment to surface, cell maturation, and dispersal. The prokaryotic genome expresses the formation of biofilm, which helps in providing a more secure environment for the microbial community. This biofilm provides a structural support and protects the microbial cells from any form of environmental stress as well as harmful effects of biocide. There are varying processes involved in the formation of biofilm, which include: organic molecule adsorption by microbial cells from the fluid phase, transport of microbial cells to the surface through chemotaxis and sedimentation, reversible followed by irreversible adhesion of cells to substratum, microbial metabolisms, growth and replication, secretion of extracellular polymetric substances (EPS), and depletion of nutrients from the environment followed by cell lysis and death (often characterized by biofilm sloughing or detachment) (Lazar, 2011).

Biofilm could be homogeneous (of a particular species) or heterogeneous (more than one species) microbial cells attached to abiotic or biotic surfaces. Nonetheless, the most predominant biofilm community in most environments are made up of more than one microbial species synergistically working together as one entity. On the contrary, biofilm made up of one type of microbial cells are often found in medical implants, where they cause infections. Although there are other single microbial species that form biofilm, however, the study of biofilm has seen *Pseudomonas aeruginosa* being a model used in understanding how other microbial biofilms form and interact with their environments. Environmental factors such as availability of nutrient velocity of bulk fluid, among many others, influences the development and quality of biofilms. Availability of nutrients ensures the continuity of microbial cells in a biofilm; otherwise, they detach themselves and assume a planktonic way of life and move on in search for greener pastures (Carlier *et al.*, 2015).

Without any external help, a planktonic cell can grow, multiply, sense, and adjust to environment changes. They can, however, intercommunicate with species of other organisms in a form of cooperation to accomplish biofilm formation, bioluminescence production, and secretion of exoenzyme, among many other cooperative activities. Microbial cells have been demonstrated to carry out a cell-to-cell mode of communication between same and different species in a mechanism known as quorum sensing (QS). QS is defined as a means of intercellular communication in microorganisms which helps regulate the expression of genes through auto-inducers (AI).

AI are extracellular molecules utilized by microorganisms in QS. This molecules are density dependent and often react when reaching a particular threshold (known as a quorum) to either suppress activate genes responsible for a particular phenotypic traits (AIs) (Carlier *et al.*, 2015; Zhang *et al.*, 2019). AIs play an important role in several physiological processes in microorganisms, which include motility, formation of biofilms, and production of antibiotics and secretion of virulence factor by bacteria (Hmelo, 2017; Zhang *et al.*, 2019). The two well reported social behaviors of microbial population are biofilms formation and QS. Some species of bacteria form surface-bound communities known as biofilms. Microbial biofilm is a community of microbial cells (predominantly bacteria) attached to a surface and embedded in matrix composed of extracellular polymeric substance produced by the cells (Lazar, 2011). These two social behaviors and their relationship have been reported by Irie and Parsek (2008), who report that biofilms formation results from gene expression and is regulated by QS.

3.2 BIOFILM FORMATION

The emerging branch of microbiology that deals with the study of the social aspects of bacterial life, which includes biofilm formation, is referred to as socio-microbiology (Parsek and Greenberg, 2005). Naturally, most bacterial cells are believe to form biofilms and these bacterial cells are usually different from their planktonic form and showcase certain differences in gene expression (Toyofuku *et al.*, 2016). Biofilm is formed through steps such as cell attachment, biofilm maturation, and biofilm dispersion (Figure 3.1).

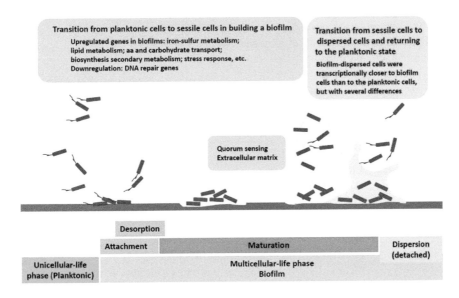

FIGURE 3.1 Stages in biofilm formation (Berlanga and Guerrero, 2016).

3.2.1 CELL ATTACHMENT

One of the most important processes in biofilm formation is surface attachment because it is a point of transition from planktonic mode to biofilm life. First, cells undergo reversible attachment to a surface (where they either attach to the surface and involve in biofilm formation or separate themselves and assume their planktonic lifestyle) using cell appendages like flagella, pili, and fimbriae. Cells undergo irreversible attachment after reversible attachment, through the extracellular secretion of surface proteins such as LapA, SadB, and EPS which helps the cells to adhere firmly to the surface. Differences in the ability to attach play a vital role in species differentiation, thus stressing how important surface attachment is in the formation of biofilm (Petrova and Sauer, 2012; Toyofuku *et al.*, 2016).

3.2.2 BIOFILM MATURATION

After bacterial cells have successfully attached firmly to a surface, they grow and undergo cell division and begin assembly to form micro-colonies. These micro-colonies continue to proliferate and produce EPS. EPS comprises lipids, nucleic acids, proteins, biopolymers, polysaccharides, and water, and is responsible for cells' surface adhesion, binding cells together, as well as maintaining the biofilm in a three-dimensional architecture. EPS also protects the bacterial cells in the biofilm against stress from the environment such as the immune response of host, metallic cations, oxidation, and effects of biocides. The EPS of biofilms also helps in housing metabolic products, extracellular enzymes, and signaling molecules (Flemming and Wingender, 2010; Drescher *et al.*, 2014).

3.2.3 BIOFILM DISPERSION

The final stage in the developmental cycle of biofilm formation in other to initiate another one is the dispersal of biofilm cells. During dispersal, cells from a matured biofilm detach and return to planktonic life, where they look for other niche surface they can colonize. Dispersal can be active or passive, and depends on two factors: degradation of EPS by waste metabolites, and physical factors (i.e. shearing force of bulk liquid medium) (Toyofuku *et al.*, 2016). Active dispersal may result from alteration of environmental conditions such as starvation, change in temperature, oxygen fluctuations, and accumulation of metabolic waste (Hong *et al.*, 2010). There are enzymes that can also cause degradation of biofilms such as alginate lyase in *P. aeruginosa*, which helps in the detachment of *P. aeruginosa* from an EPS matrix (Ansari *et al.*, 2017).

3.3 FACTORS THAT INFLUENCE BIOFILM FORMATION

The social aspect of microbial life is being affected largely by the environment. Environmental factors determine whether cells form biofilm or are dispersed and leave a biofilm (Ansari *et al.*, 2017). Other factors such as effects of types of substratum, hydrodynamics, conditioning films forming on the substratum, properties of the cell, microbial products, production of extracellular polymeric substances, and extra-cellular DNA are known to influence the processes of biofilm formation from surface colonization to growth, survival, and activities of microorganisms in the biofilm.

3.3.1　SUBSTRATUM EFFECTS

Researchers have reported that the physicochemical parameters of surfaces such as roughness, polarity, and hydrophobicity of surfaces play vital roles in the rate and depth of microbial attachment to surfaces. The moment any microbial cell is attached to a surface, the chemistry on the interface of attachment changes. Thus, influencing adhesion of cell and may confer cell survival. Colonization of sur-faces by microorganisms is influenced by roughness of surface; that is, the depth of microbial colonization increases as surface roughness increases. This is due to the fact that increases of roughness on surfaces result in increases in the surface area available for microbial action and a decrease in shear force. Again, micro-organisms adhere more rapidly to nonpolar and hydrophobic surfaces like plastic and Teflon as compared to glass, metals, and other hydrophilic materials (Donlan, 2002; Choudhary *et al.*, 2020). Microorganisms that attach to metal surfaces mainly depend on some parameters such as the growth medium, characteristic of substra-tum, and cell surface. Some of those metals include stainless steel, aluminum, and copper. However, some metals are toxic to bacteria, such as lead acetate, to which the high lead content in disinfectants and antiseptics is attributed (Donlan, 2002; Ansari *et al.*, 2017).

3.3.2　HYDRODYNAMICS

Hydrodynamic layers are affected by the type of interaction between microbial cells and the substratum. The zone found adjacent to the liquid/substratum interface where the velocity flow of liquid is negligible is known as the hydrodynamic boundary layer. The thickness of this layer is inversely proportional to the linear velocity. The characteristic velocity of the liquid controls the interaction between the submerged surface and the microbial cells. As such, at a very low linear velocity, microbial cells need to maneuver pass the hydrodynamic layer. The cell sizes of microbial cells and possession of motility structures help in controlling the interaction with the surface layer. An increase in velocity leads to a decrease in the hydrodynamic layer, expos-ing the cells to increasingly greater mixing and turbulence. Therefore, a more rapid association between microbial cells and surfaces is ensured due to linear velocities. However, this could be hampered when the velocity of bulk liquid increases, result-ing in enough shear forces capable of causing detachment of adhered microbial cells (Donlan, 2002; Choudhary *et al.*, 2020).

3.3.3　CONDITIONING FILMS FORMING ON THE SUBSTRATUM

Exposing any material to any aqueous medium often results in surface condition-ing of that material by polymers from the liquid medium. This often results in chemical modification of that material, which affects to a great extent the depth and rate of microbial attachment to that particular surface. The thin films formed have been demonstrated to be organic in nature and form when the surface is exposed for some minutes and grows continuously for some hours. In the human host, however, the nature of film conditioning is quite different. An example is the acquired pellicle, a proteinaceous conditioning film often found on enamel of teeth. This pellicle is made up of lipids, phosphoproteins, glycoproteins, albumin, lysozyme, and crevice fluid of gingival. Pellicle-conditioned surfaces become colonized within a few hours of exposure to oral cavity bacteria. Some conditioning films produced by the host – like urine, tears, blood, saliva, respiratory secretions, and intervascular fluid – influence bacterial attachment to biomaterials (Donlan, 2002).

3.3.4 PROPERTIES OF THE CELL

Hydrophobicity of the cell surface; the presence of cell appendages like flagella, pili, and fimbria; and EPS production affect the rate and extent of microbial cells adhesion to any surface. Hydrophobicity increases with an increase in the non-polarity of one or both of the microbial cell surface and the substratum surface. Fimbriae proffer cell surface hydrophobicity because of the proportion of hydrophobic amino acid residues, and they play a role in cell surface hydrophobicity and attachment by overcoming the initial electrostatic repulsion barrier that exists between the cell and substratum. Motile strains of *P. fluorescens* have been reported to attach more and against the flow faster than non-motile strains, and non-motile strains do not recolonize vacant areas on the substratum as evenly as motile strains, resulting in slower biofilm formation (Donlan, 2002).

3.3.5 MICROBIAL PRODUCTS

Biofilm formation is known to be influenced by one or more exometabolites, especially in the dispersal of biofilm cells. Accumulation of microbial wastes, along with other factors, stimulates the active dispersal of cells. Examples of bacterial exometabolites include but are not limited to siderophores, pigments, and antibiotics. Some of these microbial exometabolites may trigger the degradation of EPS. Antibiotics produced mostly by soil microorganisms can alter the population density in the biofilm. Extracellular proteases secreted into the biofilm matrix also cause cell dispersal, through the disruption and lysis of microbial cells, thus impairing the function of biofilm. The production and secretion of molecules responsible for QS by microorganisms also affects the rate of biofilm formation; likewise surfactin (Ansari *et al.*, 2017).

3.3.6 PRODUCTION OF EXTRACELLULAR POLYMERIC SUBSTANCES

As earlier stated, EPS is made up of polysaccharides, proteins, DNA, and phospholipids. The gel-like network formed by EPS acts as cementing material in binding microbial cells together through hydrophobic interactions and actions of multivalent cations. Microorganisms are packed together and adhere to surfaces with the help of

EPS, while granulation and flocculation protects microbial cells from external conditions and allow accumulation of nutrients within the EPS matrix. Different biofilms produce different amounts of EPS (Singh *et al.*, 2020).

3.3.7 EXTRACELLULAR DNA

Extracellular DNA (eDNA) is a major component of numerous single and multispecies biofilms. It plays an important role in several steps of biofilm development, such as attachment and microcolony formation. It provides the ability to protect the biofilms from abiotic stress, antibiotics, and chemicals, and as well assists as a basis of nutrients for biofilm development (Singh *et al.*, 2020).

3.3.8 ENVIRONMENTAL FACTORS

The various environmental factors which affect the formation of biofilm on both biotic and abiotic surfaces include nutrient availability, presence of oxygen, environmental pH, temperature and moisture content, and salinity.

3.3.8.1 Nutrient Availability

Nutrient availability plays an essential role in influencing the survival, growth, and metabolisms of microorganisms in any habitat. An increase in nutrient concentration is proportional to an increase in the number of attached bacterial cells (Donlan, 2002). Nutrients such as carbon source, amount of nitrate, sucrose, phosphate, and calcium enhance biofilm formation as their concentrations increase. Bacteria in biofilms may acquire nutrients by using different enzymes to break down food supplies and concentrate trace organic substances on surfaces, thus allowing them to utilize waste products from surrounding microbes. Due to the negatively charged biofilm, most cationic nutrients drawn to the surfaces of biofilms are able to exchange ions from nutrients to microbes, thereby ensuring the availability of nutrient for microbial growth and metabolism (Prakash *et al.*, 2003; Ansari *et al.*, 2017). Reports have been made that availability of carbohydrates such as trehalose and mannose stimulate biofilm formation of *Listeria monocytogenes* (Choudhary *et al.*, 2020).

3.3.8.2 Presence of Oxygen

Biofilm does not develop in absence of sufficient oxygen, which impairs bacterial attachment to surfaces of substrate. The presence of oxygen has been reported to regulate formation of biofilm by *Escherichia coli*.

3.3.8.3 Environmental pH

The effects of environmental pH have been reported on *Vibrio cholerae*. Studies have revealed that pH between 7 and 8.2 is optimum for growth and activities of *V. cholerae*. Effects of acidic pH (≤ 5) have been demonstrated to impair the biofilm-forming ability of some bacterial cells due to loss of mobility. This is not the case with *Escherichia coli* and *Streptococci epidermidis*, which can grow and form biofilms in acidic environments (i.e. urethral catheters pH of urine is acidic) (Choudhary *et al.*, 2020).

Change in temperature drastically affects biofilm formation. Active dispersal is triggered by change in temperature and other environmental factors such as nutrient availability, oxygen deficiency, and accumulation of metabolites (McDougald *et al.*, 2012). The morphology, cell density, and thickness of biofilm is affected by temperature since it goes a long way limiting the rate of EPS production. For example, the growth of spore-forming gram-positive *Clostridium perfringens* is affected by increase in temperature, which in turn affects the thickness, cell density, and morphology of the biofilm; likewise, *L. monocytogenes*, with which biofilm formation is impaired by increases in temperature (Choudhary *et al.*, 2020). This indicates that the formation of biofilm is temperature dependent, affecting the production of EPS (Toyofuku *et al.*, 2016).

Another factor that affects biofilm formation is water availability. The availability of water in some environments, such as soil, is dependent on the features of the soil, as well as the dissolved sites. At high temperatures, organic substances tend to be dissolved in water and increase in dissolution reduces the water available for microbial activities, thus impairing metabolism and biofilm formation (Ansari et al., 2017).

3.3.8.4   Salinity

Increase in salt concentration of a medium often translates to an equal increase in osmolytes accumulation and biofilm formation. Some halotolerant bacterial strains such as *Staphylococcus saprophyticus* and *Oceanobacillus profundus* increase accumulation of endogenous osmolytes (i.e. betaine, glycine, and proline), and EPS production, with subsequent increases in the rate of biofilm formation (Ansari *et al.*, 2017).

3.3.9   MICROBIAL QUORUM SENSING

QS is a mechanism known for cell-to-cell communication among microbial species (i.e. bacteria and fungi) through the production of small signal molecules known as AIs. During growth, bacterial cell often secrete these AIs, which diffuse and accumulate in the environment. Passive diffusion is the mechanism used by microbes to transport AIs across cell membranes with the help of specific transporters and efflux pumps. The amount of AIs accumulated in the environment is being sensed by all individual cells. As such, once a certain threshold of AIs is reached, the individual cells act in synchronicity to up-regulate or down-regulate gene expression responsible for a particular trait (Bandara *et al.*, 2012; Padder *et al.*, 2018). This QS phenomenon was first studied using *Alivibrio fischeri*, a bioluminescence bacteri which controls its light illumination through an integrated system of a synthase (LuxI) and a sensor receptor (LuxR), which function in producing QS molecules (i.e. N-acyl-himoserine-lactones) (Barriuso *et al.*, 2018; Padder *et al.*, 2018). Aside from the roles played QS in signaling and channeling QS molecules, they also function in social communication among the same and different microbial species in a given environment. Examples can be seen in the chemotaxis of marine diatoms toward signals of N-acylhomoserine lactone (AHL) and *Burkholderia cepacia* gene regulation produced by another bacterium (Lewenza *et al.*, 2002; Williams, 2007).

3.4   QUORUM SENSING IN BACTERIA

Acyl homoserine lactones (AHL) signaling has been described in gram-negative bacteria species. AHL signals consist of a homoserinelactone moiety that is linked by an amide bond to an acyl side chain. AHL synthesis is primarily catalyzed by a single enzyme belonging to the LuxI family, while signals are catalyzed by a single enzyme belonging to the LuxI family and sensed by cytoplasmic DNA-binding regulatory proteins belonging to the LuxR family. The first described acylhomoserine lactone (AHL) QS system was in *Vibrio fischeri*, a marine species that displays bioluminescence in the light organs of various marine animals, like the Hawaiian bobtail squid *Euprymnas colopes*. *V. fischeri* was found to be bioluminescent at high cell densities in liquid batch culture due to an accumulation of the AHL signal 3-oxo-hexanoyl homoserine lactone (Irie and Parsek, 2008; Zang *et al.*, 2019).

3.4.1   PEPTIDE AUTO-INDUCERS

Many gram-positive species have been studied to utilize peptides for QS. Streptococcal species frequently utilize competence signal peptides (CSP) among other quorum sensing molecules (QSMs). Accumulation of CSP induces autolysis, whereby chromosomal DNA is released into the environment and thereafter is taken up by neighboring cells to promote horizontal gene transfer. CSPs and other peptide-based QSMs have also been found to regulate other group-associated behaviors such as biofilm formation and bacteriocin production in different gram-positive species (Irie and Parsek, 2008).

3.4.2   AUTO-INDUCER 2

Auto-inducer 2 (AI-2) is a QS signal produced by some gram-positive and gram-negative bacterial species. The AI-2 structures have been found in *Vibrio harveyi* and *Salmonella enterica* serovar Typhimurium. While the two structures are distinct, possibly allowing for species specificities, cross-species signaling appears to be prevalent. A key step in AI-2 synthesis is catalyzed by a highly conserved enzyme LuxS, and the LuxS gene is found in a vast number of bacterial species, making it important in regulating different functions (Irie and Parsek, 2008).

3.5   QUORUM SENSING IN FUNGI

The mechanisms of QS in fungi have been reported to regulate some processes, including – but not limited to – production of secondary metabolites, transition of morphology, sporulation, and secretion of enzymes (Barriuso *et al.*, 2018). Like in bacteria, fungi QS is also population dependent in causing virulence/pathogenesis, as well as the regulation of biofilm formation. Various intraspecific – as well as interspecific – communications of QS and their nature, diversity, and ways they act in phenotypic expression of a particular trait has been discussed (Tarkka *et al.*, 2009; Padder *et al.*, 2018). An important discovery was made over 15 years ago on the control of filamentation by farnesol in dimorphic *C. albicans*, stressing the important role of QS in fungi. Farnesol, a sesquiterpene alcohol, has been demonstrated to prevent/reduce the probability of dense culture of *C. albicans* from switching to hyphal mode from yeast. Thus, farnesol inhibits formation of hyphae in fungi (Ramage *et al.*, 2002; Shirtliff *et al.*, 2009). Aside from having harmful effects on host tissues and other microbes, farnesol also plays an important role in the physiology of *C. albicans* since it serves as signaling molecules (Albuquerque and Casadevall, 2012; Padder *et al.*, 2018). Another QS molecule in fungi is the aromatic alcohol tyrosol, which plays important role in the regulation of biofilm morphogenesis and formation, as well as growth of *C. albicans*. In *Saccharomyces cerevisiae*, tryptophol and 1-phenylethanol regulate act as QS molecules and regulate biofilm morphogenesis in nitrogen-depleted environments (Albuquerque and Casadevall, 2012). The evidence of roles of QS in fungi being density dependent has been documented, although research of QS is still at a low level among fungal species (Albuquerque and Casadevall, 2012; Wongsuk *et al.*, 2016).

3.6 MECHANISMS OF MICROBIAL QUORUM SENSING

QS mechanisms are based on microbial cell-to-cell communication mediated by self secreted small molecular compounds, known as autoinducers (AIs), which are found in gram-positive and gram-negative bacteria, as well some fungal species (Irie and Parsek, 2008).

3.6.1 MECHANISMS OF QUORUM SENSING IN BACTERIA

In bacteria, QS mechanisms usually involve the synthesis and release of enzyme-specific classes of signal molecules in their environment. At a certain threshold concentration of the signal molecules, receptor proteins recognize them either directly or indirectly and coordinate bacterial group behaviors beneficial to the entire population. In gram-negative bacteria, the commonly used auto-inducer is AHLs (N-acyl-homoserine lactones) for intra-species communication, and more than 25 different kinds of gram-negative bacteria have been reportedly regulated by AHLs (Sharma *et al.*, 2020).

The mechanism involved in the regulatory role of a typical AHLs auto-inducer requires the production of AHLs and binding of LuxI (AHLs signal synthase) to LuxR (AHLs signal receptor) regulatory proteins within the cell. AHLs produced are permeable to the cell membrane and capable of diffusing randomly through the membrane into the environment and accumulates. When a certain threshold concentration is reached, they diffuse through the cell membrane and bind to the amino terminus of LuxR receptor proteins in the cytoplasm to form the LuxI/LuxR protein complexes to coordinate the expressions of certain functional genes. These LuxI/LuxR protein complexes also have a feedback regulatory effect on the production of AHLs signal molecules and their receptor proteins. Aside AHLs, other auto-inducers with different chemical structures have been discovered in gram-negative bacteria, which includes AI-2, PQS (Pseudomonas quinolone signal), indole, pyrones, DARs (Dialkylresorcinols), and CHDs (cyclohexanediones), among others (Zang *et al.*, 2019; Sharma *et al.*, 2020).

Mechanisms of QS in gram-positive bacteria are essentially similar to that in gram-negative bacteria. The main difference is that gram-positive bacteria utilize auto-inducing peptides (AIPs) as signaling molecules in exchanging information among bacterial cells. Unlike AHLs that are able to diffuse freely across cell membrane, AIPs are transported across membranes with the help of membrane proteins such as adenosine triphosphate (ATP)-binding cassette transporter (ABC transporter). The AIPs are sensed by this two-component signal transduction system (TCSTS) made up of transmembrane sensor kinase AgrC and response regulatory protein AgrA. When AIPs reach the threshold concentration, they bind to receptors on the surface of the cell, activating the two-component phospho-kinase system (TCS) that initiates the corresponding signal transduction and finally initiates gene transcription (Singh and Ray, 2014; Zang *et al.*, 2019).

Pheromones secretion in the environment is usually detected by an extracellular and intracellular pathway. Extracellular pathways involve a two-component signal transduction system whose histidine kinase binds with the pheromones in the surface of the bacteria that subsequently leads to phosphorylation of regulators and the eventual expression of target genes. However, in the intracellular pathway, an oligopeptide transport system transports pheromones into the bacteria, which leads to the activation of receptors (transcriptional regulators), and subsequently expressing the target gene. In *Bacillus* species, QS proteins bind directly to the corresponding signaling peptides, which consist of some neutral protease regulator, aspartyl phosphate phosphatases, and phospholipase C regulator. These QS systems control several microbial processes, like sporulation, virulence, biofilm formation, conjugation, and production of extracellular enzymes (Sharma *et al.*, 2020).

3.6.2 Mechanisms of Quorum Sensing in Fungi

There are many compounds that form an integrated system of QS in fungi. Since study of QS is still in its infancy, other compounds could still be discovered later. Until then, peptides (pheromones), alcohol (tryptophol, farnesol, 1-phenylethanol, tyrosol), acetaldehydes, lipids (oxylipins), and other volatile compounds have been reported to be part of molecules responsible for QS in fungi. These molecules mediate and regulate functions such as filamentation, pathogenesis, and biofilm formation, among many others (Hirota *et al.*, 2017; Padder *et al.*, 2018). Roles of QS in formation of biofilms have been reported in various fungal species, including – but not limited to – genus of *Saccharomyces*, *Aspergillus*, and *Candida* (Hornby *et al.*, 2001). Generally, fungi species such as *C. solani*, *C. zeylanoides*, *C. krusei*, *C. stellata*, *C. tenuis*, *C. intermedia*, *C. utilis*, *C. albicans*, *Histoplasma capsulatum*, *Aspergillus niger*, *Aspergillus fumigates*, *Ustilagomaydis*, and *Ceratocystisulmi* – among many others – have been reported one or more of the aforementioned molecules used in fungi QS (Albuquerque and Casadevall, 2012; Padder *et al.*, 2018).

In parasitic *H. capsulatum*, α-(1, 3)-glucan control the transition from a filamentous to yeast form, and vice versa. *H. capsulatum* in the soil exist as free-living filamentous fungi with a saprophytic mode of nutrition. However, once it gets into the system of animals through inhalation, they switch from a filamentous cell to a yeast by forming a cell wall of α-1,3-glucan (polysaccharide) required for pathogenicity in a density-dependent manner. The biosynthesis of α-1,3-glucan has been known to be peculiar factor for *H. capsulatum* virulence. Aside from playing roles virulence, α-1,3-glucan also protects cells of *H. capsulatum* from macrophages and within the phagolysosomes, as well as latency establishment intracellularly (Romani, 2011; Padder *et al.*, 2018).

Two QS signaling molecules (tyrosol and farnesol) have been reported in *C. albicans*. The farnesol molecules are sesquiterpene alcohols made up of 12 carbon and three isoprene units (3,7,11-trimethyl-2,6,10-dodecatriene-1-ol). This farnesol is produced as an intermediate compound during biosynthesis of sterol (Hornby *et al.*, 2001) and acts in preventing differentiation from yeast to hyphae (Hornby *et al.*, 2001), but enhances the switch from hypha to yeast form (Lindsay *et al.*, 2012). On the flip side, the lag time of cells of *C. albicans* is shortened by tyrosol accelerating the development of hyphae from germ tubes formed. Tyrosol is regarded as a minor QS molecule of morphogenesis in *C. albicans*, because, its effects only come into existence when molecules of farnesol are limited or absent from the environment (Padder *et al.*, 2018). QS molecules of *C. albicans* have also been reported to play an active role in the regulation of structures of biofilms and their dispersal, making these molecules essential in pathogenesis (Barriuso *et al.*, 2018).

Fungal pheromones play important roles in fungal reproduction, since they serve as informative molecules in karyogamy and plasmogamy by identifying sexual partners that are compatible. Volatile compounds of fungi also affect their growth. For example, colonies of *S. cerevisiae* produce volatile ammonia, which are turbid on agar. In *Trichoderma* spp., volatile compounds have been demonstrated to induce the formation of conidia (Padder *et al.*, 2018).

3.6.3   ROLE OF QUORUM SENSING IN MICROBIAL BIOFILMS FORMATION

Biofilm formation is made up of three stepwise stages: surface adhesion, followed by bacterial replication and production of exopolysaccharides matrix, and finally, disassembly/dispersion of cells. In finding out the role that QS plays in biofilm development, the first approach is to know the step at which bacterial density reaches the threshold level that allows QS signal molecules to participate in biofilm regulation. During biofilm formation steps, QS has been observed to play an important role in biofilm development during cell replication, as well as dispersion of cells. QS is not involved in the initial attachment of cells to surfaces because it involves bacteria that are swimming freely in the medium and accumulation of QS signals is not involved. When the cells are well attached, they divide and form micro-colonies, which enable the population density to increase with an increasing level of QS signals that attain sufficient quantity to activate the maturation and disassembly of the biofilm in a coordinate manner (Solano *et al.*, 2014).

QS-regulated activities initiate biofilm formation from an inducing concentration of QS molecules, which may result from starvation and other types of stress associated with high cell density of microbial populations. In response to such types of stress and as a way of protecting themselves, bacteria may form biofilms to ensure resistance to harsh environmental conditions. QS may also function as a control to the size of microbial population in a biofilm, promoting dispersion of subpopulation of cells in order to escape the nutritional stress that may accompany the inducing concentrations of QS molecules. In some non-motile gram-positive bacteria, autolysis is initiated in response to reaching a quorum (Irie and Parsek, 2008).

In cells of a particular biofilm, QS may induce behaviors during transition from a state not induced by QS to a QS-induced state, which disrupts processes of biofilm development such as secretion of adhesions and EPS. Another way QS affects biofilm development is through repressing or inducing surface motility of microcommunities, which in turn affects the architecture of biofilm (Irie and Parsek, 2008). The importance of QS in biofilm dispersal cannot be overemphasized, as it creates avenues for colonization of new surfaces and begins a new biofilm when nutrients are depleted and toxic waste products are accumulated in the old one. QS ensures biofilm dispersal through down-regulating EPS synthesis and up-regulating the synthesis of degradative enzymes that can disrupt the EPS and other bonds binding the components of biofilm together.

In *C. albicans*, QS holds a vital role in the complete biofilm cycle since secretion of tyrosol induces the formation of hyphal at the early stages of development. At the later stages, QS down-regulates the production of tyrosol while promoting the secretion of farnesol, which induces transition of hyphal cells back to yeast, thereby promoting biofilm dispersals (Alem *et al.*, 2006; Padder *et al.*, 2018). This implies that syntheses of farnesol and tyrosol are differential all through the developmental stages of biofilm in fungi, and each cell responds adequately to the signal molecule accumulated at any particular stage of biofilm development (Barriuso *et al.*, 2018). At high cell density during biofilm formation and growth, farnesol (an acyclic sesquiterpene alcohol in the pathogenic fungi *C. albicans*) is released into the environment, which blocks the transition from yeast to filamentous fungi but cannot inhibit the elongation of already existing hyphae. It also inhibits germ tube formation and triggers the dissemination of yeast phase cells to inhabit new environments (Alem *et al.*, 2006). Farnesol also inhibits the formation of *C. parapsilosis* biofilms, but because this fungus does not form true hyphae, the mechanisms of action may be distinct from the inhibition produced on *C. albicans* biofilms. It alters the expression of oxidoreductases and genes involved in sterol metabolism (Albuquerque and Casadevall, 2012). In *H. capsulatum*, a pathogenic thermodimorphic fungus, α-(1, 3)–glucan QS molecules synthesized in the cell wall occur in response to cell density is responsible for pathogenecity, and its absence in the yeast form results in a loss of virulence by the fungus. It is also responsible for changing from filamentous to yeast form, intracellular latency establishment, protection of yeast within phagolysosomes, and the control of proliferation of yeast in host macrophages (Albuquerque and Casadevall, 2012; Padder *et al.*, 2018).

In *P. aeruginosa* biofilm formation, QS is plays an important role in synthesis of AHL signal molecules and mediates the formation of biofilms. In some other bacteria, QS may function in the dispersal of individual organisms from the biofilm. Albuquerque and Casadevall (2012) reported that AHL signaling molecules produced by *P. aeruginosa* do not inhibit the yeast form of *C. albicans*. However, in the *C. albicans* hyphal form, there is usually the inhibition of cell growth due to the AHL molecules produced by *P. aeruginosa*. As such, the moment *C. albicans* notices the presence of *P. aeruginosa* through QS molecules, there is usually a morphological switch to the yeast form as a survival strategy.

QS has played a great role in inter-kingdom interactions between bacteria and fungi. For example, *P. aeruginosa* can inhibit biofilm formation of *Aspergillus furmigatus* through secretion of a heat-stable factor in a concentration dependent fashion (Barriuso *et al.*, 2018). QS molecules of diffusible lipopeptide known as ralsolamycin are produced by *Ralstonia solanacearum*, a soil-borne pathogenic bacteria of plant. These diffusible molecules are produced by an integrated QS system of PhcBSR in a concentration-dependent manner, which enhances the hyphae of fungi to invade host tissues, thereby providing a favorable niche for bacterial colonization. However, these molecules produced by *R. solanacearum* impair fungal survival by inhibiting secretion of protective agents that help the fungal cells against oxidative stress (Li *et al.*, 2017; Khalid *et al.*, 2018); likewise, *Bacillus licheniformis* ComX pheromone, which inhibits *A. flavus* growth and metabolisms (Barriuso *et al.*, 2018).

3.6.4 ROLE OF QUORUM SENSING IN MULTISPECIES BIOFILM COMMUNITIES

Biofilms found in many environments – such as industrial, clinical, and natural habitats – are usually mixed microbial species of very high cell density, which result in high QS signal concentrations found in these communities. High concentrations of signal molecules produced maybe very vital for every species present in the environment. It may trigger their response to either competing for survival or in a way that is beneficial to themselves as a result of the sensing molecules they produce. However, some microbes do not produce but rather respond to signal molecules produced by other species as a mechanism to attain survival in the environment. Some of the interactions studied in multispecies biofilm communities may be antagonistic or synergistic interactions.

3.6.4.1 Antagonistic Interactions

Competition usually occurs among microorganisms that occupy the same niche either for nutrients or other growth requirements. In a mixed-species biofilm, there is high number of competing species that are fixed spatially at a close range to one another. *P. aeruginosa* is able to establish dominance over *Agrobacterium tumefacien* in any biofilm environment due to the leverage offered by its QS molecules. When *Staphylococcus aureus* is exposed to farnesol QS molecules produced by yeast species, they become more susceptible to antibiotics and eventually result in decreased biofilm formation. In addition to promoting biofilm dispersion, QS may sometimes mediate means by which established bacterial biofilms are prevented from predation by eukaryotes, protozoans, and predatory bacterial species. *P. aeruginosa* is capable of degrading farnesol required by *C. albicans* to transition from yeast to parasitic filamentous form and mature biofilm formation, whereas in the presence of AHL signal molecules produced by *P. aeruginosa*, *C. albicans* remain as a yeast, a form more resistant to killing by *P. aeruginosa*, thereby enhancing their survival in the environment (Irie and Parsek, 2008).

3.6.4.2   Synergistic Interactions

This form of interactions occurs in mixed-species biofilm when QS signal molecules produced by one species of microbes may stimulate the expression of a particular gene in another species present in the same biofilm. Interspecies QS communication has been reported among AHL-producing organisms. For example, in a dual species biofilm of *P. aeruginosa* and *Burkholderia cenocepacia*, AHLs produced by *P. aeruginosa* may promote the expression of AHL-regulated virulence gene by *B. cenocepacia*. This happens because *B. cenocepacia* is able to sense *P. aeruginosa* AHL signal, but *P. aeruginosa* could not perceive *B. cenocepacia* AHL. There are other multispecies biofilm systems that have triggered some gene expressions only when mixed species are present, such as those formed by bacteria isolated from marine algae (Burmølle *et al.*, 2006; Irie and Parsek, 2008).

3.7   CONCLUSION

The cell density–dependent phenomenon of quorum sensing (QS) and discoveries of molecules involved in biofilm formation are an emphatic breakthrough in understanding how microbes interact effectively with each other and their environments, using QS analysis of how microbes communicate in a density-dependent manner, both in inter-species and intra-species contexts. The diverse study of QS and the discovery of different QSMs revealed the importance and role of QS in microorganisms, as well as biofilm formation and their ability to successfully inhabit different habitats.

REFERENCES

Albuquerque P, Casadevall A (2012) Quorum sensing in fungi – a review. Medic. Mycolog. 50:337–345. https://doi.org/10.3109/13693786.2011.652201

Alem MA, Oteef MD, Flowers TH, Douglas LJ (2006) Production of tyrosol by Candida albicans biofilms and its role in quorum sensing and biofilm development. *Eukar. Cell.* 5:1770–1779. https://doi.org/10.1128/ec.00219-06

Ansari FA, Jafri H, Ahmad I, Abulreesh HH (2017) Factors affecting biofilm formation in in vitro and in the rhizosphere. Biof. Plan. Soil Healt. 275–290. https://doi.org/10.1002/9781119246329.ch15

Bandara HMHN, Lam OLT, Jin LJ, Samaranayake L (2012) Microbial chemical signaling: a current perspective. Crit. Rev. Microbiol. 38:217–249. https://doi.org/10.3109/1040841x.2011.652065

Barriuso J, Hogan DA, Keshavarz T, Martínez MJ (2018) Role of quorum sensing and chemical communication in fungal biotechnology and pathogenesis. FEMS Microbiol. Rev. 42:627–638. https://doi.org/10.1093/femsre/fuy022

Berlanga M, Guerrero R (2016) Living together in biofilms: the microbial cell factory and its biotechnological implications. Microb Cel. Fact. 15:165. https://doi.org/10.1186/s12934-016-0569-5

Burmølle M, Webb JS, Rao D, Hansen LH, Sørensen SJ, Kjelleberg S (2006) Enhanced biofilm formation and increased resistance to antimicrobial agents and bacterial invasion are caused by synergistic interactions in multispecies biofilms. *Appl Environ Microbiol.* 72:3916–3923. https://doi.org/10.1128/aem.03022-05

Carlier A, Pessi G, Eberl L (2015) Microbial biofilms and quorum sensing. In Principles of plant-microbe interactions. In: B Lugtenberg: editor: Principles of Plant-Microbe Interactions. New York City: Springer Cham: 45–52. https://doi.org/10.1007/978-3-319-08575-3_7

Carradori S, Di Giacomo N, Lobefalo M, Luisi G, Campestre C, Sisto F (2020): Biofilm and quorum sensing inhibitors: the road so far. Exp. Opin. Therap. Patent. 30:917–930. https://doi.org/10.1080/13543776.2020.1830059

Choudhary P, Singh S, Agarwal V (2020) Microbial Biofilms. In Bacterial Biofilms. IntechOpen. https://doi.org/10.5772/intechopen.90790

Donlan RM (2002) Biofilms: microbial life on surfaces. Emerg. Infec. Dis. 8:881. https://doi.org/10.3201/eid0809.020063

Drescher K, Nadell CD, Stone HA, Wingreen NS, Bassler BL (2014) Solutions to the public goods dilemma in bacterial biofilms. Curr. Biol. 24:50–55. https://doi.org/10.1016/j.cub.2013.10.030

Flemming HC, Wingender J (2010) The biofilm matrix. Nat. Rev. Microbiol. 8:623–633. https://doi.org/10.1038/nrmicro2415

Hirota K, Yumoto H, Sapaar B, Matsuo T, Ichikawa T, Miyake Y (2017) Pathogenic factors in Candida biofilm related infectious diseases. J. Appl. Microbiol. 122:321–330. https://doi.org/10.1111/jam.13330

Hmelo LR (2017) Quorum sensing in marine microbial environments. Ann. Rev. Marin. Sci. 9:257–281. https://doi.org/10.1146/annurev-marine-010816-060656,

Hong SH, Lee J, Wood TK (2010) Engineering global regulator Hha of *Escherichia coli* to control biofilm dispersal. Microb. Biotech. 3:717–728. https://doi.org/10.1111/j.1751-7915.2010.00220.x

Hornby JM, Jensen EC, Lisec AD, Tasto JJ, Jahnke B, Shoemaker R, Nickerson KW (2001) Quorum sensing in the dimorphic fungus *Candida albicans* is mediated by farnesol. Appl. Environ. Microbiol. 67:2982–2992. https://doi.org/10.1128/aem.67.7.2982-2992.2001

Irie Y, Parsek MR (2008) Quorum sensing and microbial biofilms. Bacter. Biof. 67–84. https://doi.org/10.1007/978-3-540-75418-3_4

Khalid S, Baccile JA, Spraker JE, Tannous J, Imran M, Schroeder FC, Keller NP (2018) NRPS-derived isoquinolines and lipopeptides mediate antagonism between plant pathogenic fungi and bacteria. ACS Chem. Biolog. 13:171–179. https://doi.org/10.1021/acschembio.7b00731

Lazar V (2011) Quorum sensing in biofilms – how to destroy the bacterial citadels or their cohesion/power?. Anaerob. 17:280–285. https://doi.org/10.1016/j.anaerobe.2011.03.023

Lewenza S, Visser MB, Sokol PA (2002) Interspecies communication between *Burkholderia cepacia* and *Pseudomonas aeruginosa*. Canad. J. Microbiol. 48:707–716. https://doi.org/10.1139/w02-068

Li P, Yin W, Yan J, Chen Y, Fu S, Song S, Zhang LH (2017) Modulation of inter-kingdom communication by PhcBSR quorum sensing system in *Ralstonia solanacearum* phylotype I strain GMI1000. Front. Microbiol. 8:1172. https://doi.org/10.3389/fmicb.2017.01172

Lindsay AK, Deveau A, Piispanen AE, Hogan DA (2012) Farnesol and cyclic AMP signaling effects on the hypha-to-yeast transition in *Candida albicans*. Eukary Cel. 11:1219–1225. https://doi.org/10.1128/ec.00144-12

McDougald D, Rice SA, Barraud N, Steinberg PD, Kjelleberg S (2012) Should we stay or should we go: mechanisms and ecological consequences for biofilm dispersal. *Nat Rev Microbiol*. 10:39–50. https://doi.org/10.1038/nrmicro2695

Padder SA, Prasad R, Shah AH (2018) Quorum sensing: A less known mode of communication among fungi. Microbiol. Res. 210:51–58. https://doi.org/10.1016/j.micres.2018.03.007

Parsek MR, Greenberg EP (2005) Sociomicrobiology: the connections between quorum sensing and biofilms. Trend. Microbiol. 13:27–33. https://doi.org/10.1016/j.tim.2004.11.007

Petrova OE, Sauer K (2012) Sticky situations: key components that control bacterial surface attachment. J. Bacteriol. 194:2413–2425. https://doi.org/10.1128/jb.00003-12

Prakash B, Veeregowda BM, Krishnappa G (2003) Biofilms: a survival strategy of bacteria. Curr. Sci. 1299–1307. https://doi.org/10.1177/0020294019866854

Ramage G, Saville SP, Wickes BL, López-Ribot JL (2002) Inhibition of *Candida albicans* biofilm formation by farnesol, a quorum-sensing molecule. Appl. Environ. Microbiol. 68:5459–5463. https://doi.org/10.1128/aem.68.11.5459-5463.2002

Romani L (2011) Immunity to fungal infections. Nat. Rev. Immunol. 11.275–288. https://doi.org/10.1038/nri2939

Sharma A, Singh PBK, Nandi SP (2020) Quorum sensing its role in microbial social networking. Res. Microbiol. https://doi.org/10.1016/j.resmic.2020.06.003

Shirtliff ME, Krom BP, Meijering RA, Peters BM, Zhu J, Scheper MA, Jabra-Rizk MA (2009) Farnesol-induced apoptosis in *Candida albicans*. *Antimicro Agent Chemother.* 53:2392–2401. https://doi.org/10.1128/aac.01551-08

Singh MP, Singh P, Li HB, Song QQ, Singh RK (2020) Microbial biofilms: development, structure, and their social assemblage for beneficial applications. In: New and Future Developments in Microbial Biotechnology and Bioengineering: Microbial Biofilms. Amsterdam: Elsevier: 125–138. https://doi.org/10.1016/b978-0-444-64279-0.00010-4

Singh R, Ray P (2014) Quorum sensing-mediated regulation of staphylococcal virulence and antibiotic resistance. *Fut Microbiol.* 9:669–681. https://doi.org/10.2217/fmb.14.31

Solano C, Echeverz M, Lasa I (2014) Biofilm dispersion and quorum sensing. Curr. Opin. Microbiol. 18:96–104. https://doi.org/10.1016/j.mib.2014.02.008

Tarkka MT, Sarniguet A, Frey-Klett P (2009) Inter-kingdom encounters: recent advances in molecular bacterium – fungus interactions. Curr. Gene. 55:233–243. https://doi.org/10.1007/s00294-009-0241-2

Toyofuku M, Inaba T, Kiyokawa T, Obana N, Yawata Y, Nomura N (2016) Environmental factors that shape biofilm formation. Biosci. Biotechnol. Biochem. 80:7–12. https://doi.org/10.1080/09168451.2015.1058701

Williams P (2007) Quorum sensing, communication and cross-kingdom signalling in the bacterial world. Microbiol. 153:3923–3938. https://doi.org/10.1099/mic.0.2007/012856-0

Wongsuk T, Pumeesat P, Luplertlop N (2016) Fungal quorum sensing molecules: role in fungal morphogenesis and pathogenicity. J. Bas. Microbiol. 56:440–447. https://doi.org/10.1002/jobm.201500759

Zhang J, Feng, T, Wang J, Wang Y, Zhang XH (2019) The mechanisms and applications of quorum sensing (QS) and quorum quenching (QQ) J. Ocean Uni. Chi. 18:1427–1442. https://doi.org/10.1007/s11802-019-4073-5

4 Microbial Mats Ecosystems

Oluwafemi Adebayo Oyewole, Daniel Gana, and Chinedu Enemalu

CONTENTS

4.1 INTRODUCTION

Microorganisms do not exist in isolation. They are found in active interaction with both biotic and abiotic factors of their environment. The interaction between these microorganisms may often result in the creation of microbial communities with special structures capable of attaching to solid material, forming an intricate ecological entity in various environments across the world (Davey and O'Toole, 2001). Adherence to a surface is an adaptive mechanism developed by microbes several million years ago. This ability gave them a better chance at survival and evolution in communities, while allowing them to withstand severe climatic and environmental factors. Microbial interactions include a simple biofilm non-specific in nature composed of a single species of microorganism, and an intricate microbial mats which comprise vast array of microorganisms (Bonilla *et al.*, 2012).

According to Ruvindy *et al.* (2016), communities of microorganisms arising from different species attach to the surface of solid materials, which aids the formation

DOI: 10.1201/9781003184942-5

65

of complex ecological entities in various environments around the world. The technique adopted by microorganisms for over a million years is the attachment to a solid surface, which aids survival and evolution of organisms in various communities and also enhances their adaptation to the abiotic factors surrounding them, although some conditions may impose stress on them. Bolhuis *et al.* (2014) and Wong *et al.* (2015) reported that microbial mats are commonly found on the seafloor occurring in different vertical layers, each with distinct communities, that are usually formed at the boundary between the solid and liquid components of various environments. Unlike what is obtainable in biofilms, a microbial mat consists of several million species of microorganisms. (Ruvindy *et al.*, 2016) reported that all microorganisms in a microbial mat are in constant interaction and signal exchange, and to ensure swifter flow of resources and energy to enable continuous survival of the community, they are surrounded by a matrix of exopolysaccharide and nutrients. There are limitations to the type of interactions commonly encountered in microbial mats, with symbiosis being the most prominent, thus giving them relatively selective advantage (Al-Thani *et al.*, 2014).

Furthermore, according to Ruvindy *et al.* (2016), microbial mats comprise diverse species amounting to several millions of microorganisms, which are in continuous interaction and signal relay lodged in a matrix of exopolysaccharide and nutrients to aid increase in resources and energy transmission for the community of different microorganisms to survive the adverse conditions in the environment. According to Nutman *et al.* (2016), the existence of microbial mats on earth dates back to the Paleolithic era, with the most ancient being found in some classes of sedimentary rocks, 3.7 GA west of Australia and 3.4 GA South Africa from the Archean era, although the Proterozoic era has been shown to contain the highest abundance of microbial mats (2.5–0.57 GA) and it is found in different regions of the world. The detailed study of the fossils proposes that there is stability and flexibility in the different communities found in microbial mats in adapting to the constantly changing conditions of the environment (Revsbech *et al.*, 2016).

Persistence of these communities is commonly observed in hypersaline ponds, hot springs, and sulfur springs, among other extreme environments where growth and proliferation of some multicellular organisms and eukaryotes are limited and restricted by some environmental conditions (Revsbech *et al.*, 2016). Microbial mats have been shown to play many important roles in different environments ranging such as modifying atmospheric composition, producing hydrogen and oxygen, as well as methane, through the process of photosynthesis and other anaerobic degradation processes, and also represent the earliest ecosystem, along with stromatolites (Bolhuis *et al.*, 2014). Microbial mat is a unique ecosystem whereby such processes as microbial diversity (structure and patterns of the community), evolutionary trends, and their adaptation to extreme conditions of the environment can be studied, making it a natural laboratory (Klatt *et al.*, 2013).

4.1.1 STRUCTURE OF MICROBIAL MATS

Microbial mats can be seen with unaided eyes, unlike biofilms. Biofilms are mainly composed of single species that form a mass on electrode surfaces, whereas microbial mats exhibit a high diversity of species with different levels of metabolism taking place simultaneously. Biofilms are observed actively covering solid surfaces, while microbial mats are seen covering sediments from a few cell layers to a few millimeters in thickness but ranging from < 1 mm to several centimeters depending on the communities involved, as seen in Figure 4.1. Microbial mats are often formed via the combination of several biofilms of organisms, which are embedded in a matrix of exopolysaccharides. The energy flow in biofilm is strictly heterotrophic and reliant on the provision of substrate, whereas microbial mats can exist in different forms ranging from photoautotrophic to chemoautotrophic levels. These organisms serve as the primary producers in the first few millimeters of the mats.

Mats exist in vertical fashion due to the physical gradients which form the major factor of biological biodiversity in the mats. The biological processes and physical gradients observed in microbial mats make available the needed microenvironment and explicate the functional roles of the microorganisms in the microbial mat ecosystem (Bolhuis *et al.*, 2014). The organisms found in microbial mats are usually bacterial species with some archaea and eukarya. Studying microbial mats requires a

FIGURE 4.1   Structure of microbial mats (Prieto-Barajas *et al.*, 2018).

combination of chemical and physical parameters with a sound knowledge of biological interactions. Chemical parameters like saline concentration, redox potential, pH, presence of oxygen, electron donor and electron acceptor, and the availability of various chemical species are inherent to studying microbial mats. Physical parameters such as pressure, temperature, and light must be closely monitored. It is pertinent to note that processes like metal reduction, nitrogen fixation, photosynthesis, denitrification, reduction in sulfate compounds, and methanogenesis, among other processes, are key in describing the microbial mat performance (Woebken *et al.*, 2015).

4.1.2 Microbial Communities Found in Microbial Mats

The biological communities commonly found in microbial mats include the cyanobacteria, proteobacteria (purple bacteria), green sulfur bacteria (*Chlorobi*), anoxygenic photosynthetic bacteria, aerobic and anaerobic heterotrophs, methanogenic archaea, sulfur oxidizing, and sulfate-reducing bacteria (SRB), as shown in Figure 4.2 (Klatt *et al.*, 2016).

Microbial mats consist of many microbial groups that are closely compacted into mats of various biological activities and processes such as sulfate reduction, photosynthesis, metal reduction, nitrogen fixation, methanogenesis, and denitrification. Because

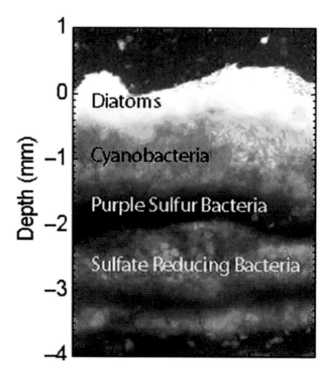

FIGURE 4.2 Some biological communities found in a microbial mat (Prieto-Barajas *et al.*, 2018).

of the myriads of microbial communities in microbial mats, several ecological relationships exist within the microbial communities. Interactions in microbial mats could be beneficial or detrimental to all participating species, or they could be neutral with species neither benefited nor harmed. Mutualism, commensalism, and proto-cooperation are the various forms of beneficial interaction occurring among the participating species while amensalism, antagonism, predation, parasitism, and competition are the detrimental associations among the various species in a microbial mat. Some photosynthetic activities of the photosynthetic group, mainly cyanobacteria, may create an anoxic environment, especially in the upper layer. This may create an excellent environment for the anaerobes such as the SRB at the lower layer (Rich and Maier, 2015).

4.1.3 ENERGY TRANSFER IN MICROBIAL MATS

Although there are possible cases where non-photosynthetic mats exist, photosynthesis is the major source of energy in microbial mats. The primary producers or cyanobacteria perform the photosynthetic activity. Various biogeochemical cycles, coupled with biochemical processes, are associated with all microbial mats which function as a consortium. The immediate interaction among different species in the microbial mat enables the utilization of the metabolic product of one group by other groups of microorganisms (Severin *et al.*, 2010). The filamentous cyanobacteria communities found in microbial mats carry out the process of nitrogen fixation. However, SRB also play vital roles in the biological process. The flow of nutrients in a microbial mat is summarized in Figure 4.3.

Ecological succession brings about the formation of complex communities in a mat, with cyanobacteria serving as the colonizers and modifiers of microenvironment for the later colonization by specialized bacteria with greater and more specific environmental requirements (Boomer *et al.*, 2009). Microbial mats also serve as dynamic communities, allowing change in position of motile organisms in the mat to achieve a stable and favorable environmental conditions like redox potential and luminous intensity.

Since the discovery of mats, geologists and microbiologists have continued to explore the best ways to understand the physiology of mats. It was quite impossible to investigate microbial processes taking place in mats due to their sizes (about 1 mm thick). This narrative changed when the applications of microelectrodes and microlight sensors became widespread. These instruments allow researchers to measure microbial activity at tens-of-micrometer scales, thus revealing a detailed high-resolution spatial and temporal information on photosynthesis, redox, pH, sulfide, oxygen, and light levels in the mats. Even with this technological advancement, the activities of microorganisms in mats are difficult to measure or observe. It was the application of genetic techniques that led to exploring the information of organisms found in mats. The cultivation and microscopy of communities in mats show that they exhibit a high level of diversity with different metabolic pathways, depending on the energy need of the organisms. Since organisms existing in mats depend on each other for survival, the waste product of one organism usually serves as the energy source of the next – and the chain continues (Stal, 2012).

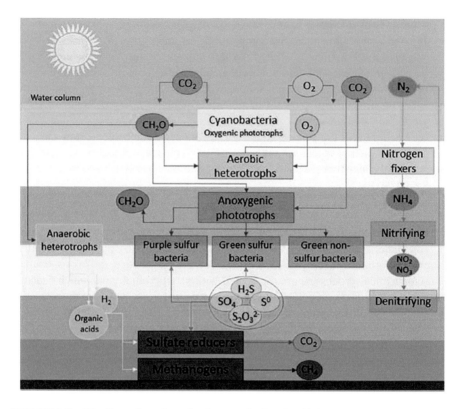

FIGURE 4.3 The flow of nutrients in photosynthetic microbial mats (Prieto-Barajas *et al.*, 2018).

4.2 TYPES OF MICROBIAL MATS

According to Drewniak *et al.* (2016), microbial mats are commonly found in the biosphere, although they thrive and prosper in extreme environments. Microbial mats exhibit high diversity of species, described in the following sections; it is the various microbial diversity that characterizes the different microbial mats, which are mostly phototrophic with a significant photosynthetic components making them light dependent.

4.2.1 Hypersaline Mats

Hypersaline mats are commonly found in areas of saline lakes (Figure 4.4). These mats are characterized by extreme conditions such as high salt content (salinity), high temperature, and high radiation energy. These factors impede the formation of complex microbial communities (Charlesworth and Burns, 2016). Various molecular techniques have been studied extensively to provide adequate information for the description of the community in mats; common among the techniques include

FIGURE 4.4 Hypersaline mat (Ruvindy *et al.* (2016).

microbial culture, amplification, cloning, and sequencing of 16S rRNA and metagenomics. Bacteria are the dominant organism in this community, with an approximate 90% bacteria, 9% archaea, and 1% eukaryotes (Tripathi *et al.*, 2016). The distribution of bacterial species in hypersaline mats is determined by the presence of light, oxygen, H_2S (hydrogen sulfide), and the formation of spatiotemporal chemical gradients (Kunin *et al.*, 2008). The distribution of the different groups of bacteria into vertical sections is a function of the presence of oxygen, light, and hydrogen sulfide, and the structure of the microbial community is strongly affected by the formation of spatiotemporal chemical gradient (Feller, 2013).

4.2.2 COASTAL MATS

The two most biologically diverse microbial mats with extensive coastal distribution are the coastal and hypersaline mats. These mats are commonly found in irregular flood plains, in intertidal coastal zones with fluctuations in high salt content and intense temperature. Cyanobacteria are the most commonly encountered species inhabiting the mats, with other bacterial groups like bacteroidetes and proteobacteria (Bolhuis *et al.*, 2014). According to Mahajan and Balachandran (2017),

Schiermonnikoog Island off the coast of the Netherlands measuring 300 m in width and 5 km in length has a green bench with a massive strip of microbial mats. The dominant bacteria divisions were identified using sequence analysis of the 16S rRNA gene and were shown to be cyanobacteria, proteobacteria of the order rhodobacteriales and sphingomonadales, chromatiales, S-proteobacteria of the order desulfobacteriales and desulfovibrionales, and actinobacteria. important archean elements as euryarchaeota (particularly methanogens) are crenarchaeota are also scarcely present (Feller, 2013).

Mahajan and Balachandran (2017), reported that the most abundant divisions in the great Sippewisset swamp (Massachusetts, USA) are the cyanobacteria, proteobacteria, and chloroflexi, with traces of spirochetes, verrucomicrobia, acidobacteria, caldithirix, and actinobacteria. The vast majority of microbial mats have shown similar trends in organizational levels and structural complexity, although large number of eukaryotes have been discovered in the coastal mats, mainly diatoms of the genera *Navicula sp., Amphora sp., Diploneis sp.*, and *Cylindrotheca*, and the algae of the genera *Chlorophyta* and *Enteromorpha*.

4.2.3 MICROBIAL MATS IN OLIGOTROPHIC ENVIRONMENTS

Al-Thani *et al.* (2014) reported the detailed study of the features of the oligotrophic mats of Cuatro Cienegas present in the Coahuila desert of northern Mexico. These mats are extremely low in concentration of phosphorus. Phosphorus in form of phosphate is one of the major constituents of the DNA, proteins, and other energy molecules, making it an essential limiting factor upon which all life forms depend (Thiel *et al.*, 2017). The diversity of community largely depends on the degree of exposure and disturbance; mats with little exposure to constant disturbances have high diversity of community with no dominant groups, with the most diverse groups as proteobacteria, cyanobacteria, and bacteriodetes alongside 16 other divisions and 28 bacterial orders (Bolhuis *et al.*, 2014).

4.2.4 ACID MICROBIAL MATS

Vast majority of microbial mats form primarily at alkaline pH, with few being developed in acidic environments. A number of microbial mats lack photosynthetic microbial groups, and as such, the predominant metabolism in the communities being oxidation of iron and sulfur compounds. (Bolhuis *et al.*, 2014) observed that pyrite, chalcopyrite, arsenopyrite, FeS_2, FeAsS, and $CuFeS_2$ – among other sulfur minerals – are present in acid mine drainage which is strongly acidic with pH values between 0.77 and 1.21, characterized by high concentration of toxic metals. The rate of propagation of microbial diversity of a mat is usually very minimal due to the effects of environmental conditions, archaea and bacteria being the predominant organisms with oxidation and reduction of iron sulfate reduction as the metabolism. Firmicutes, S-proteobacteria, actinobacteria, and nitrospirae are the various phyla in acidic microbial mats, others being *Ferromicrobium acidophilus, Leptospirillum, Acidomicrobium*, and Thermoplasmales (Baker and Banfield, 2003).

The pH value of acidic springs ranges between 3 and 3.5, and the mats from Yellowstone National Park (USA) are the most studied mats. Metabolism of iron and sulfur compounds are crucial for a dynamic community in acid mines. *Hydrogenobacculum* sp., *Metallosphaera yellowstonensis*, heterotrophic archaea, and members of a new Geoarchaeota archaea division are present in springs as One Hundreds Spring Plain and Beowulf (Beam *et al.*, 2016).

4.2.5 PSYCHROPHILE MICROBIAL MATS

Psychrophile microbial mats are shown in Figure 4.5. According to Drewniak *et al.* (2016), a vast array of cold environments occupies the largest proportion of earth, with low temperature of less than 5°C found in oceans, caves, polar regions, and Alpine areas. Polar region mats in Antarctic and Arctic shelters serve as hot spots of ecological diversity and primary production (Brito *et al.*, 2014). These environments include Alpine areas, caves, oceans, and polar regions. Locations like Artic shelter and Antarctica represent the best locations for studying the ecological diversity and primary production in extremely cold mat. The photosynthetic mats in cold

FIGURE 4.5   Psychrophile microbial mat (Nutman *et al.*, 2016).

regions are dominated by filamentous cyanobacteria and various species of diatoms, algae, nematodes, protozoans (ciliates and flagellates), rotifers, and microinverte-brates (Siddiqui *et al.*, 2004). Filamentous cyanobacteria of the orders Dichothrix, Nostocales-Tolypothrix, and Oscillatoriales-Tychonema dominate the photosynthetic mats of these poles; they help in producing wide variety of polysaccharides matrix that afford protection to other organisms such as ciliates, diatoms, rotifers, algae, flagellates and microinvertebrates with lower tolerance that make up the community (Drewniak *et al.*, 2016).

Charlesworth and Burns (2016) reported various growth limiting factors as low temperature, drought, selective pressure, high solar radiation, prolonged winter dark-ness, nutrient deficiency, and freezing, as well as thawing cycles, are imposed by the psychrophilic conditions. Heterotrophic bacteria play vital role in nutrients cycling in Antarctica with major bacterial phyla as proteobacteria, actinobacteria, firmicutes, bacteroides and Deinococcus–Thermus (Charlesworth and Burns, 2016). The domi-nant group of the photosynthetic bacteria in the mat are the cyanobacteria with the orders Nostocales and Oscillatoriale, which are filamentous in nature and the most abundant. Chloroflexi and Chlorobi, among other photosynthetic groups, are also present in trace amount (Brito *et al.*, 2014).

4.2.6 HOT SPRINGS MICROBIAL MATS

A hot springs microbial mat is shown in Figure 4.6. They are associated with high temperatures and extreme environments such as hot springs and geysers. The limit-ing factors that affect the development of life and diversity of community are temper-ature, sulfur concentration, and pH, with temperature serving as the most essential factor for the distinctive modeling of the communities (Cardoso *et al.*, 2017). Within phototrophs, hot spring microbial mats have the lowest diversity, however, the study of the most ancient communities on earth are best understood using these mats (Klatt *et al.*, 2016).

Thermophilic communities may be associated with water columns, sediments, streams, and microbial mats (Wong *et al.*, 2015). The vital role of metabolic dynam-ics of the community in microbial mats is carried by photosynthetic organisms. There are two major problems associated with photosynthesis in heated water: high temperature decreases the concentration of dissolved gases such as CO_2 and O_2, and it also denatures the proteins and their biomolecules; as a result, high temperature limits the activity and proliferation of photosynthetic bacteria which is normally 75°C, as chlorophyll degrades at that temperature (Cardoso *et al.*, 2017).

Cyanobacteria are major and one of the most abundant groups in these mats. Cyanobacteria function as nitrogen and carbon fixers in the community. *Synechococcus*, among other unicellular cyanobacteria, are commonly associated with springs with thermal water temperature exceeding 55°C (Woebken *et al.*, 2015). Various thermophilic microbial mats have been discovered in different regions of the world, with the Yellowstone National Park in the USA being the most studied; others include Thailand, Tibet, Romania, Patagonia, the Philippines. The geothermal zones determine the geographical distribution of the mats (Thiel *et al.*, 2017).

FIGURE 4.6   Hot spring microbial mat (Charlesworth and Burns, 2016).

4.2.7   QUORUM SENSING IN BIOFILM: MAT PRODUCTION

A cell-to-cell communication process in bacteria involving the detection of signal, generation of response to extracellular signaling, or messenger molecules called autoinducers (AIs) is termed as quorum sensing. An increase in the bacterial population density results in accumulation of the autoinducers, and bacteria detect changes in their cell number, thereby collectively altering the gene expression by monitoring the information. A compound is detected when a certain threshold concentration of the signal is attained, translating into activation or repression of the target gene. N-acyl-homoserine lactones (AHL) are the most studied bacterial autoinducers, and they are synthesized by LuxI-type proteins (Boyer and Wisniewski-Dye, 2009). Certain genes that regulate the beneficial activities of group of bacteria acting in synchrony are controlled by quorum sensing. Antibiotics production, plant-microbe interaction, bioluminescence, biofilm formation, and virulence factor secretion are the various processes controlled by quorum sensing (Williams and Camara, 2009).

Biofilms are multicellular sessile communities formed by bacteria in which the cells of a community are surrounded by a matrix of secreted extracellular polymeric substances. Quorum sensing and the formation of biofilms are fundamental and mostly the interrelated features that characterize bacterial social life (Li and Tian, 2012). The development of various microbial communities in a mat is described as biofilm formation, while the various checkpoints of different phases in biofilm progress is referred to as quorum sensing (Irie and Parsek, 2008). It may be difficult to ascertain the differences between the primary quorum sensing system responsible in biofilm-related phenotype; however, quorum sensing–related mechanisms are found to regulate many steps in biofilm development. Involvement of quorum regulatory units is observed in the formation and maturation of biofilms; this was revealed by studies of different quorum sensing mutant strains (Kjelleberg and Molin, 2002).

Biofilms are assemblages of small groups of microorganisms that are attached to one another and fastened to a surface, and are embedded in a matrix secreted by the organisms themselves. The vast majority (99.9%) of microorganisms have the capability to form a biofilm. Various benefits bacteria derive from forming biofilms include protection against antibiotics and other disinfectants, bacterial tolerance to harsh environmental conditions are enhanced, and it prevents the washing away of bacteria by water flow or bloodstream. Also, bacterial motility is hampered by the formation of biofilm, and it also increases the cell density, thereby providing a suitable environment for easy transfer of plasmid among organisms in a community through the process of conjugation. An antibiotics resistance gene may be carried in some of these plasmids to enable them withstand various environmental stresses. Quorum sensing is a unique chemical communication technique used be bacteria in a biofilm to relay information among themselves.

The formation of biofilm takes place in the following stages: (i) the organisms are accumulated or adsorbed on the surface of aggregator, i.e. substrate, this is known as deposition; (ii) formation of polymer bridges upon desegregation of the layer between the organisms and the aggregator; (iii) multiplication or proliferation of the organisms on the surface of the aggregator; (iv) formation and maturation of biofilm; and (v) dispersal or detachment. High cell concentration in biofilm is characterized by the involvement of a cascade of cell signaling mechanisms. Gene expression is triggered by the autoinducers or signaling molecules such as homoserine lactones and small peptides through enzymatic processes for the formation and maturation of the biofilm (Subramani and Jayaprakashvel, 2019).

The functioning of quorum sensing signal molecules in a biofilm and how often quorum sensing is activated in a biofilm remains largely unknown. Biofilm formation is promoted by some quorum sensing systems, while others are involved in the dispersion of biofilms. Biofilm formation is seen as series of stages involving continuous attachment of microbial communities to aggregator surfaces, while quorum sensing is described as checkpoint for restarting the cycle by aiding dispersion or dissolution of a subpopulation of cells. In this case, the nutritional stress that accompanies inducing concentrations of a quorum sensing signal is overcome by the dispersing cells. The population density of non-motile species in a biofilm is regulated using different mechanisms. The mechanistic pathway of biofilm development such as the production of exopolysaccharide and other secreted factors is altered as quorum sensing induce behaviors in the biofilm cells (as they transition from

QS-uninduced state to QS-induced state) (Irie and Parsek, 2008; Li and Tian, 2012; Munir *et al.*, 2020).

Group activities such as surface motility might be induced or repressed by quorum sensing, which could have a profound impact on the structure of biofilm. The production of vps exopolysaccharide in *Vibrio cholerae* is regulated by quorum sensing. Biofilm formation was observed to be promoted by the expression of vps (Silva and Benitez, 2016). Alem *et al.* (2006) reported that at the initial stages of biofilm development, hyphal formation is promoted or enhanced by the quorum sensing signal tyrosol, while in the later stages, dispersal of yeast cells from the biofilm is promoted by farnesol, another quorum sensing signal. This implies that tyrosol and farnesol syntheses are differential throughout biofilm development, and as such, the quorum sensing signal that accumulates at specific stages of biofilm development determines the cells' response.

The role of the regulatory protein las quorum sensing is the formation of biofilm of *Pseudomonas aeruginosa*. Quorum sensing in the biofilm of this organism in very intricate and has two interconnected N-acyl homoserine lactone-dependent regulatory circuits; these are acted upon by regulators at translational and post-translational levels, thus modifying them. It was discovered that the lasI mutant, which lacks the capacity to produce autoinducer N-3-oxo-dodecanoyl-L-homoserine lactone (3-oxo-C12-HSL), formed biofilm cell clusters with thickness of 20% of the wild-type biofilm and they were found to be highly sensitive to detergent removal (Venturi, 2006). The lasI mutant was found to regain the ability to form structured biofilm when autoinducer 3-oxo-C12-HSL was added to the system. The study reveals the important role played by quorum sensing in biofilm development, and most significantly, it establishes an inextricable link between quorum sensing and biofilm formation. *Staphylococcus aureus*, and a wide variety of *Streptococcus* sp. among other gram-positive bacteria, make use of signal peptide-mediated system for quorum sensing (Rutherford and Bassler, 2012). The main causative agent of nosocomial infection worldwide is *S. aureus*, and it causes diseases ranging from mild skin infection to potentially fatal systemic disorders. Infections like endocarditis, osteomyelitis, and foreign body–related infections caused by *S. aureus* are caused by biofilms and not by free-living cells. Various aspects of the biofilm phototype may be influenced by the agr phototype locus in *S. aureus* and pattern of expression from the accumulated evidence. These include cells attachment to surfaces, dispersal of biofilm, and – to some extent – the chronic nature of biofilm-associated infections (Tong *et al.*, 2015)

Bacteria in liquid cultures are believed to be physiologically similar, and thus, the rate of signal molecules production is the same. However, complications might be encountered in quorum sensing and signal transduction in biofilms; these are attributed to the range of chemical, physical, and nutritive factors influencing signal production, stability, distribution, and efficiency to interact with their cognate receptors in the biofilm (Li and Tian, 2012). N-acyl-L-homoserine lactones (ALHs) are observed to have little problem to reach their target receptors through free diffusion in the biofilm matrix; this is due to the free diffusion of these molecules across the cell membrane. However, the physical, chemical, and biological factors within a biofilm likely influence the signaling peptides produced by gram-positive bacteria because of the interaction of the small peptides with the charged molecules. Presently, there is little knowledge of the chances of signal peptides being affected by the limit of diffusion or by non-specific binding to polysaccharides, proteins, DNA, and the components of the cell wall within the biofilm (Horswill *et al.*, 2007; Roya *et al.*, 2018).

The cost of producing an active signal peptide by a gram-positive bacterium is a very expensive process. The estimated cost for production of signal peptide in *S. aureus* by Keller and Surette is 184 ATP (Li and Tian, 2012), but is only 8 ATP for ALH in *P. aeruginosa*. Generally, in gram-positive bacteria, the cost of producing a signal peptide is much more expensive. It is therefore necessary to assume that the signal peptide–mediated quorum sensing and activities in gram-positive biofilms are influenced by important factors as nutrient and energy source. Theoretically, the concentration of signal molecules affects the signal molecules which are used to estimate the population density. Signal molecules could also be affected by other factors as the limit of diffusion, accessibility to the receptor, degradation, and synthesis of the same autoinducer, such as AI-2 by third parties, intentionally or by chance. Since activation of quorum sensing depends on the diffusion of a signal molecule to and interact with the cognate receptor, quorum sensing could be regarded as diffusion sensing (DS) (Moghaddam *et al.*, 2014; Radlinski *et al.*, 2017; Hotterbeekx *et al.*, 2017).

4.2.8 Ecology and Functional Diversity of Microbial Mats

In microbial mats, different communities of microorganisms are found, and these organisms carry out different biological processes. The microenvironment required by the organisms and the functional roles of microorganisms with specific needs are provided by the biological processes and physical gradients (Sauder *et al.*, 2017). The communities that make up these microbial mats are mostly bacteria, although other less abundant domains such as archaea and eukarya with low diversity are found in microbial mats. Chemical parameters such as pH, presence of oxygen, redox potential, presence of electron donor and acceptor compounds, saline concentration, and diversity of chemical species are considered for studying microbial mats, while light, temperature, and pressure are important physical factors also to be considered (Adessi *et al.*, 2017).

The biological interactions that exist in microbial mats include symbiotic, neutralism, and amensalism, while essential processes for efficient performance of microbial mats include nitrogen fixation, photosynthesis, metal reduction, iron and sulfate reduction, denitrogenation, and methanogenesis (Langille *et al.*, 2013). The basic functional groups of microbial mats are cyanobacteria, green sulfate bacteria (Chlorobi), proteobacteria (purple bacteria), anoxygenic photosynthetic bacteria (known as non-sulfur green bacteria of the Chloroflexi division), aerobic and anaerobic heterotrophs, SRB, sulfur oxidizing bacteria, and methanogenic archaea (Mobberley *et al.*, 2017).

Photosynthesis is the primary source of energy and nutrition in microbial mats, although non-photosynthetic mats exist (Nelson *et al.*, 2015). Photosynthesis is the first step for survival of this tropical network in typical mats; it is a process in which the primary producers cyanobacteria transform the inorganic carbon (CO_2) to organic carbon ($[CH_2O]n$) using light energy with the release of oxygen (Tripathi *et al.*, 2016). According to Pearl *et al.* (2000), the close interaction between the biogeochemical cycle and various biological processes in the consortium enables the utilization of the product of metabolism of one group by other microorganisms. Filamentous cyanobacteria and unicellular organisms carry out nitrogen fixation, while SRB are those group of organisms that have the ability to reduce sulfates to sulfur, and also oxidized organic matter as well in the process of obtaining energy (Sauder *et al.*, 2017).

Vital processes such as calcium precipitation and lithification of mats are carried out by the SRB; therefore, they are responsible for the preservation of mats in fossil records (Sauder *et al.*, 2017). Complex communities are formed by a process of ecological succession whereby cyanobacteria serve as colonizers and modifiers of microenvironments for subsequent colonization by more specialized bacteria with higher and more specific environmental requirements (Langille *et al.*, 2013).

4.2.9 SIGNALING SYSTEM IN MICROBIAL MATS

The high microbial diversity exhibited by mats is an indication that several signals may exist in them. The organisms present in mats exhibit diversity with different functional groups, enclosed in a matrix of extracellular polymeric substances. The laminated sedimentary biofilms can provide a perfect system for examining the environmental effects of quorum sensing. This is largely because a wide range of extractable quorum sensing signals are produced by mats, which possess an extensive small-scale horizontal gradient of physicochemical conditions that may vary dramatically over a diel cycle (Jorgensen, 1983).

Chemical signals between microbes in a mat can operate within short range (usually tens of micrometers). This is sponsored by the diffusion constraints. Moving chemical signals over long distances may take too long. The chemical signals exchanged by microbes in a mat can be utilized as extracellular sensors, providing information to cells about the properties of their proximal environment such as local diffusivity. The process is known as "efficiency sensing". For instance, the potential benefits of contributing to extracellular processes can be assessed using important determinants such as gauging the relative diffusivity of molecules released by cells into the environment.

Thus, microbial mats have shown to play many important roles from modification of atmospheric compositions, producing H_2, O_2, and CH_4. Microbial mats, along with stromatolites, also represent the first ecosystem. In addition, vital processes such as microbial diversity, evolutionary trends, and their adaptation to extreme environments can be efficiently studied, making microbial mats a natural laboratory.

REFERENCES

Adessi A, Corneli E, De Philippis, R (2017) Photosynthetic purple non sulfur bacteria in hydrogen producing systems: new approaches in the use of well known and innovative substrates. In *Modern topics in the phototrophic prokaryotes*. Springer, 321–350.

Alem MAS, Oteef MDY, Flowers TH, Douglas LJ (2006). Production of Tyrosol by Candida albicans Biofilms and Its Role in Quorum Sensing and Biofilm Development. *Eukaryotic Cell, 5*(10), 1770–1779. https://doi.org/10.1128/ec.00219-06

Al-Thani R, Al-Najjar MAA, Al-Raei AM, Ferdelman T, Thang NM, Shaikh IA, Al-Ansi M, & de Beer D (2014). Community Structure and Activity of a Highly Dynamic and Nutrient-Limited Hypersaline Microbial Mat in Um Alhool Sabkha, Qatar. *PLoS ONE, 9*(3), e92405. https://doi.org/10.1371/journal.pone.0092405

Baker BJ, Banfield JF (2003). Microbial communities in acid mine drainage. FEMS *Microbiology Ecology, 44*(2), 139–152. https://doi.org/10.1016/s0168-6496(03)00028-x

Beam, J. P., Bernstein, H. C., Jay, Z. J., Kozubal, M. A., Jennings, R. D., Tringe, S. G., & Inskeep, W. P. (2016). Assembly and Succession of Iron Oxide Microbial Mat Communities in Acidic Geothermal Springs. *Frontiers in Microbiology, 7.* https://doi.org/10.3389/fmicb.2016.00025

Bolhuis H, Cretoiu MS, Stal LJ. (2014) Molecular ecology of microbial mats. *FEMS Microbiol Ecology.* 90(2):335–350.

Bonilla-Rosso, G., Peimbert, M., Alcaraz, L. D., Hernández, I., Eguiarte, L. E., Olmedo-Alvarez, G., & Souza, V. (2012). Comparative Metagenomics of Two Microbial Mats at Cuatro Ciénegas Basin II: *Community Structure and Composition in Oligotrophic Environments. Astrobiology,* 12(7), 659–673. https://doi.org/10.1089/ast.2011.0724

Boomer SM, Noll KL, Geesey GG, Dutton BE (2009). Formation of Multilayered Photosynthetic Biofilms in an Alkaline Thermal Spring in Yellowstone National Park, Wyoming. *Applied and Environmental Microbiology, 75*(8), 2464–2475. https://doi.org/10.1128/aem.01802-08

Boyer M., Wisniewski-Dye F. (2009) Cell-cell signaling in bacteria: not simply a matter of quorum. *FEMS Microbiological Ecology.* 70:1–19.

Brito EM., Villegas-Negrete N, Sotelo-González IA, Caretta CA, Goñi-Urriza M, Gassie C, Piñón-Castillo HA *et al.* (2014) Microbial diversity in Los Azufres geothermal field (Michoacán, Mexico) and isolation of representative sulfate and sulfur reducers. *Extremophiles.* 18(2):385–398.

Cardoso DC, Sandionigi A, Cretoiu MS, Casiraghi M, Stal L, Bolhuis H. (2017) Comparison of the active and resident community of a coastal microbial mat. *Sci Rep.* 7(1):1–10.

Charlesworth J, Burns B (2016). Extremophilic adaptations and biotechnological applications in diverse environments. *AIMS Microbiology, 2*(3), 251–261. https://doi.org/10.3934/microbiol.2016.3.251

Drewniak L, Krawczyk PS, Mielnicki S, Adamska D, Sobczak A, Lipinski L, Sklodowska A *et al.* (2016) Physiological and metagenomic analyses of microbial mats involved in self-purification of mine waters contaminated with heavy metals. *Front Microbiol.* 7:1252.

Feller G. (2013) Psychrophilic enzymes: from folding to function and biotechnology. *Scientifica.* 512840. https://doi.org/10.1155/2013/512840

Horswill A.R., Stoodley, P., Stewart, P.S., Matthew R. Parsek, M.R. (2007) The effect of the chemical, biological, and physical environment on quorum sensing in structured microbial communities. Anal Bioanal Chem (2007) 387:371–380 DOI 10.1007/s00216-006-0720-y

Hotterbeekx A, Kumar-Singh S, Goossens H, Malhotra-Kumar S (2017) In vivo and In vitro Interactions between Pseudomonas aeruginosa and Staphylococcus spp. *Front Cell Infect Microbiol.* 7,106. doi: 10.3389/fcimb.2017.00106

Irie Y, Parsek MR (2008) Quorum sensing and microbial biofilms. *Curr Top Microbiol Immunol.* 67–79.

Jorgensen BB. *et al.* (1983) Photosynthesis and structure of benthic microbial mats: microelectrode and SEM studies of four cyanobacterial communities. *Limnol Oceanogr.* 28:1075–1093.

Kjelleberg S, Molin S, (2002) Is there a role for quorum sensing signals in bacterial biofilms? *Curr Opin Microbiol.* 5:254–258.

Klatt CG, Inskeep WP, Herrgard MJ, Jay ZJ, Rusch DB, Tringe S, Niki Parenteau M, Ward DM, Boomer SM, Bryant DA, & Miller S. (2013). Community Structure and Function of High-Temperature Chlorophototrophic Microbial Mats Inhabiting Diverse Geothermal Environments. *Frontiers in Microbiology*, 4. https://doi.org/10.3389/fmicb.2013.00106

Klatt JM, Meyer S, Hausler S, Macalady JL, De Beer D., Polerecky L. (2016) Structure and function of natural sulphide-oxidizing microbial mats under dynamic input of light and chemical energy. *ISME J.* 10(4):921–933.

Kunin V, Raes J, Harris JK, Spear JR, Walker JJ, Ivanova N, von Mering C, Bebout BM, Pace NR, Bork P, Hugenholtz P (2008). Millimeter-scale genetic gradients and community-level molecular convergence in a hypersaline microbial mat. *Molecular Systems Biology*, 4(1), 198. https://doi.org/10.1038/msb.2008.35

Langille MG, Zaneveld J, Caporaso J G, McDonald D, Knights D, Reyes JA., Beiko RG *et al.* (2013) Predictive functional profiling of microbial communities using 16S rRNA marker gene sequences. *Nature Biotechnol.* 31(9):814.

Li YH., Tian X. (2012) Quorum Sensing and Bacterial Social Interactions in Biofilms. *Sensors.* 12: 2519–2538.

Mahajan GB, Balachandran L. (2017) Sources of antibiotics: Hot springs. *Biochemical Pharmacology.* 134:35–41.

Mobberley JM, Lindemann SR, Bernstein HC, Moran JJ, Renslow RS, Babauta J, Hu D, Beyenal H, Nelson W. C (2017). Organismal and spatial partitioning of energy and macronutrient transformations within a hypersaline mat. *FEMS Microbiology Ecology*, 93(4). https://doi.org/10.1093/femsec/fix028

Moghaddam MM, Khodi S, Mirhosseini A (2014) Quorum Sensing in Bacteria and a Glance on *Pseudomonas Aeruginosa Clin Microbial.* 3:156. doi:10.4172/2327-5073.1000156

Munir S, Shah AA, Shahid M, Manzoor I, Aslam B, Rasool MH, Saeed M, Ayaz S, Khurshid M (2020, August 31) Quorum Sensing Interfering Strategies and Their Implications in the Management of Biofilm-Associated Bacterial Infections. *Brazilian Archives of Biology and Technology.* 63:e20190555. https://doi.org/10.1590/1678-4324-2020190555

Nutman AP, Bennett VC, Friend CRL, van Kranendonk MJ, Chivas AR (2016). Rapid emergence of life shown by discovery of 3,700-million-year-old microbial structures. *Nature*, 537(7621), 535–538. https://doi.org/10.1038/nature19355

Paerl HW, Pinckney JL, Steppe TF (2000). Cyanobacterial-bacterial mat consortia: examining the functional unit of microbial survival and growth in extreme environments. *Environmental Microbiology*, 2(1), 11–26. https://doi.org/10.1046/j.1462-2920.2000.00071.x

Prieto-Barajas CM, Valencia-Cantero E, Santoyo G (2018). Microbial mat ecosystems: Structure types, functional diversity, and biotechnological application. *Electronic Journal of Biotechnology, 31*, 48–56. https://doi.org/10.1016/j.ejbt.2017.11.001

Radlinski L, Rowe SE, Kartchner LB, Maile R, Cairns BA, Vitko NP, *et al.* (2017) *Pseudomonas aeruginosa* exoproducts determine antibiotic efficacy against *Staphylococcus aureus.* *PLoS Biol.* 15(11):e2003981. https://doi.org/10.1371/journal.pbio.2003981

Revsbech, NP, Trampe E, Lichtenberg M, Ward DM., Kahl M. (2016) In situ hydrogen dynamics in a hot spring microbial mat during a diel cycle. *Appl Environ Microbiol.* 82(14):4209–4217.

Rich VI, Maier RM. (2015) Aquatic Environments. In Environmental Microbiology: Third Edition. Elsevier Inc. 2015. p. 111–138 https://doi.org/10.1016/B978-0-12-394626-3.00006-5

Roya R, Tiwaria M, Donellib G., Tiwaria V (2018) Strategies for combating bacterial biofilms: A focus on anti-biofilm agents and their mechanisms of action. *Virulence.* 9(1):522–554 https://doi.org/10.1080/21505594.2017.1313372

Rutherford ST, Bassler BL (2012) Bacterial Quorum Sensing: Its Role in Virulence and Possibilities for Its Control. Cold Spring Harb Perspect Med. 2(11):a012427

Ruvindy R, White III RA, Neilan BA, Burns BP. (2016) Unravelling core microbial metabolisms in the hypersaline microbial mats of shark bay using high-throughput metagenomics. *ISME J.* 10(1):183–196.

Sauder LA, Albertsen M, Engel K, Schwarz J, Nielsen PH, Wagner M, Neufeld JD. (2017) Cultivation and characterization of Candidatus Nitrosocosmicus exaquare, an ammonia-oxidizing archaeon from a municipal wastewater treatment system. *ISME J.* 11(5):1142–1157.

Severin I, Acinas SG, Stal LJ (2010). Diversity of nitrogen-fixing bacteria in cyanobacterial mats. *FEMS Microbiology Ecology*, no. https://doi.org/10.1111/j.1574-6941.2010.00925.x

Siddiqui KS, Poljak A, Cavicchioli R (2004). Improved activity and stability of alkaline phosphatases from psychrophilic and mesophilic organisms by chemically modifying aliphatic or amino groups using tetracarboxy-benzophenone derivatives. *Cellular and Molecular Biology (Noisy-Le-Grand, France), 50*(5).

Silva AJ, Benitez JA (2016) *Vibrio cholerae* Biofilms and Cholera Pathogenesis. *PLoS Negl Trop Dis.* 10(2):e0004330. doi:10.1371/journal. pntd.0004330

Stal, LJ (2012) Cyanobacterial mats and stromatolites. In Whitton BA (Ed.), *The ecology of cyanobacteria II: Their diversity in space and time.* Springer, 65–125.

Subramani R, Jayaprakashvel M (2019) Bacterial quorum sensing: biofilm formation. *Survival Behav Antibio Res.* 3:21–33.

Thiel V, Hügler M, Ward DM, Bryant DA (2017). The Dark Side of the Mushroom Spring Microbial Mat: Life in the Shadow of Chlorophototrophs. II. Metabolic Functions of Abundant Community Members Predicted from Metagenomic Analyses. *Frontiers in Microbiology, 8.* https://doi.org/10.3389/fmicb.2017.00943

Tong SYC, Davis JS, Eichenberger E, Holland TL, Fowler VG Jr. (2015) *Staphylococcus aureus* infections: epidemiology, pathophysiology, clinical manifestations, and management. *Clin Microbiol* Rev doi:10.1128/CMR.00134-14

Tripathi C, Mahato NK, Rani P, Singh Y, Kamra K, Lal R. (2016) Draft genome sequence of Lampropedia cohaerens strain CT6 T isolated from arsenic rich microbial mats of a Himalayan hot water spring. *Standards Genomic Sciences.* 11(1):64.

Venturi V (2006) Regulation of quorum sensing in *Pseudomonas, FEMS Microbiol Reviews.* 30(2):274–291. https://doi.org/10.1111/j.1574-6976.2005.00012.x

Williams P, Camara M (2009) Quorum sensing and environmental adaptation in Pseudomonas aeruginosa: a tale of regulatory networks and multifunctional signal molecules. *Curr Opin Microbiol.* 12:182–191.

Woebken D, Burow LC, Behnam F, Mayali X, Schintlmeister A, Fleming ED, Pett-Ridge J *et al.* (2015) Revisiting N 2 fixation in Guerrero Negro intertidal microbial mats with a functional single-cell approach. *ISME J.* 9(2):485–496.

Wong HL, Smith DL, Visscher PT, Burns BP (2015) Niche differentiation of bacterial communities at a millimeter scale in Shark Bay microbial mats. *Sci Rep.* 5:15607.

5 Succession of Bacterial Communities in Environmental Biofilm Structures

Fiorela Nievas, Walter Giordano and Pablo Bogino

CONTENTS

5.1 INTRODUCTION

Biofilms, a very common community in nature, are composed of microorganisms associated in a self-produced matrix of extracellular polymeric compounds. Bacterial biofilms are made up of either a single species – or, more often in nature, multiple species. This collective lifestyle has made it possible for most microorganisms to live in natural, clinical and industrial locations. In fact, most bacteria appear to form biofilms in different environments, and biofilms work not only as a protective mechanism against hostile conditions but also as functional blocks with crucial ecological roles (Armbruster and Parsek, 2018). Their formation depends on bacteria's ability to adhere to different surfaces (i.e. biotic or abiotic), colonize a niche and develop a well-organized community. Sessile cells form a multispecies consortium immersed into a matrix of organic exopolymeric compounds (EPS), which include exopolysaccharides, extracellular DNA and proteins (Jamal *et al.*, 2018; Bogino *et al.*, 2013). This is effectively a dynamic three-dimensional structure in which the activities of its members and the role of EPS are well-defined, so that specific functions may be fulfilled for the whole community (Flemming and Wingender, 2010). Resembling multicellular assemblies, environmental bacterial biofilms are characterized by task

DOI: 10.1201/9781003184942-6

division among members, functional complexity and high taxonomical diversity (Shi *et al.*, 2015; Jackson, 2003). Thus, both succession of species and succession of activities are expected to occur within them.

The theory of ecological succession was among the first to be defined in the field of ecology (Odum, 1969). Applied mainly to plant communities, this theory serves to explain how the diversity and structure of an ecosystem change. A very common example occurs after a forest fire: succession theory can describe the patterns followed by plants which are able to colonize the habitat in the immediate aftermath of the event and also in the following generations, as some initial colonizers become locally extinct while others come to dominate or are joined by immigrant species. Researchers attempted to observe similar trends in microbial systems, but the process of succession in these communities turned out not to fit the same definition (Connell and Slatyer, 1977) due to the differences between superior eukaryotic organisms and microorganisms. Predictions about the succession in bacterial biofilm communities are conditioned by multiple factors (Lyautey *et al.*, 2005; Rickard *et al.*, 2003) which are difficult to measure, since they are associated with environmental conditions that are ever-changing, often at a microscale. This was confirmed by several studies that observed bacterial community succession under varying conditions in different habitats, such as plant rhizospheres (Tkacz *et al.*, 2015). Although there seems to be a certain common order characterized by the recruitment of new organisms and the loss of others, this is only a very small part of the whole process (Zhao *et al.*, 2019). For instance, one gram of beech forest and one of field soil have been estimated to contain 10^5 and 10^6 different bacterial species, respectively. The majority of these bacteria are likely to be present in biofilms, which offer several advantages compared to a planktonic lifestyle in the soil (Burmølle *et al.*, 2007).

The advent of molecular methods has revolutionized microbial ecology, and the taxonomic characterization of microbial communities has become widespread (Woodcock and Sloan, 2017). Naturally occurring microbial biofilms that contain not only bacteria but also yeasts, fungi, algae and protozoa feature diverse assemblages at both taxonomic and functional levels. Due to this complexity, succession patterns in biofilms do not match those of simple microbial communities (Jackson *et al.*, 2001). Although bacteria are well-known to play essential roles in the functioning of different ecosystems, the mechanisms underlying those roles would be better understood if we could learn more about species succession and its related physiological changes in bacterial biofilm communities. Several questions remain to be answered about the role and progression of different microbiomes, and this is why microbial community succession remains an active topic of research (Brislawn *et al.*, 2019). This chapter reviews the available knowledge about some aspects of microbial succession in biofilm communities in different environments.

5.2 MODELS TO DESCRIBE BACTERIAL COMMUNITY SUCCESSION IN BIOFILM STRUCTURES

Ecological succession is an early theory, traditionally applied to superior organisms to describe structural changes in the communities within an ecosystem (Odum, 1969). Its principles could not be applied to the microbial world until recent decades, when molecular techniques fostered the emergence of microbial ecology (Margulies *et al.*, 2005; Fierer *et al.*, 2010). Since most bacterial life develops within biofilm structures, the study of bacterial succession often centers around that occurring in communities associated as biofilm. In recent years, some crucial characteristics of the development of these communities have been observed, and they have served as the basis for conceptual models of microbial community succession.

5.2.1 NICHE-BASED MODEL

Initial conceptual models for bacterial succession were based on observations of how the communities dynamically varied from their initial colonizers to the members in mature structures. This eventually led to the niche-based model, which posits that the structural and functional patterns of the community are essentially governed by ecological processes related to interspecies interactions and competition for resources (Jackson *et al.*, 2001). At later stages of biofilm succession, bacterial interactions and predation are believed to be important factors (Hibbing *et al.*, 2010; Zhou and Ning, 2017; Kurm *et al.*, 2019). Although such deterministic models are useful to understand the concept of microbial succession, they do not provide mathematical explanations through measurable parameters.

5.2.2 NEUTRAL COMMUNITY MODEL

The neutral community model, based on stochastic ecological factors, offers mathematical explanations of the patterns of succession in biofilms (Woodcock *et al.*, 2007; Chase and Myers, 2011). It asserts the competitive equivalence of individuals, the use of finite resources and spaces, and events of replacement triggered by deaths and replication or immigration. Application of this model was shown to be better than niche-based ones, since it reproduces common features found in different studies (Woodcock and Sloan, 2017). Successional events described by the neutral model are organized into three stages of development. The first stage involves: (i) colonization of the surface (or substrate) by any individual cells able to occupy available sites; (ii) no competition between individuals; and (iii) initial growth of species whose richness progressively increases. After the substrate is covered by different cells, growth and biodiversity decline during the second phase due to new cells competing for the vacancies left by dead cells. The arrival of other taxa is hindered by replicative cells already present in the community. The last stage is characterized by a second growth phase and a new increase in species richness, thanks to the biofilm maturing into a bigger structure able to hold new cell arrivals.

5.2.3 TOWARD A FUNCTIONAL MODEL?

The formation of biofilm involves three main events: attachment to the surface, growth and detachment. Succession takes place in all three. During attachment, more adhesive strains or species are at an advantage to efficiently colonize the substrate. Biofilm growth is dominated by cell division, the synthesis of exopolymers and task division. The result is a mature structure whose characteristics depend on the colonized surface, the prevalent species and the ecological role of the whole community. Finally, the detachment of certain members also contributes to succession, since they look for new colonization niches and the composition of the biofilm changes.

Such simplified assumptions can become more complex by taking into account the physiological roles of the members, their appearance, and the generation or obstruction of localized niches (spatial heterogeneity). The primary colonizers might play a major role in the fate of succession by modifying localized niches through functional processes (primary productivity, nitrogen fixation, inter-kingdom interactions, etc.). This would mean that similar phylogenetic taxa are more likely to be recruited due to analogous habitat requirements and activities (Martiny *et al.*, 2013). Succession patterns may also be linked to members having different ecological roles, and prokaryotes may be included, as well as eukaryotes (Bernstein *et al.*, 2017). This would make sense since biofilms are multispecies communities with well-defined task division among members (van Gestel *et al.*, 2015; Momeni, 2018).

5.2.4 Usefulness of Models to Study Bacterial Biofilm Succession

The models described are relevant and useful under certain considerations. They explain the changes observed in species diversity in a range of long-term biofilm studies in aquatic systems, but they are not applicable to all cases of microbial community succession in natural environments. The environment might therefore be the main determinant in the settlement and progression of a given microbial community. In just a few very different environments, such as soil, marine water or industrial pipeline systems, variability is readily introduced by many factors specific to each setting. The situation can be further complicated by the existence of superior organisms in the same niche (plant rhizosphere or phylosphere, dental plaque, etc.). All things considered, a unified observable model remains elusive for biofilm community succession, and particular models or theories offer better explanations.

5.3 COMMON FEATURES OF SUCCESSIONAL EVENTS IN BACTERIAL BIOFILM COMMUNITIES

Bacteria face different challenges imposed by the environment, such as variable amounts of nutrients, exposure to stressors and the presence of other organisms in the same niche. Several reports have described patterns of bacterial biofilm succession in different environments, ranging from natural habitats – such as marine waters, plant rhizospheres and dental plaque – to abiotic surfaces such as plastic (Table 5.1).

In general, succession dynamics are associated with the settlement of a pioneering community by primary colonizers, some of which will persist while others will be replaced over time. Different processes might take place during this initial colonization (e.g. microenvironment modification, generation of new niches, use of resources, and production of different exopolymeric compounds, communicative signals, and substances harmful to other organisms). All of this profoundly changes the initial environment into somewhere more or less accessible to different microorganisms, and opens up the possibility of various interactions such as collaboration or

TABLE 5.1

Summary of Representative Studies on Succession of Bacterial Biofilm Communities in Different Sources or Environments

Source/ Environment	Initial Taxa (Primary or Pioneer Colonizers)	Taxa Replacement/ Co-Existence (Secondary Colonizers)	Overall Predominant Taxa	Remarks	Reference
Tidal creek in a salt marsh	Gammaproteobacteria (*Alteromonas*).	Gammaproteobacteria → Alphaproteobacteria.	Alphaproteobacteria (Roseobacter subgroup).	Microorganisms of Roseobacter subgroup as rapid and ubiquitous colonizers of surfaces in coastal environments.	Dang and Lovell, 2000
Coastal marine water	Gammaproteobacteria (*Alteromonas, Pseudomonas, Acinetobacter*, uncultured).	Gammaproteobacteria → Alphaproteobacteria.	Alphaproteobacteria (*Methylobacterium, Loktanella, Pelagibacter*, and uncultured).	Gammaproteobacteria species were the pioneering population on three kinds of surfaces (acrylic, glass and steel).	Lee *et al.*, 2008
Coastal marine water	Gammaproteobacteria (genus *Oleibacter*) and Bacteirodetes (Flavobacteriia).	Gammaproteobacteria → Alphaproteobacteria (Rhodobacteraceae).	Alphaproteobacteria and Bacteirodetes (Flavobacteriia).	Marine biofilms were shown to be highly dynamic, with a predominance of inter-specific cooperative and mutualistic interactions rather than competitive.	Pollet *et al.*, 2018
Seawater	PVC group: gammaproteobacteria (Alteromonadaceae, Cellvibrionaceae and Oceanospirillaceae). Other plastics: Cyanobacteria and Flavobacteriaceae and Oleiphilaceae families.	Hyphomonadaceae, Flavobacteriaceae, Rhodobacteraceae, Planctomycetaceae families dominating all surfaces. Cyanobacteria remains as a main group on polypropylene surfaces.	High relative abundance on all surfaces of Flavobacteriaceae, Rhodobacteraceae, Planctomycetaceae and Phyllobacteriaceae families.	Specific bacterial taxa were found at all stages of biofilm development on specific surfaces (PVC vs. other plastics). Differences were greater after short periods (one week) than at later stages of incubation.	Pinto *et al.*, 2019

(Continued)

TABLE 5.1 *(Continued)*

Source/ Environment	Initial Taxa (Primary or Pioneer Colonizers)	Taxa Replacement/ Co-Existence (Secondary Colonizers)	Overall Predominant Taxa	Remarks	Reference
Marine fouling	Flavobacteria, a and gammaproteobacteria.	Primary colonizers → Actinobacteria and Planctomycetia.	Actinobacteria, Planctomycetia and Cyanobacteria.	Biofouling was driven by attachment of free-living seawater microorganisms to the surfaces (recruitment of primary colonizers). Then, they were replaced by more competitive specialist taxa.	Raeid *et al.*, 2019
Marine fouling	Spring: gammaproteobacteria (*Oleibacter* genus) and Mollicutes. Winter: alphaproteobacteria (Rhodobacterales). Also, early strong presence of Firmicutes and Actinobacteria.	Actinobacteria (*Propionibacterium*), alphaproteobacteria (*Loktanella*, *Octadecabacter*) and gammaproteobacteria (*Pseudoalteromonas*).	Overall dominance of the orders Rhodobacterales (Alphaproteobacteria) and Flavobacteriales (Bacteroidetes) and late dominance of photoautotrophic groups (mostly diatoms).	Distinct temporal patterns were determined in different seasons (winter – spring) at early stages. Later, biofilm communities appeared to converge toward a similar composition independently of the season.	Antunes *et al.*, 2020
Drinking water	Gammaproteobacteria (*Serratia* and *Pseudomonas*) Bacteroidetes (*Cloacibacterium*) and betaproteobacteria (*Diaphorobacter*).	Initial community → Gammaproteobacteria (Chromatiales) and Mycobacterium-related species.	Mycobacterium-like phylotypes.	Succession explained by a combination of different ecological interactions (facilitation and competition).	Revetta *et al.*, 2013
Chlorinated drinking water distribution system	Bacteria: Gamma and betaproteobacteria. Fungi: Ascomycota (unspecified class), Sordaryomicetes, Dothideomycetes and Saccharomycetes.	Bacteria: Gamma, beta and alphaproteobacteria. Fungi: Ascomycota (unspecified class) and Sordaryomicetes.	Bacteria of the genera *Pseudomonas* (gammaproteobacteria), *Massillia* (betaproteobacteria) and	Bacterial community grew, was enriched and became more diverse over time compared to fungal community.	Douterelo *et al.*, 2018

			Sphingomonas (alphaproteobacteria); and fungi *Acremonium* and *Neocosmopora* (Sordaryomicetes) were increasingly present over time.		
Wastewater distribution systems	Alphaproteobacteria (*Pedomicrobium, Sphingomonas, Sphingobium, Blastomonas*) and betaproteobacteria (*Azospira, Cupriavidus, Sideroxidans, Zooglea*) were primary colonizers regardless the pipe material or disinfection system. Other pioneer taxa were Sphingobacteriia and Actinobacteria.	Pioneer species created adaptive habitats for Nitrospira, Acidobacteria, and other genera of alpha and betaproteobacteria.	Proteobacteria, Nitrospirae, Bacteroidetes, Acidobacteria, Planctomycetes, Actinobacteria, and Verrucomicrobia comprised the dominant phyla in biofilm.	A primary succession pattern was followed in the biofilm communities. Bacteria were primarily selected by abiotic factors (disinfection and properties of pipe materials). Established bacteria could modify the niche (adaptation, alteration of microenvironment) to enable the settlement of more diverse microbiota.	Zhang *et al.*, 2019
Activated sludge in flow cells fed with different substrate concentrations	*Enterobacter* and *Acinetobacter.*	Persistence of primary colonizers. Appearance of other genera (*Bacillus, Comamonas, Elizabethkingia, Stenotrophomonas*) mainly under reduced substrate concentration.		Gammaproteobacteria were the dominant bacterial taxa.	The succession dynamics of dominant (Gammaproteobacteria) and other taxa (Bacilli, Betaproteobacteria, Flavobacteria) were dependent on nutrient availability.	Yuan *et al.*, 2020

(Continued)

TABLE 5.1 (Continued)

Source/Environment	Initial Taxa (Primary or Pioneer Colonizers)	Taxa Replacement/Co-Existence (Secondary Colonizers)	Overall Predominant Taxa	Remarks	Reference
Supragingival plaque	*Streptococcus mitis* and *Neisseria mucosa* (one day).	*Capnocytophaga gingivalis*, *Eikenella corrodens*, *Veillonella parvula* and *Streptococcus oralis* (1–4 days).	*Campylobacter rectus*, *Campylobacter showae*, *Prevotella melaninogenica* and *Prevotella nigrescens* (4–7 days).	Succession was similar in patients with periodontitis and healthy individuals.	Teles *et al.*, 2012
Subgingival plaque	*Streptococcus mitis* (one day).	*Veillonela parvula* and *Capnocytophaga gingivalis* (1–4 days).	*Capnocytophaga sputigena* and *Prevotella nigrescens* (4–7 days).		
Root of lettuce grown aeroponically	Mainly beta and alphaproteobacteria. Presence of gammaproteobacteria (12 days).	Mainly beta and alphaproteobacteria (19 days).	Mainly alphaproteobacteria (26 days).	Water delivery method and plant development were associated with successional patterns and dynamism of bacterial communities associated with plant roots.	Edmonds *et al.*, 2019
Rhizosphere of *Avena fatua*	The relative abundance of several bacterial phyla was assessed for 12 weeks in two different seasons. Members of the Proteobacteria phylum (mainly alpha and betaproteobacteria) and Bacteroidetes became enriched in the proximity of plant roots, whereas Actinobacteria, Acidobacteria, Firmicutes, Planctomycetes, Gemmatimonadetes, and Chloroflexi decreased in the rhizosphere of *A. fatua*.			Roots selectively stimulated the bacterial phyla associated with them. A pattern of temporal succession characterized by a decrease in phylogenetic diversity was determined over different seasons.	Shi *et al.*, 2015
Rhizospheric soil of *Medicago sativa*	Gammaproteobacteria (Xanthomonadales).	Gammaproteobacteria (Xanthomonadales and Enterobacterales).	Alphaproteobacteria and Actinobacteria.	Gammaproteobacteria were identified as highly adhesive strains and the primary colonizers of developing biofilm associated with roots. Mature biofilm in the rizosphere was mainly formed by α-proteobacteria and Actinobacteria.	Nievas *et al.*, 2021

cooperation and co-existence, competition and replacement, competition and persistence, etc. Secondary colonizers then join the community and modify its initial structure and functioning. The whole structure acquires more biomass and integrates different predominant taxa that fulfills specific tasks. The ecological roles of the community are also determined by successional changes in functionality.

This kind of progression (pioneering colonizers, secondary colonizers, overall predominant taxa) has been described for both biotic and abiotic surfaces. Some common features have been identified in both surfaces (Nemergut *et al.*, 2013): (i) appearance or loss of taxa (or functions) that modifies the local habitat; leading to (ii) a change in niche availability during community succession (Jones *et al.*, 1997); (iii) introduction of new taxa and co-existence or replacement (and loss of taxa) in a particular niche, which defines mechanisms of turnover and nestedness (Baselga, 2012); (iv) abundance of species distribution congruent with low abundance for most taxa; (v) high abundance of minor taxa related to better adaptation to environmental factors (Delgado-Baquerizo *et al.*, 2018); and (vi) influence of stochastic and deterministic factors (strong environmental features and selection forces) (Stegen *et al.*, 2012). In spite of these generalizations, and as expressed in (vi), successional dynamics are markedly influenced by the characteristics of each environment (biotic and abiotic factors, type of surface, presence of superior organisms, physical variations, time of observation, anthropogenic activities, etc.). Many ecosystems contain numerous microhabitats suitable for both microbial and superior organisms. In the soil ecosystem and its different microenvironments (from soil particles to the rhizosphere of plants), bacterial succession in biofilm communities is influenced by soil properties (composition, organic matter content, availability of mineral nutrients, porosity, etc.), plant species, seasonal changes and crop rotation, among other factors (Zhao *et al.*, 2019; Micallef *et al.*, 2009). The rhizosphere – the portion of soil directly influenced by plant root secretions – works as an interface between soil, microorganisms and plants. Crucial ecological events taking place in this environment modify the nutritional and physicochemical properties of the surrounding soil, provide protection to roots growing under abiotic stresses and improve the soil's moisture and porosity (Khan *et al.*, 2020). Microbial rhizospheric succession is of particular importance in agriculture, since the close bidirectional interaction between the microbial community (microbiota) and the host plant is an essential determinant of crop health and yield. The succession of biofilm-like bacterial structures in plant rhizospheres can provide clues about the mechanisms behind microbial adaptability, crop growth and their responses to agricultural practices (Ramakrishna *et al.*, 2019; Tkacz *et al.*, 2015).

Succession in the alfalfa rhizosphere depends on the quick colonization of the roots by strains (mainly gammaproteobacteria) with a high capacity to develop biofilm on inert surfaces (i.e. highly adhesive strains), followed by their replacement with strains less able to produce biofilm in vitro but probably more stable when associated with roots (alphaproteobacteria and actinobacteria) (Nievas *et al.*, 2021). Thus, the mature root-associated biofilm community is composed of less adhesive taxa incapable of initial colonization, but better suited for interaction with the plant. The settlement of such a stable interactive community depends on how well the niches created by the pioneering colonizers enable fast diversification later on (Jones *et al.*, 2007; Dang *et al.*, 2008; Briand *et al.*, 2017). The key role of gammaproteobacteria as initial colonizers, as well as the change toward other taxa when the community starts to mature, have also been identified for biofilms in drinking and marine water (Revetta *et al.*, 2013; Pollet *et al.*, 2018).

5.4 CONCLUDING REMARKS

Primary ecological succession involves the colonization of available habitats by certain species. These initial species can be progressively replaced by others, or new species can join the community, as allowed by mechanisms of facilitation, tolerance, inhibition and competition. The study of these processes, which has been extensive in superior organisms such as plants or animals, requires different tools in the case of microbial communities living in biofilms. The whole process could be depicted as similar to the one observed for superior organisms in terms of mechanisms, but distinctive succession models have come up to address the unique properties of bacterial biofilm communities. Nevertheless, not enough is known about the determinants affecting their settlement, successional progression and functional progressive activities. A more in-depth understanding of the role of bacterial biofilms in different ecological niches would have a significant impact on human, animal, plant and ecological healthcare, agricultural and antifouling practices, and the economic aspects of many other activities.

REFERENCES

Antunes JT, Sousa AGG, Azevedo J, Rego A, Leão PN, Vasconcelos V (2020) Distinct Temporal Succession of Bacterial Communities in Early Marine Biofilms in a Portuguese Atlantic Port. *Front Microbiol.* 11:1938. doi: 10.3389/fmicb.2020.01938.

Armbruster CR, Parsek MR (2018) New insight into the early stages of biofilm formation. *Proc Natl Acad Sci USA.* 115(17):4317–4319. doi: 10.1073/pnas.1804084115.

Baselga A (2012) The relationship between species replacement, dissimilarity derived from nestedness, and nestedness. *Glob Ecol Biogeogr.* 21:1223–1232. https://doi.org/10.1111/j.1466-8238.2011.00756.x.

Bernstein HC, Brislawn CJ, Dana K, Flores-Wentz T, Cory AB, Fansler SJ, Fredrickson JK, Moran JJ (2017) Primary and heterotrophic productivity relate to multi-kingdom diversity in a hypersaline mat. *FEMS Microbiol Ecol.* 93(10):fix121. https://doi.org/10.1093/femsec/fix121.

Bogino P, Oliva M, Sorroche F, Giordano W (2013) The role of bacterial biofilms and surface components in plant-bacterial associations. *Int J Mol Sci.* 14:15838–15859. doi: 10.3390/ijms140815838.

Briand J-F, Barani A, Garnier C, Réhel K, Urvois F, LePoupon C, Bouchez A, Debroas D, Bressy C (2017) Spatio-temporal variations of marine biofilm communities colonizing artificial substrata including antifouling coatings in contrasted French coastal environments. *Microb Ecol.* 74:585–598. doi: 10.1007/s00248-017-0966-2.

Brislawn CJ, Graham EB, Dana K, Ihardt P, Fansler SJ, Chrisler WB, Cliff JB, Stegen JC, Moran JJ, Bernstein HC (2019) Forfeiting the priority effect: turnover defines biofilm community succession. *ISME J.* 13:1865–1877. doi: 10.1038/s41396-019-0396-x.

Burmølle M, Hansen LH, Sørensen SJ (2007) Establishment and early succession of a multi-species biofilm composed of soil bacteria. *Microb Ecol.* 54(2):352–362. DOI: 10.1007/s00248-007-9222-5.

Chase JM, Myers JA (2011) Disentangling the importance of ecological niches from stochastic processes across scales. *Philos Trans R Soc Lond B Biol Sci.* 366(1576):2351–2363. doi: 10.1098/rstb.2011.0063.

Connell JH, Slatyer RO (1977) Mechanisms of succession in natural communities and their role in community stability and organization. *Am Nat.* 111:1119–1144. https://www.jstor.org/stable/2460259?seq=1.

Dang H, Li T, Chen M, Huang G (2008) Cross-ocean distribution of Rhodobacterales bacteria as primary surface colonizers in temperate coastal marine waters. *Appl Environ Microbiol.* 74(1):52–60. doi: 10.1128/AEM.01400-07.

Dang H, Lovell CR (2000) Bacterial primary colonization and early succession on surfaces in marine waters as determined by amplified rRNA gene restriction analysis and sequence analysis of 16S rRNA genes. *Appl Environ Microbiol.* 66(2):467–475. doi: 10.1128/AEM.66.2.467-475.2000.

Delgado-Baquerizo M, Oliverio AM, Brewer TE, Benavent-González A, Eldridge DJ, Bardgett RD, Maestre FT, Singh BK, Fierer N (2018) A global atlas of the dominant bacteria found in soil. *Science.* 359(6373):320–325. doi: 10.1126/science.aap9516.

Douterelo I, Fish KE, Boxall JB (2018) Succession of bacterial and fungal communities within biofilms of a chlorinated drinking water distribution system. *Water Res.* 141:74–85. doi: 10.1016/j.watres.2018.04.058.

Edmonds JW, Sackett JD, Lomprey H, Hudson HH, Moseret DP (2019) The aeroponic rhizosphere microbiome: community dynamics in early succession suggest strong selectional forces. *Ann van Leeu.* 113(1):83–99. https://doi.org/10.1007/s10482-019-01319-y.

Fierer N, Nemergut D, Knight R, Craine JM (2010) Changes through time: integrating microorganisms into the study of succession. *Res Microbiol.* 161:635–642. doi: 10.1016/j.resmic.2010.06.002.

Flemming HC, Wingender J (2010) The biofilm matrix. *Nat Rev Microbiol.* 8(9):623. doi: 10.1038/nrmicro2415.

Hibbing ME, Fuqua C, Parsek MR, Peterson SB (2010) Bacterial competition: surviving and thriving in the microbial jungle. *Nat Rev Microbiol.* 8(1):15–25. doi: 10.1038/nrmicro2259.

Jackson CR (2003) Changes in community properties during microbial succession. *Okios.* 101:444–448. doi:10.1034/j.1600-0706.2003.12254.x.

Jackson CR, Churchill PF, Roden EE (2001) Successional changes in bacterial assemblage structure during epilithic biofilm development. *Ecology.* 82:555–566. https://doi.org/10.1890/0012-9658(2001)082[0555:SCIBAS]2.0.CO;2.

Jamal M, Ahmad W, Andleeb S, Jalil F, Imran M, Nawaz MA, Hussain T, Ali M, Rafiq M, Kamil MA (2018) Bacterial biofilm and associated infections. *J Chin Med Assoc.* 81:7–11. doi: 10.1016/j.jcma.2017.07.012.

Jones CG, Lawton JH, Shachak M (1997) Positive and negative effects of organisms as physical ecosystem engineers. *Ecology.* 78:1946–1957. https://doi.org/10.1890/0012-9658(1997)078[1946:PANEOO]2.0.CO;2.

Jones P, Cottrell M, Kirchman D, Dexter SC (2007) Bacterial community structure of biofilms on artificial surfaces in an estuary. *Microb Ecol.* 53(1):153–162. doi: 10.1007/s00248-006-9154-5.

Khan N, Ali S, Tariq H, Latif S, Yasmin H, Mehmood A, Shahid MA (2020) Water conservation and plant survival strategies of rhizobacteria under drought stress. *Agronomy.* 10:1683. https://doi.org/10.3390/agronomy10111683.

Kurm V, van der Putten WH, Weidner S, Geisen S, Snoek BL, Bakx T, Gera Hol WH (2019) Competition and predation as possible causes of bacterial rarity. *Environ Microbiol.* 21(4):1356–1368. doi: 10.1111/1462-2920.14569.

Lee JW, Nam JH, Kim YH, Lee KH, Lee DH (2008) Bacterial communities in the initial stage of marine biofilm formation on artificial surfaces. *J Microbiol.* 46(2):174–182. doi: 10.1007/s12275-008-0032-3.

Lyautey E, Jackson CR, Cayrou J, Rols JL, Garabetian F (2005) Bacterial community succession in natural river biofilm assemblages. *Microb Ecol.* 50(4):589–601. https://doi.org/10.1007/s00248-005-5032-9.

Margulies M, Egholm M, Altman WE, Attiya S, Bader JS, *et al.* (2005) Genome sequencing in microfabricated high-density picolitre reactors. *Nature.* 437(7057):376–380. doi: 10.1038/nature03959.

Martiny AC, Treseder K, Pusch G (2013) Phylogenetic conservatism of functional traits in microorganisms. *ISME J* 7(4):830–838. doi: 10.1038/ismej.2012.160.

Micallef SA, Channer S, Shiaris MP, Colón-Carmona A (2009) Plant age and genotype impact the progression of bacterial community succession in the *Arabidopsis* rhizosphere. *Plant Signal Behav.* 4(8):777–780. https://doi.org/10.4161/psb.4.8.9229.

Momeni B (2018) Division of Labor: How microbes split their responsibility. *Curr Biol.* 28(12):R697–R699. doi: 10.1016/j.cub.2018.05.024.

Nemergut DR, Schmidt SK, Fukami T, O'Neill SP, Bilinski TM, Stanish LF, Knelman JE, Darcy JL, Lynch RC, Wickey P, Ferrenberg S (2013) Patterns and processes of microbial community assembly. *Microbiol Molec Biol Rev.* 77(3):342–56. doi: 10.1128/MMBR.00051-12.

Nievas F, Primo E, Foresto E, Cossovich S, Giordano W, Bogino P (2021) Early succession of bacterial communities associated as biofilm-like structures in the rhizosphere of alfalfa. *Appl Soil Ecol.* 157:103755. https://doi.org/10.1016/j.apsoil.2020.103755.

Odum EP (1969) The strategy of ecosystem development. *Science.* 164(3877):262–270. doi: 10.1126/science.164.3877.262.

Pinto M, Langer TM, Hüffer T, Hofmann T, Herndl GJ (2019) The composition of bacterial communities associated with plastic biofilms differs between different polymers and stages of biofilm succession. *PLoS One.* 14(6):e0217165. https://doi.org/10.1371/journal.pone.0217165.

Pollet T, Berdjeb L, Garnier C, Durrieu G, Le Poupon C, Misson B, Jean-François B (2018) Prokaryotic community successions and interactions in marine biofilms: the key role of Flavobacteriia. *FEMS Microbiol Ecol.* 94(6). doi:1093/femsec/fiy083.

Raeid MMA, Al Fahdi D, Muthukrishnan T (2019) Short-term succession of marine microbial fouling communities and the identification of primary and secondary colonizers. *Biofouling.* 35(5):526–540. doi: 10.1080/08927014.2019.1622004.

Ramakrishna W, Yadav R, Li K (2019) Plant growth promoting bacteria in agriculture: two sides of a coin. *Appl Soil Ecol.* 138:10–18. https://doi.org/10.1016/j.apsoil.2019.02.019.

Revetta RP, Gomez-Alvarez V, Gerke TL, Curioso C, Santo Domingo JW, Ashbolt NJ (2013) Establishment and early succession of bacterial communities in monochloramine-treated drinking water biofilms. *FEMS Microbiol Ecol.* 86:404–414. https://doi.org/10.1111/1574-6941.12170.

Rickard AH, Gilbert P, High NJ, Kolenbrander PE, Handley PS (2003) Bacterial coaggregation: an integral process in the development of multi-species biofilms. *Trends Microbiol.* 11(2):94–100. https://doi.org/10.1016/s0966-842x(02)00034-3.

Shi S, Nuccio E, Herman DJ, Rijkers R, Estera K, Li J, da Rocha UN, He Z, Pett-Ridge J Brodie EL, Zhou J, Firestone M (2015) Successional trajectories of rhizosphere bacterial communities over consecutive seasons. *mBio.* 6(4):e00746. doi: 10.1128/mBio.00746-15.

Stegen JC, Lin X, Konopka AE, Fredrickson JK (2012) Stochastic and deterministic assembly processes in subsurface microbial communities. *ISME J.* 6(9):1653–1964. doi: 10.1038/ismej.2012.22.

Teles FR, Teles RP, Uzel NG, Song XQ, Torresyap G, Socransky SS, Haffajee AD (2012) Early microbial succession in redeveloping dental biofilms in periodontal health and disease. *J Periodontal Res.* 47(1):95–104. https://doi.org/10.1111/j.1600-0765.2011.01409.x.

Tkacz A, Cheema J, Chandra G, Grant A, Poole PS (2015) Stability and succession of the rhizosphere microbiota depends upon plant type and soil composition. *ISME J.* 9(11):2349–2359. doi: 10.1038/ismej.2015.41.

van Gestel J, Vlamakis H, Kolter R (2015) Division of labor in biofilms: the ecology of cell differentiation. *Microbiol Spectr.* 3(2). MB-0002-2014. doi: 10.1128/microbiolspec. MB-0002-2014.

Woodcock S, Sloan WT (2017) Biofilm community succession: a neutral perspective. *Microbiol.* 163:664–668. doi: 10.1099/mic.0.000472.

Woodcock S, van der Gast CJ, Bell T, Lunn M, Curtis TP, Head IM, Sloan WT (2007) Neutral assembly of bacterial communities. *FEMS Microbiol Ecol.* 62:171–180. doi: 10.1111/j.1574-6941.2007.00379.x.

Yuan S, Yu Z, Pan S, Huang J, Meng F (2020) Deciphering the succession dynamics of dominant and rare genera in biofilm development process. *Sci Total Environ.* 739:139961. doi: 10.1016/j.scitotenv.2020.139961.

Zhang G, Li B, Guo F, Liu J, Luan M, Liu Y, Guan Y (2019) Taxonomic relatedness and environmental pressure synergistically drive the primary succession of biofilm microbial communities in reclaimed wastewater distribution systems. *Environ Int.* 124:25–37. doi: 10.1016/j.envint.2018.12.040.

Zhao M, Yuan J, Shen Z, Dong M, Liu H, Wen T, Li R, Shen Q (2019) Predominance of soil vs root effect in rhizosphere microbiota reassembly. *FEMS Microbiol Ecol.* 95(10):139. https://doi.org/10.1093/femsec/fiz139.

Zhou J, Ning D (2017) Stochastic community assembly: does it matter in microbial ecology? *Microbiol Mol Biol Rev.* 81(4):e00002–e00017. doi: 10.1128/MMBR.00002-17.

6 Oral Biofilm of Hospitalized Patients

Thainah Bruna Santos Zambrano, Naga Raju Maddela, Rubén Jaime Szwom, and Ricardo Sergio Couto de Almeida

CONTENTS

6.1 INTRODUCTION

This chapter presents aspects focused on oral biofilm in hospitalized patients and its relationship with respiratory diseases, from the medical and practical point of view, for the creation of awareness across dental management and practices in the hospital environment to improve the buccal health and life of our patients.

Throughout history, it is observed that towns have developed intending to improve the quality of life of their population. Among the significant advances are hospitals, health aspects, and the appearance of practices carried out by professionals.

DOI: 10.1201/9781003184942-7

As diseases and calamities affected humanity, sometimes originating from human degradation itself, health professionals have been forced to seek new practices or techniques to minimize the suffering of their patients and the cure of their ills. Even with scientific and technological advancements, the process of change will always face new challenges.

The World Health Organization (WHO) declares the infections associated with healthcare like a severe problem for the security of the patient. Their impact has apparent effects on the hospitalization time and some kinds of impairment and aftermath, increasing resistance of pathogens, health expenditures, and morbidity and mortality (World Health Organization, 2004).

Some hospitalized patients have poor buccal health, and they do not have any way to report their status and their disagreement; the professionals who care for them are responsible for caring about maintenance of their health and life, which must be as complete as possible. Also, the critical patient's dental attention contributes to preventing hospitalization infections, mainly the respiratory, amid nosocomial pneumonia, which is one of the central infections in hospitalized patients in intensive care units (ICUs) helped by microorganisms which proliferate in the oropharynx. Its recurrence is worrying because it is prevalent in this kind of patient, causing an actual number of deaths, extending patient stays, and requiring more care and medicines.

This chapter reminds of the necessity of promotion and prevention measures for maintaining oral health and the stomatognathic system during a patient's stay, controlling oral biofilms, preventing caries, periodontal disease, periimplantitis infections, and some buccal problems such as malignant and premalignant injuries. It is added, in addition, that the dental care of the critic patient helps in the prevention of some hospital infections, mainly the respiratory diseases, including nosocomial pneumonia, which is one of the central infections in ICUs helped by microorganisms which grow in the oropharynx. Its recurrence is worrying because it is widespread in these types of patients, causing a significant number of deaths, increasing the time of stays in hospitals and use of medications and care.

6.2 BUCCAL BIOFILM

Biofilms are the community of microbes attached to the surface, notably in medical, industrial, and natural environments. In fact, life in one biofilm probably represents the predominant growth of the microbes in almost every environment (O'Toole, 2010).

In the oral cavity, biofilm is formed not only in natural teeth but also in restorative materials, prosthetic constructions, and dental implants (Eick, 2020), in which there are consortia of bacteria, viruses, and fungi. Most of the microorganisms found in oral biofilm are harmless natural inhabitants, but some have the potential to cause mineralized tissue damage or soft tissue infections (Øilo and Bakken, 2015).

Biofilm formation is a process of various lapses and is extremely hard. There is a requirement of some factotums and relevant conditions which shall exist in the oral cavity to guarantee the flow of the whole process (Krzysciak et al., 2016). Thus, Figure 6.1 illustrates all stages of dental biofilm formation, from initial colonization of the tooth surface by *Streptococcus* sp. to maturation, with acid production and demineralization of tooth enamel. The oral biofilm is unique among the different kinds of biofilms because it usually requires salivary glycoproteins of the guest to stay attached. The first stage of oral biofilm formation is the binding of acquired films, a very slim film containing proteins derived from salivary glycoproteins to a clean tooth surface (Figure 6.1C). The mechanism of the formation of the acquired film is based on the Gibbs law of free enthalpy in the sense of a better release of energy if the glycoproteins stay attached to the teeth surface. They have interaction between the different glycoproteins, other salivary components, and the teeth surface. The forces of these interactions can be roughly divided into three types: Long-range forces (50–100 nm between the two interacting components), midrange force (10–50 nm), and short-range forces (less than 5 nm) (Huang *et al.*, 2011). The long-range forces include Coulomb's interactions, van der Waals forces, and dipole–dipole interactions; the midrange strength forces include hydrophobic interactions; and the short-range strength forces include covalent connections, electrostatic interactions,

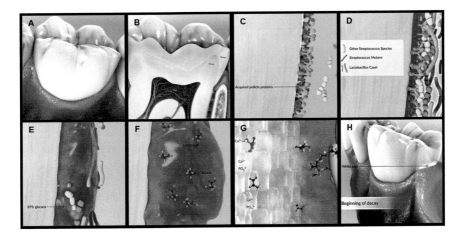

FIGURE 6.1 Stages of dental biofilm development: A – Illustrated image of a mandibular molar. B – Illustrated image of inner structure of the tooth. C – The acquired pellicle contains salivary proteins and cell debris adhered to the surface of the dental enamel. D – *Streptococcus* species make up the first colonizers, which will adhere to the tooth surface through the acquired pellicle and produce extracellular polysaccharides (in red). *Streptococcus mutans* and *Lactobacillus casei* are cariogenic species that can colonize the formed biofilm. E – The extracellular polymeric substances (EPS) produced by bacteria is mainly composed of glucans. As long as the bacteria get oxygen for cellular respiration, fermentation will not take place, and consequently there will be no acid production. F – After the biofilm reaches a certain thickness, the lack of oxygen forces the bacteria to carry out acid fermentation; the main acid produced is lactic acid. G – Lactic acid accumulation at the interface between dental enamel and biofilm causes the pH to drop below 5.5, dissociating hydroxyapatite from enamel prisms into Ca^{2+} and PO_4^{3-}. H – The initial erosion of the enamel surface causes the appearance of a white spot, called reversible caries. The figures were kindly provided by the Company Angelus S/A, Brazil.

hydrogen connections, ionic interactions, and Lewis acid–base interactions (Hannig and Hannig, 2009). With these forces, the proteins are absorbed and reorganized, some conformational changes take place, and some kinds of films are ready for the adhesion of pioneer bacteria.

About 1,000 different bacteria species have been identified in oral biofilm. The adaptation to a biofilm lifestyle implies regulating genes set, and then the micro-organisms may optimize the phenotypic properties for the particular environment. The biofilm structure provides various customs life environments (with different pH, nutrient, and oxygen availability) (Saini et al., 2011).

Despite the different harmless bacterial communities found in the biofilm, it has been shown that oral biofilms are the leading cause of oral diseases, like caries and periodontitis, which are caused by dysbiosis and imbalance of biofilm composition. The oral bacteria communities are influenced by physic and biological factotums, nutrient availability, pH potential, the oxy-reduction, diet, smoking and drinking habits, and general health and oral hygiene of the patient. These factotums may change the bacteria growth and the biofilm characteristics (Gedif Meseret, 2021).

Epidemiological studies, laboratory tests, and animal testing verify the asso-ciation between *Streptococcus mutans* with dental caries and bad hygiene in the buccal cavity (Figure 6.1D). *S. mutans* is a bacteria located in the buccal cavity making part of the dental biofilm, is gram-positive morphologically similar to the coconut, with optional anaerobic respiration, and presents lactic acid as the final product fermentation. The excessive production of this acid for this coconut in the presence of carbohydrates causes the pH to fall (under 5.5) and the demineraliza-tion of dental tissues, letting cavities develop (Figure 6.1F–H). This sickness has a multifactorial and global grasp and affects kids and adults. This sickness in some places of the world is rife because of the minimum hygiene habits in the popula-tion and bad local environment (unfluoridated water, lack of appropriate den-tal medication, and cariogenic diet) (Akthar et al., 2014; Dalmasso et al., 2015). *S. mutans* also has connections with non-oral infections like bacterial endocar-ditis. Therefore, it is crucial to get appropriate hygiene for reducing the bacterial amount in the dental biofilm, preventing bacteremia apparition and endocarditis.

For efficient biofilm control, mainly in special requirements, patients in hospital beds and ICUs use mechanical nature methods (toothbrush and floss) to encounter some problems that the patient has (Sharma and Galustians, 1994) (Figure 6.2).

6.3 INTENSIVE CARE UNIT

The origin of the ICU was at the start of the 20th century in the Johns Hopkins Hospital (USA) with the creation of 'recuperation halls' following neurosurgery (Pereira Júnior et al., 1999).

The complexity of the healthcare required by critical patients and some of its spe-cial characteristics – such as greater susceptibility, exposure to cross-transmission, and the need for invasive procedures – make the ICU, together with the emergency services and operating rooms, places where there is a high rate of incidents related to healthcare.

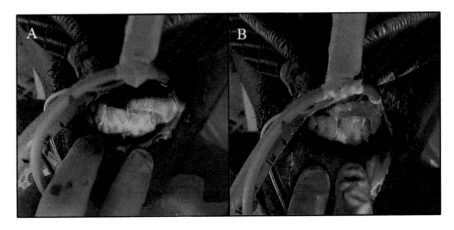

FIGURE 6.2   Dental biofilm of patient in ICU: A – Dental biofilm before oral care. B – Reduction of dental biofilm after oral hygiene (Santos Zambrano *et al.*, 2020).

The ICU is destined for critical patients with survival probabilities, requiring constant monitoring (24 hours a day) and more complex care than other patients. They currently play a decisive role in the probability of survival of critically sick patients, whether they are victims of trauma or any life threat. This situation has been increasing due to the number of new cases, and the quality of life support has accompanied the technology evolution.

Some of the equipment located in the ICU are thermometers, pulse oximeters (which measure the oxygen in patient tissue), blood pressure monitors, heart monitors, nasoenteral tubes (for feeding patients), urinary catheters (to collect urine), mask and oxygen catheters (to help with breathing), catheters (devices to help in the blood circulation), orotracheal tubes (which help with patient breathing), and mechanical fans (which help to inject and remove air when the patient cannot do it alone) (Celik and Eser, 2017).

The objective of the ventilatory prosthesis wear is to reduce the respiratory muscle work and maintain the gaseous changes, allowing the patient to recover their normal physiological state, capable of maintaining adequate levels of O_2 and CO_2 in the blood. The necessity of ventilator support may be the cause of the patient's placement in the ICU or be necessary for the course of treatment or in a postoperative state when the patient suffers from respiratory insufficiency, hypoxemia, and hypercapnia. Such ventilator support can be partial or complete; it depends on the patient's situation and the cause of their placement in the ICU (Jones *et al.*, 2010).

Mechanical ventilation is a life support method, and its primary function is to replace or help the respiration temporally. The objectives of mechanical ventilation can be broken down into physiological and clinical. The physiological ones seek to maintain or improve gas exchange, increase lung volume, and reduce the work of the respiratory muscles. The clinical objective of mechanical ventilation is to reverse hypoxemia, reverse respiratory acidosis, relieve respiratory effort, prevent or reverse

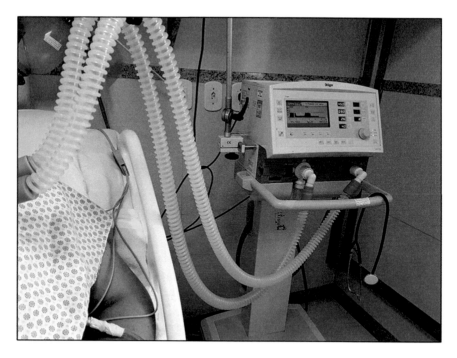

FIGURE 6.3   Mechanical artificial ventilation apparatus.

atelectasis, reverse respiratory muscle fatigue, allow the use of sedation and muscle relaxation, decrease systemic oxygen consumption and myocardial, reduce intracranial pressure, and stabilize the chest wall (Slutsky, 1993) (Figure 6.3).

6.4   HOSPITAL DENTISTRY

Hospital dentistry is a dental branch that seeks attention and buccal procedures in the hospital environment. The first dental services established occurred in 1901, in the Philadelphia General Hospital (USA), and its function was the dental care of patients and the training of students in this area (Willis, 1965).

The development of hospital dentistry arose in the United States after the middle of the 19th century with the efforts of Drs. Simon Hullihen and James Garretson. Hospital dentistry encompasses actions that go beyond imagined proportions and are attributed by society; since the procedures performed do not refer only to surgical interventions, the patient's approach must be taken into account and not only in the aspects related to the care of the oral cavity. As a harmony status, normality, or stiffness, buccal health has a meaning if it goes hand in hand with the patient's general healthcare (Cillo, 1996).

Hospital dentistry may be defined as the odonatological practice performed in a hospital environment to include the dental surgeon (DDS, for doctor of dental surgery) in the multi-professional team to give attention to hospitalized patients and improve the patients' general health and life quality (Figure 6.4).

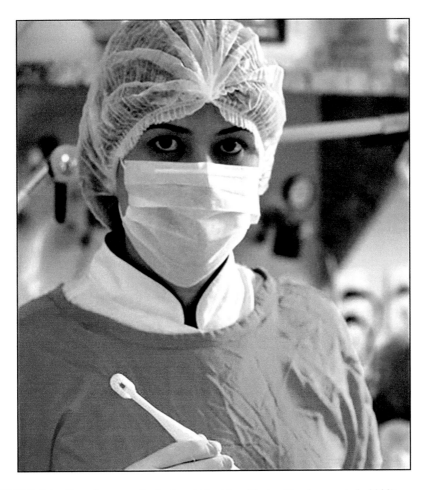

FIGURE 6.4 Dental surgeon in the hospital setting (Santos Zambrano *et al.*, 2020).

After creating the family health program, the DDS – a professional long seen as isolated from other medical health professionals – joined the multidisciplinary health team. The dental surgeon's performance gained space in a hospital environment, and the DDS started attending to hospitalized patients, wrecking every barrier and stereotype from the culture among the health professionals and nursing directly linked with this service.

6.5 RISK FACTORS FOR INFECTIONS ASSOCIATED WITH INTENSIVE THERAPY UNITS

Infection control practices based on the evidence are essential but represent a challenge for health institutions. Their achievement depends on the contextual factotums of the health system.

The infections associated with sanitary assistance represent an opposite effect more frequently, affecting hospitalized patients, resulting in increased morbidity and mortality, and hospitalization time and imbalances (Sax *et al.*, 2013).

ICU infections can be prevented by some caring methods, like hand washing, raising the head of the bed, constant absorption of oral secretions, thrombosis prevention, and caring for the medical legs and tubes. Regarding expiration, pneumonia associated with mechanical ventilation (VAP) can be prevented with buccal hygiene protocols and odonatological procedures for the remoting of oral infectious foci to reduce the number of buccal microorganisms, which can require the increasing of VAP – implying more time for the patient in the ICU as also the cost of such care (Franco *et al.*, 2018) (Figure 6.5).

The risk of contracting infections related to sanitary assistance is especially significant in the ICU. European data reveal that 19.5% of infections contracted during hospitalization were contracted while patients were in ICU compared to the 5.2% of acquired infections in other hospital units. Thus, the use of antimicrobials was high and, therefore, necessary in 56.5% of patients in ICUs. The WHO mentions that the result is unfavorable: around 30% of the patients affected by one or more healthcare-related infection contracted their infections in intensive therapy environments (Allegranzi *et al.*, 2010).

Healthcare-related infections pose the greatest threat to patient safety. Recent estimates of global morbidity and mortality associated with healthcare-related infections show evidence of a significant global public health problem (Haque *et al.*, 2018).

There are also indications for interventions and recommendations that can substantially reduce the incidence of healthcare-related infections. It is highlighted that 20–30% of these infectious complications are preventable (Magill *et al.*, 2014). On the face of it, healthcare-related infection control represents the key to improving the quality of care and the opportunity to save lives and reduce budgets (Zimlichman *et al.*, 2013).

Infection surveillance analysis is a critical prerequisite for quality care and the prevention of healthcare-related infections. It is recognized that routine surveillance

FIGURE 6.5   Oral conditions of patients prone to aspiration of pathogenic microorganisms.

for these infections can reduce their incidence. However, in developing countries, due to the lack of formal surveillance, the rate of healthcare-related infections is high, which leverages the relevance of the issue (Datta *et al.*, 2014).

Some patient-related factors (age, nutritional status, pre-existing infections, and comorbidity, procedures, and technical difficulties) and long durations of surgery, preoperative preparation, and inadequate sterilization of surgical instruments may worsen the risk of infection significantly. In addition, the virulence and the invasiveness of the organism involved, the physiological state of the patient, and the immunological integrity of the host also represent some of the elements that may or may not occur in infection (Negi *et al.*, 2015).

The occurrence of infections related to healthcare in patients hospitalized in ICUs contributes to increased hospitalization time, higher mortality, and increased expenses for medicines and materials (Hu *et al.*, 2013).

These infectious complications can represent a patient burden, expressed in the increased burden of disease and the establishment of associated septic symptoms (Rodríguez *et al.*, 2011).

6.6 INFECTIONS ASSOCIATED WITH DENTAL BIOFILMS IN A HOSPITAL SETTING

Patients admitted to a hospital who cannot perform their hygiene self-care due to their systemic condition are more likely to colonize the oral biofilm by pathogenic microorganisms, especially respiratory pathogens. This can increase the risk of developing nosocomial pneumonia, which often occurs when aspirating the mucous membrane content present in the mouth and pharynx, mainly in newborns.

6.6.1 PNEUMONIA ASSOCIATED WITH MECHANICAL VENTILATION

Ventilator-associated pneumonia (VAP) is an acute respiratory infection that causes inflammation of the lung parenchyma. It develops in the patient 48–72 hours from the start of mechanical ventilation, either through an endotracheal tube or a tracheostomy tube, or 48–72 hours after extubation (American Thoracic Society, 2005).

Pneumonia is the second most common infectious complication in the hospital environment and occupies the first place in intensive medical services. Eighty percent of nosocomial pneumonia episodes occur in patients with an artificial airway, which is the most frequent cause of mortality among nosocomial infections in the ICU, mainly due to *Pseudomonas aeruginosa* and *Staphylococcus aureus* resistance to the medicine (MRSA) (Diaz *et al.*, 2010). In addition, it increases the days of mechanical ventilation and the mean stay in the hospital and ICU. Despite the available tests, the diagnosis of a PAV remains clinical. The presence of opacity in the chest radiograph with purulent tracheal secretions is necessary conditions for its diagnosis. In addition, we will have to evaluate their status and the risk factors for pathogens that are difficult to treat. If the PAV is early and these risk factors do not exist, most of the empirical guidelines present a correct coverage of the flora that we will find (Figure 6.6).

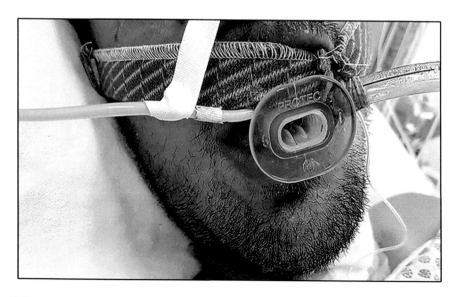

FIGURE 6.6   Patient with mechanical ventilation.

Pneumonia is a prevalent infectious disease that accounts for approximately 15% of all hospital infections (Sikora and Zahra, 2021). VAP is one of the most frequent complications in ICUs, with a mortality rate between 25% and 75%, and it has been recognized that 60% of all deaths from nosocomial infections are caused by this infection. Ventilator patients are at higher risk because respiratory assistance devices disrupt their defenses and normal secretion clearance mechanisms. High ventilation-associated pneumonia creates the need to review practices to improve the quality of care constantly. A prevention practice that must be incorporated is the complete oral care protocol in institutional practices for reducing VAP, which may save lives (Munro *et al.*, 2009).

6.6.2 NOSOCOMIAL PNEUMONIA OR HOSPITAL-ACQUIRED PNEUMONIA

Nosocomial pneumonia or hospital-acquired pneumonia (HAP) is described as an illness that presents in 48–72 hours after hospitalization, provided that an infectious pulmonary process present or in the incubation period at the time of admission and that pneumonia that occurs within seven days after discharge from hospital has been excluded. It is the second most common cause of hospital infections after urinary tract infections and the first cause of infections in ICUs. The incidence varies depending on the age group, being 5 cases per 1,000 hospitalized with age younger than 35 and rising to over 15 cases per 1,000 in older than 65. This incidence multiplies by 20 in patients connected to invasive mechanical ventilation (IMV), for whom mortality rates can go up to around 50% (Cantón-Bulnes *et al.*, 2019; Rodríguez and Díaz, 2020; Picazo *et al.*, 2020). The delay in an adequate antibiotic treatment for the critic nosocomial pneumonia is accompanied by a worse prognostic. The entry of germs to the respiratory tract is the aspiration of oropharyngeal secretions in most cases, so the etiology of the HAP will depend on the colonizing microorganisms.

The most frequent pathogens are the enteric gram-negative bacilli (GNB) (not *Pseudomonas*), sensitive *Haemophilus influenzae, Staphylococcus aureus* methicillin, and *Streptococcus pneumoniae*. They can be polymicrobial, especially HAP associated with ventilation. If HAP occurs early, the microbial spectrum is similar to community-acquired pneumonia (CAP), while if the hospitalization time grows, the colonizing flora of the oropharynx changes, then the GNB etiology predominates. The diagnosis of pneumonia, in general, is syndromic in the presence of suggestive symptoms and a radiological infiltrate. However, the specificity of these data in the HAP is low, especially in VAP, in which it may become necessary to use microbiological and pathological methods to arrive at the diagnosis. Also, in this VAP, modifying an inappropriate antibiotic therapy once the microorganism causing the infection has been isolated does not significantly improve the initial poor outcome. The election of an adequate empirical antibiotic therapy is one of the factotums of the modification of the HAP. Another factor that has been associated with the prognosis of HAP is the etiology itself, mortality being higher when the infection is caused by *Pseudomona aeruginosa* and lower when the isolated germ is *H. influenzae* or gram-positive cocos (Cantón-Bulnes *et al.*, 2019; Rodríguez and Díaz, 2020; Picazo *et al.*, 2020).

6.6.3 EARLY-START VAP

Early-start VAP appears in patients who have been on a ventilator between 48 and 96 hours. Patients are often suffering from an emergency or traumatic intubations, chest surgery, or neurosurgery. Pathogens are more common, sensitive to antibiotics, e.g. *Methicillin-sensitive Staphylococcus aureus (MSSA), Haemophilus influenza*, or *Streptococcus pneumoniae* (Butcher *et al.*, 2018).

6.6.4 LATE-INITIATION VAP

Late-initiation VAP appears in patients who have been on a ventilator for five days or more. These patients have chronic obstructive pulmonary diseases (COPD) or cardiogenic pulmonary edema. The most common pathogens are the most strong bacterial strains like *Methicillin-resistant Staphylococcus aureus (MRSA)* or gram-negative bacteria like *Pseudomonas aeruginosa, Acinobacter, Enterobacter, Klebsiella*, or *Serratia* (Butcher *et al.*, 2018).

6.6.5 INFECTIONS IN NEWBORNS

VAP in newborns is 50% when they are less than 25 weeks old, or their weight is lower than 750 g, or when they are inversely proportional to infection rates (Stoll *et al.*, 2002; Tripathi *et al.*, 2010).

The bacteria that are in the oral cavity in the newborns in the hospital are *Klebsiella pneumoniae (K. pneumoniae), Burkholderia cepacia (B. cepacia), Enterobacter cloacae (E. cloacae), Enterococcus, Escherichia coli (E. coli)*, and *Pseudomonas aeruginosa (P. aeruginosa)* (Lee *et al.*, 2017).

6.7   ORAL HYGIENE CARE AS A REDUCTION FACTOR OF ORAL BIOFILM IN HOSPITALIZED PATIENTS

Dental treatment for hospitalized patients, as health promotion actions, contribute to the prevention and improvement of the patient's systemic condition, improving the respiratory infections rate because of the reduction of the need for systemic antibiotics – and as a consequence, diminishing of the mortality rate — and this results in an economic impact.

The concern about oral infections as the main focus of systemic infections in patients dependent on hospital care in ICUs, although it is little documented, has been relevant in the discussions of interdisciplinary teams. The measures to reduce oral infections include local hygiene care and techniques.

Dental biofilm is the precursor to many oral diseases (like gingivitis, periodontitis, and cavities), and therefore, its removal and control is one of the important aspects of oral hygiene, mainly in the hospital environment (de Carvalho Baptista *et al.*, 2018; Zambrano *et al.*, 2018) (Figure 6.7).

Oral healthcare in patients in the ICUs includes tooth brushing and antiseptics in the oral cavity (Munro *et al.*, 2009). This idea is reinforced by the US Centers for Disease Control and Prevention (CDC) (Iosifidis *et al.*, 2016) guidelines on

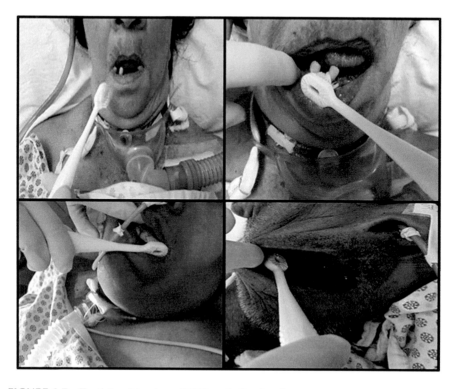

FIGURE 6.7   Tooth brushing in an ICU hospitalized patient.

implementing comprehensive oral hygiene preventive programs that include the use of an antiseptic for patients at risk of acquiring nosocomial pneumonia.

In a revision of seven randomized controlled trials (RTC) and a meta-analysis with the sampling of 3,370 patients, Halm and Armola (2009) conclude that ineffective oral hygiene is associated with increased dental plaque and oropharyngeal colonization, and respiratory infections. In the last two decades, similar etiological studies can be found in the literature that relate to VAP with the dental plaque or the colonization of the oropharyngeal mucosa, since both act as reservoirs of microorganisms that – together with macro and micro aspirations – favor the appearance of the VAP in the critic patient.

One of the most commonly used antiseptic agents in oral hygiene interventions is chlorhexidine; described as the gold standard due to its great antimicrobial power, it does not generate resistance, it is effective at low concentrations, and the trends of studies and reviews suggest its use in patient oral hygiene (Torres *et al.*, 2009; Khezri *et al.*, 2013).

According to Santos Zambrano *et al.* (2020), to have adequate sanitation, it is necessary to use a brush connected to the hospital suction system, and it is introduced into the patient's mouth first for aspiration of saliva and debris present. It is necessary to use chlorhexidine 0.12% gel in small portions on the bristles of the brush to start brushing correctly; in addition, the brush head must be gently placed in the region between the free gum and the tooth, forming an angle of 45°. With soft vibratory movements, the bristles of the brush should be gently pressed toward the gum to achieve penetration into the vestibular gingival sulcus of the teeth, in addition to cleaning the vestibular surface. Next, it is pertinent to carry out sweeping movements in the gingival-dental direction, smoothly and repeatedly, approximately fewer than five repetitions, and involving 2–3 teeth, in the dental hemiarchs, through the buccal and lingual region (Santos Zambrano *et al.* 2020).

For the occlusal face, hygiene of the lower and upper teeth should use sway movements and a soft tongue, palate, and internal mucosa of the cheek brushing. In ICU patients, it is necessary to externally clean the tube and catheter. At the end of this protocol, the need to aspire the buccal cavity surface with the brush and sterile gauze soaked in 0.2% aqueous chlorhexidine is emphasized for cleaning the external buccal cavity (Santos Zambrano *et al.*, 2020).

The apparition of the VAP in the neonatal intensive care unit (NICU) may be reduced through the application of effective prevention strategies and proper oral hygiene. Currently, the use of water-soaked swabs –sterile and applied to the palate, tongue, oral mucosa, and upper and lower gingival rollers – are found in the literature (Berry *et al.*, 2011; Fernandez Rodriguez *et al.*, 2017).

6.8 FREQUENT BUCCAL ALTERATIONS IN THE HOSPITAL ENVIRONMENT

Infection is a frequent complication with a high death rate in hospitalized patients. Infections can be divided into exogenous (pathogen native of the outside environment) or endogenous (pathogen belonging to the patient's microbial flora). However, after exogenous colonization, there is an alteration in the endogenous flora. Taking into account that almost half of the microbiota present in the human body are found in the oral cavity, as it is constantly colonized and presents an adequate environment to favor the proliferation of microorganisms, in addition to the use of associated drugs in the treatment of patients hospitalized, they may be located in different oral alterations such as candidiasis, mucositis, xerostomia, and trismus (Colombo and Guimarães, 2003).

6.8.1 Oral Candidiasis

The oral cavity constitutes a favorable environment for the colonization of opportunistic microorganisms such as fungi of the genus *Candida*; however, they are not pathogenic because the normal bacterial flora and the immune system limit their growth and slow down their excessive proliferation, thus maintaining a balance (Pardi and Cardozo, 2002).

In cancer patients, this microbial ecological balance is generally altered as a result of the tumor disease itself or of the therapeutic strategies used to treat it (chemotherapy and/or radiotherapy), functional alterations of cellular and humoral immune system, alterations in physical barriers, and changes in endogenous microflora, among other factors. Therapeutic interventions such as chemotherapy and radiotherapy cause – as the main adverse effect – generalized immunosuppression, with neutropenia being the leading risk factor for severe infection (Carratalà *et al.*, 1998).

Chemotherapy reduces the number of leukocytes and alters the function of polymorphonuclear leukocytes (PML), which facilitates the proliferation of *Candida*. On the other hand, radiotherapy – considered an effective instrument to control and cure head and neck tumors – produces undesirable effects, such as xerostomia, leading to oral colonization or infection, mainly by *C. albicans* (Panizo and Reviákina, 2001).

C. albicans is the principal etiological agent of oral candidiasis and has acquired clinical importance due to its high number of isolates in cancer patients due to the disease itself or the therapeutic strategies used to treat it (Costa *et al.*, 1999).

In this type of immunocompromised patients, oral candidiasis can spread through the bloodstream or upper gastrointestinal tract, spreading to other parts of the body and causing a more severe infection, significantly increased morbidity, and mortality (Birman *et al.*, 1997).

6.8.2 Mucositis

Oral mucositis is an inflammation and ulceration of oral mucosa with pseudomembrane formation and possible deadly infection source. Its initial manifestation is the erythema, continued with the growth of scaly white plaques that are sensitive to contact (Biron *et al.*, 2000). Epithelial crusts and fibrinous exudate carry pseudomembrane formation and ulceration, representing the most pronounced form of mucositis. The patients invariably have painful symptoms. The most severe mucositis form is represented by the exposition of the stroma of the richly innervated connective tissue because of the loss of epithelial cells, which usually occurs between five and seven days after the drug administration.

The ulceration produces so much intense pain that the patient's diet has to be changed and parenteral administration of narcotics for palliation started. In myelo-suppression patients, ulcerous mucositis may be useful as a vehicle for systemic invasion of bacteria or bacterial cell wall products (Volpato *et al.*, 2007).

Oral mucositis is a significant complication of mucotoxic cancer therapy that affects 5–40% of patients receiving standard-dose chemotherapy and more than 75% of patients receiving high doses with stem cell transplantation or with radiation therapy for head and neck cancer (Niikura *et al.*, 2020). Therapeutic drugs and chemotherapy drugs used to treat cancer can cause oral mucositis, causing pain and other symptoms, making it difficult for the patient to ingest orally, and this may result in malnutrition. Likewise, oral mucositis can also lead to an increase in bacteria inside the mouth, which can lead to pneumonia (Niikura *et al.*, 2020).

6.8.3 XEROSTOMIA

Xerostomia is a subjective sensation of dry mouth, resulting from the salivary glands, with changes in the quantity and the quality of saliva. It is a frequent symptom in palliative care patients, and its prevalence is 60–88% in progressive and advanced cancer (Werfalli *et al.*, 2020). Xerostomia has physical consequences, but also psychological and social consequences. Saliva plays an important role in maintaining normal physiological conditions of the tissues of the mouth. In addition to humidifying the oral cavity tissues, the lubricating property of saliva helps the formation and swallowing of the bolus, facilitates phonetics, and prevents tissue damage caused by mechanical agents or microorganisms. Xerostomia results from three basic causes: factors that affect the salivary center, factors that alter autonomic stimulation, and changes in gland function. The diagnosis is fundamentally clinical (Johansson *et al.*, 2020).

The state of the mouth and the actual functional situation should be evaluated in detail, and quantitative methods can be used to determine salivary secretion at rest or by stimulation when the situation warrants. Treatment should be guided by knowing the etiology and the repercussions that dry mouth has on the loss of comfort and quality of life of the patient. Therefore, we must monitor the use of xerogenic drugs and treat the underlying disease, if possible, and if it contributes to xerostomia, promote hydration and take measures for symptomatic control. Symptomatic treatment is divided into three areas of action: increasing saliva production through mechanical, gustatory, or pharmacological stimulation; use of saliva substitutes when stimulation is not possible; and actions to promote oral health. The mechanical stimulation of saliva secretion is done with chewing gum. Taste stimulation is carried out, for example, by taking vitamin C tablets. The pilocarpine is the available medicine for stimulating saliva secretion (Pappa *et al.*, 2020). A soft diet should also be recommended, avoiding very hard or dry foods and the use of tobacco and alcoholic beverages. Health technicians must teach patients with xerostomia the best way to obtain relief, which must be taken to prevent complications that could seriously compromise their quality of life (Figure 6.8).

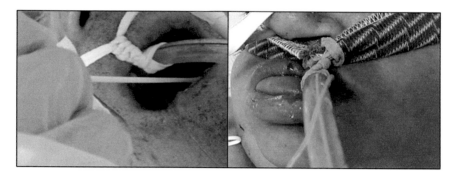

FIGURE 6.8   Images of hospitalized patients with clinical features of xerostomia.

6.8.4   TRISMUS

Surgical resection, radiation therapy, and chemoradiotherapy are the primary modalities for treating head and neck cancer. Trismus, or lockjaw, can occur as a pre-existing problem or develop as a complication in treating this type of cancer.

Trismus, or hypomobility of the mouth, is the result of restriction and limitation of the normal range of motion of the mouth. It is defined as a mouth opening of 35 mm or less. Trismus causes significant implications on health and activities of daily living, as it is associated with pain, speech problems, difficulty swallowing, malnutrition, dehydration, difficulty with oral hygiene, social isolation, and lower quality of life (Lee *et al.*, 2015).

There is no evidence on how it should be prevented or limited, or for treatment protocols, but there is information on therapeutic techniques that can help treat it. There is surgical management for severe trismus; for this, a comprehensive evaluation and management led by a specialist is necessary to avoid comorbidities that may compromise the patient's health (Loh *et al.*, 2017).

Early treatment has the potential to prevent or minimize many of the consequences of this condition. A pre-treatment assessment of the patient's condition, teaching and educating the patient about possible therapeutic techniques that can be implemented to avoid its establishment, and symptoms and warning signs that require prompt attention, are necessary. Although the clinical examination was done after treatment and reveals a decrease in mouth opening, treatment should begin as soon as possible. If the restriction becomes more severe and probably irreversible, the need for treatment is urgent (Stubblefield *et al.*, 2010).

6.8.5   ORAL MANIFESTATIONS IN HUMAN PAPILLOMAVIRUS PATIENTS

Human papillomavirus (HPV) constitutes a heterogeneous viral group capable of producing hyperplastic, papillomatous, and verrucous lesions in the skin and mucosa, demonstrating an essential role in carcinogenesis (Robles *et al.*, 2020).

HPV infection may be a cofactor in oral carcinogenesis, although the presence of the virus would not be sufficient to cause malignant transgression, probably requiring additional genetic changes for progression to a neoplastic stage. This infection continues to increase in the population; hence, the need to highlight the importance of making an early diagnosis of benign lesions in the mouth, which would allow adequate preventive treatment of the lesion, avoiding its transformation and progression to becoming a premalignant or malignant lesion (Santos-Zambrano *et al.*, 2019). HPV is sexually transmitted in most cases, with a variable incubation period ranging from three weeks to an imprecise time.

HPV spreads from many benign papillary growths on the oral mucosa, presenting clinically as single or multiple areas of thickened epithelium, often with a papillary surface, and may be pedunculated or flat and diffuse on a sessile base. Most are whitish, but flatter, broad-based lesions may be reddish or have the pink color of the normal oral mucosa (Aguiar *et al.*, 2019).

Periodic evaluation of this type of lesions should be maintained, regardless of the size of the lesion, and even if it is asymptomatic, since a large percentage of deaths

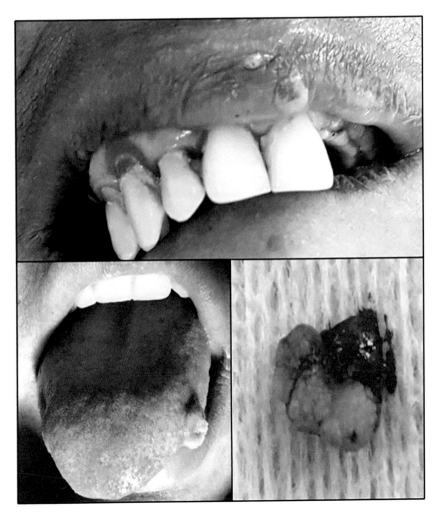

FIGURE 6.9 Clinical characteristics of oral HPV lesions (Santos-Zambrano *et al.*, 2019).

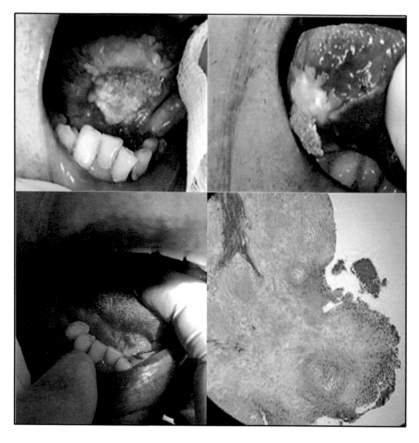

FIGURE 6.10 Clinical characteristics of verrucous leukoplakia (Santos-Zambrano *et al.*, 2020).

due to tumor lesions are due to treatment being performed at an advanced stage due to deficiencies in diagnosis, follow-up and consequently not being performed in a timely manner (Figure 6.9).

6.8.6 ORAL MANIFESTATIONS IN HIV/AIDS-SEROPOSITIVE PATIENTS

Acquired immunodeficiency syndrome (AIDS) is an infectious disease caused by the human immunodeficiency virus (HIV) which leads to progressive loss of immunity. The disease is a syndrome characterized by a set of signs and symptoms resulting from the drop in the rate of our defense mechanisms such as CD4 lymphocytes, cells that are very important in the body's immune defense. When the disease increases, the more it compromises the immune system and, consequently, the ability of the carrier to defend against infections.

Today, the AIDS patient can lead a life of considerable quality concerning oral health and is healthy enough to tolerate most surgical and dental procedures.

In addition, several studies have indicated that patients with HIV are no longer susceptible to complications, regardless of CD4 cell count.

In the later stages, opportunistic diseases are common, caused by agents such as other viruses, bacteria, fungi, and parasites, mainly affecting the oral cavity, such as:

Candidiasis – a white lesion on the tongue, inner cheek wall, and palate. The patient complains of burning and decreased taste.

Herpes labialis – small, grouped vesicles, mainly in the mucosa and skin transition area. These lesions are preceded by itching, burning, needling, and tingling.

Hairy leukoplakia – a whitish lesion most common on the lateral edges of the tongue.

HIV-related periodontal disease – people with HIV would be subject to a faster rate of progression of periodontal disease (Figure 6.10).

REFERENCES

Aguiar, Carmen, Alejandra Arriola, Macarena Vassel, Sylvia Jaumandreu, and María Carmen Rodríguez. (2019) "Presentación Clínica Infrecuente de Lesión VPH Anorrectal." *Acta Gastroenterológica Latinoamericana*. 49(1):65–68.

Akthar, MS, Degaga B, Azam T (2014, January) Antimicrobial activity of essential oils extracted from medicinal plants against the pathogenic microorganisms: a review. *Issues Bio Sci Pharma Res*. 2:1–7.

Allegranzi, Benedetta, Hugo Sax, Loséni Bengaly, Hervé Riebet, Daouda K Minta, Marie-Noelle Chraiti, Fatoumata Maiga Sokona, Angele Gayet-Ageron, Pascal Bonnabry, and Didier Pittet. (2010) "Successful Implementation of the World Health Organization Hand Hygiene Improvement Strategy in a Referral Hospital in Mali, Africa." *Infection Control & Hospital Epidemiology*. 31(2):133–141.

American Thoracic Society. (2005) "Guidelines for the Management of Adults with Hospital-Acquired, Ventilator-Associated, and Healthcare-Associated Pneumonia." *American Journal of Respiratory and Critical Care Medicine*. 171(4):388.

Berry, A M, Patricia M Davidson, Janet Masters, K Rolls, and R Ollerton. (2011) "Effects of Three Approaches to Standardized Oral Hygiene to Reduce Bacterial Colonization and Ventilator Associated Pneumonia in Mechanically Ventilated Patients: A Randomised Control Trial." *International Journal of Nursing Studies*. 48(6):681–688.

Birman, Esther Goldenberg, Sergio Kignel, F R X Da Silveira, and C R Paula. (1997) "Candida Albicans: Frequency and Characterization in Oral Cancer(Stage I) from Smokers and Drinkers." *Revista Iberoamericana de Micologia*. 14:101–103.

Biron, P, C Sebban, R Gourmet, G Chvetzoff, I Philip, and J Y Blay. (2000) "Research Controversies in Management of Oral Mucositis." *Supportive Care in Cancer*. 8(1):68–71.

Butcher HK, Bulechek, GM, McCloskey Dochterman, JM, Wagner, CM (2018) *Nursing interventions classification (NIC)*. e-Book. Elsevier Health Sciences.

Cantón-Bulnes, María Luisa, María Ascensión González-García, Manuela García-Sánchez, Ángel Arenzana-Seisdedos, and José Garnacho-Montero. (2019) "Estudio Caso-Control Del Impacto Clínico de La Traqueobronquitis Asociada a La Ventilación Mecánica En Pacientes Adultos, Que No Desarrollan Neumonía Asociada a Ventilación Mecánica." *Enfermedades Infecciosas y Microbiología Clínica*. 37(1):31–35.

Carratalà, Jordi, Beatriz Rosón, Alberto Fernández-Sevilla, Fernando Alcaide, and Francesc Gudiol. (1998) "Bacteremic Pneumonia in Neutropenic Patients with Cancer: Causes, Empirical Antibiotic Therapy, and Outcome." *Archives of Internal Medicine*. 158(8):868–872.

de Carvalho Baptista, Ivany Machado de, Frederico Canato Martinho, Gustavo Giacomelli Nascimento, Carlos Eduardo da Rocha Santos, Renata Falchete do Prado, and Marcia Carneiro Valera. (2018) "Colonization of Oropharynx and Lower Respiratory Tract in Critical Patients: Risk of Ventilator-Associated Pneumonia." *Archives of Oral Biology.* 85:64–69.

Celik, Gul Gunes, and Ismet Eser. (2017) "Examination of Intensive Care Unit Patients' Oral Health." *International Journal of Nursing Practice.* 23(6):1–9. https://doi.org/10.1111/ijn.12592.

Cillo Jr, J E. (1996) "The Development of Hospital Dentistry in America – the First One Hundred Years (1850–1950)" *Journal of the History of Dentistry.* 44(3):105–9.

Colombo, Arnaldo Lopes, and Thaís Guimarães. (2003) "Epidemiologia Das Infecções Hematogênicas Por Candida Spp." *Revista Da Sociedade Brasileira de Medicina Tropical.* 36:599–607.

Costa, Lino João da, Esther Goldenberg Birman, Sidney Hartz Alves, and Arlete Emily Cury. (1999) "Antifungal Susceptibility of Candida Albicans Isolated from Oral Mucosa of Patients with Cancer." *Revista de Odontologia Da Universidade de São Paulo.* 13:219–23.

Dalmasso, Marion, Eric De Haas, Horst Neve, Ronan Strain, Fabien J. Cousin, Stephen R. Stockdale, R. Paul Ross, and Colin Hill. (2015) "Isolation of a Novel Phage with Activity against Streptococcus Mutans Biofilms." *PLoS ONE* 10 (9): 1–18. https://doi.org/10.1371/journal.pone.0138651.

Datta, Priya, Hena Rani, Rajni Chauhan, Satinder Gombar, and Jagdish Chander. (2014) "Health-Care-Associated Infections: Risk Factors and Epidemiology from an Intensive Care Unit in Northern India." *Indian Journal of Anaesthesia.* 58(1):30.

Diaz E, L Lorente, J Valles, and J Rello. (2010) "Mechanical ventilation associated pneumonia." *Medicina Intensiva.* 34(5):318–24.

Eick S (2020) Biofilms. *Monographs in Oral Science.* 1–11. https://doi.org/10.1159/000510184.

Fernandez Rodriguez, Beatriz, Lorena Pena Gonzalez, Maria Cruz Calvo, Fernando Chaves Sanchez, Carmen Rosa Pallas Alonso, and Concepción de Alba Romero. (2017) "Oral Care in a Neonatal Intensive Care Unit." *The Journal of Maternal-Fetal & Neonatal Medicine.* 30(8):953–57.

Franco, Juliana Bertoldi, Sumatra Melo da Costa Pereira Jales, Camila Eduarda Zamboni, Fabio José Condino Fujarra, Márcio Vieira Ortegosa, Priscila Fernandes Ribas Guardieiro, Diogo Toledo Matias, and Maria Paula Siqueira de Melo Peres. (2018) "Higiene Bucal Para Pacientes Entubados Sob Ventilação Mecânica Assistida Na Unidade de Terapia Intensiva: Proposta de Protocolo/Oral Hygiene for Intubated Patients Assisted with Mechanical Ventilation in Intensive Care Unit: Proposal Protocol." *Arquivos Médicos Dos Hospitais e Da Faculdade de Ciências Médicas Da Santa Casa de São Paulo.* 59(3):126–31.

Gedif Meseret, Abebe. (2021) "Oral Biofilm and Its Impact on Oral Health, Psychological and Social Interaction." *International Journal of Oral and Dental Health.* 7(1). https://doi.org/10.23937/2469-5734/1510127.

Halm, Margo A, and Rochelle Armola. (2009) "Effect of Oral Care on Bacterial Colonization and Ventilator-Associated Pneumonia." *American Journal of Critical Care.* 18(3):275–78.

Hannig, Christian, and Matthias Hannig. (2009) "The Oral Cavity – A Key System to Understand Substratum-Dependent Bioadhesion on Solid Surfaces in Man." *Clinical Oral Investigations.* 13(2):123–39. https://doi.org/10.1007/s00784-008-0243-3.

Haque, M, Sartelli M, McKimm J, Abu Bakar M (2018) Health care-associated infections – an overview. *Infect Drug Resist.* 11:2321.

Hu, Bijie, Lili Tao, Victor D Rosenthal, Kun Liu, Yang Yun, Yao Suo, Xiandong Gao, Ruisheng Li, Danxia Su, and Hungmei Wang. (2013) "Device-Associated Infection Rates, Device Use, Length of Stay, and Mortality in Intensive Care Units of 4 Chinese Hospitals: International Nosocomial Control Consortium Findings." *American Journal of Infection Control.* 41(4):301–6.

Huang, Ruijie, Mingyun Li, and Richard L. Gregory. (2011) "Bacterial Interactions in Dental Biofilm." *Virulence*. 2(5):435–44. https://doi.org/10.4161/viru.2.5.16140.

Iosifidis, Elias, Elpis Chochliourou, Asimenia Violaki, Elisavet Chorafa, Stavroula Psachna, Afroditi Roumpou, Maria Sdougka, and Emmanuel Roilides. (2016) "Evaluation of the New Centers for Disease Control and Prevention Ventilator-Associated Event Module and Criteria in Critically Ill Children in Greece." *Infection Control & Hospital Epidemiology*. 37(10):1162–66.

Johansson, Ann-Katrin, Anders Johansson, Lennart Unell, Gunnar Ekbäck, Sven Ordell, and Gunnar E Carlsson. (2020) "Self-reported Dry Mouth in 50-to 80-year-old Swedes: Longitudinal and Cross-sectional Population Studies." *Journal of Oral Rehabilitation*. 47(2):246–54.

Jones, Deborah J., Cindy L. Munro, Mary Jo Grap, Todd Kitten, and Michael Edmond. (2010) "Oral Care and Bacteremia Risk in Mechanically Ventilated Adults." *Heart and Lung: Journal of Acute and Critical Care*. 39 (6): S57–65. https://doi.org/10.1016/j.hrtlng.2010.04.009.

Khezri, Hadi Darvishi, Mohammad Ali Haidari Gorji, Ali Morad, and Heidari Gorji. (2013) "Comparison of the Antibacterial Effects of Matrica & Persica™ and Chlorhexidine Gluconate Mouthwashes in Mechanically Ventilated ICU Patients: A Double Blind Randomized Clinical Trial." *Rev Chilena Infectol* 30 (4): 368–73.

Krzysciak, Wirginia, Anna Jurczak, and Jakub Piątkowski. (2016) "The Role of Human Oral Microbiome in Dental Biofilm Formation." *Microbial Biofilms – Importance and Applications*, no. July 2016. https://doi.org/10.5772/63492.

Lee, Li-Yun, Shu-Ching Chen, Wen-Cheng Chen, Bing-Shen Huang, and Chien-Yu Lin. (2015) "Postradiation Trismus and Its Impact on Quality of Life in Patients with Head and Neck Cancer." *Oral Surgery, Oral Medicine, Oral Pathology and Oral Radiology* 119 (2): 187–95.

Lee, Pei-Lun, Wei-Te Lee, and Hsiu-Lin Chen. (2017) "Ventilator-Associated Pneumonia in Low Birth Weight Neonates at a Neonatal Intensive Care Unit: A Retrospective Observational Study." *Pediatrics & Neonatology* 58 (1): 16–21.

Loh, Sook Y, Robert W J Mcleod, and Hassan A Elhassan. (2017) "Trismus Following Different Treatment Modalities for Head and Neck Cancer: A Systematic Review of Subjective Measures." *European Archives of Oto-Rhino-Laryngology* 274 (7): 2695–2707.

Magill, Shelley S, Jonathan R Edwards, Wendy Bamberg, Zintars G Beldavs, Ghinwa Dumyati, Marion A Kainer, Ruth Lynfield, Meghan Maloney, Laura McAllister-Hollod, and Joelle Nadle. (2014) "Multistate Point-Prevalence Survey of Health Care – Associated Infections." *New England Journal of Medicine* 370 (13): 1198–1208.

Munro, Cindy L, Mary Jo Grap, Deborah J Jones, Donna K McClish, and Curtis N Sessler. (2009) "Chlorhexidine, Toothbrushing, and Preventing Ventilator-Associated Pneumonia in Critically Ill Adults." *American Journal of Critical Care* 18 (5): 428–37.

Negi, Vikrant, Shekhar Pal, Deepak Juyal, Munesh Kumar Sharma, and Neelam Sharma. (2015) "Bacteriological Profile of Surgical Site Infections and Their Antibiogram: A Study from Resource Constrained Rural Setting of Uttarakhand State, India." *Journal of Clinical and Diagnostic Research: JCDR* 9 (10): DC17.

Niikura, N, Nakatukasa K, Amemiya T, Watanabe KI *et al.* (2020) Oral care evaluation to prevent oral mucositis in estrogen receptor-positive metastatic breast cancer patients treated with everolimus (oral care-BC): a randomized controlled phase III trial. *Oncolog.* 25(2):e223.

Øilo, Marit, and Vidar Bakken. (2015) "Biofilm and Dental Biomaterials." *Materials* 8 (6): 2887–2900. https://doi.org/10.3390/ma8062887.

O'Toole, George A. (2010) "Microtiter Dish Biofilm Formation Assay." *Journal of Visualized Experiments*, no. 47: 10–11. https://doi.org/10.3791/2437.

Panizo, M M, and V Reviákina. (2001) "Candida Albicans y Su Efecto Patógeno Sobre Las Mucosas." *Revista de La Sociedad Venezolana de Microbiología* 21 (2): 38–45.

Pappa E, Vougas K, Zoidakis J, Vastardis H (2020) Proteomic advances in salivary diagnostics. *Biochim Biophys Acta (BBA)-Proteins Proteomics*. 140494.

Pardi, G, Cardozo EI (2002) Algunas consideraciones sobre candida albicans como agente etiológico de candidiasis bucal. *Acta Odontol Venez*. 9–17.

Pereira Júnior, Gerson Alves, Francisco Antônio Coletto, Maria Auxiliadora Martins, Flávio Marson, Rosana Claudia Lovato Pagnano, Maria Célia Barcellos Dalri, and Anibal Basile-Filho. (1999) "Papel Da Unidade De Terapia Intensiva No Manejo Do Trauma." *Medicina (Ribeirao Preto. Online)* 32 (4): 419. https://doi.org/10.11606/issn.2176-7262. v32i4p419-437.

Picazo L, Gracia Arnillas MP, Muñoz-Bermúdez R, Durán X, Álvarez Lerma F, Masclans JR (2020) La humidificación activa en ventilación mecánica no se asocia con un aumento de complicaciones infecciosas respiratorias en un estudio cuasi-experimental pre-postintervención. *Medicina Intensiva*. 285–292.

Robles, JLM, Martin Moya LA, Mendoza NB, Alcívar Cedeño V, Santos Zambrano TB (2020) Leucoplasia verrugosa con asentamiento del virus papiloma humano subtipo 33: Reporte de un caso clínico. *Revista San Gregorio*. 1:38.

Rodríguez, Ferney, Lena Barrera, Gisela De La Rosa, Rodolfo Dennis, Carmelo Dueñas, Marcela Granados, Dario Londoño, Francisco Molina, Guillermo Ortiz, and Fabián Jaimes. (2011) "The Epidemiology of Sepsis in Colombia: A Prospective Multicenter Cohort Study in Ten University Hospitals." *Critical Care Medicine* 39 (7): 1675–82.

Rodríguez RG, Barcón Díaz L (2020) Modos de ventilación mecánica no invasiva en una unidad de cuidados intensivos. *Rev Cub Med Int Emerg*. 19:1.

Saini, Rajiv, Santosh Saini, and Sugandha Sharma. (2011) "Biofilm: A Dental Microbial Infection." *Journal of Natural Science, Biology and Medicine* 2 (1): 71–75. https://doi.org/10.4103/0976-9668.82317.

Santos-Zambrano, TB, Barreiro-Mendoza N, Alarcón-Barcia N, Ramos-León MV *et al.* (2019) Verrugas vulgares bucales múltiples: reporte de un caso clínico. *Paideia XXI*. 9(2).

Santos-Zambrano, TB, A C Poletto, B Gonçalves Dias, R Guayato Nomura, W Junior Trevisan, and R S Couto de Almeida. (2020) "Evaluación de Un Protocolo de Cepillado Dental Con Aspiración En Pacientes Hospitalizados En La Unidad de Cuidados Intensivos Utilizando Análisis de Imagen y Microbiología: Estudio Piloto." *Medicina Intensiva* 44 (4): 256–59.

Sax, Hugo, Lauren Clack, Sylvie Touveneau, Fabricio da Liberdade Jantarada, Didier Pittet, and Walter Zingg. (2013) "Implementation of Infection Control Best Practice in Intensive Care Units throughout Europe: A Mixed-Method Evaluation Study." *Implementation Science* 8 (1): 1–11.

Sharma, N C, and J Galustians. (1994) "Efeitos Clínicos Sobre a Placa Dental e a Gengivite Obtidos Com o Uso de Quatro Escovas Dentais Manuais de Desenho Complexo Por Período de Três Meses." *J Clin Dentistr* 5: 114–18.

Sikora, A, and F Zahra. (2021) Nosocomial infections. *StatPearls*. https://www.ncbi.nlm.nih.gov/books/NBK559312/

Slutsky, Arthur S. (1993) "Mechanical Ventilation." *Chest* 104 (6): 1833–59.

Stoll, Barbara J, Nellie Hansen, Avroy A Fanaroff, Linda L Wright, Waldemar A Carlo, Richard A Ehrenkranz, James A Lemons, Edward F Donovan, Ann R Stark, and Jon E Tyson. (2002) "Late-Onset Sepsis in Very Low Birth Weight Neonates: The Experience of the NICHD Neonatal Research Network." *Pediatrics* 110 (2): 285–91.

Stubblefield, Michael D, Laura Manfield, and Elyn R Riedel. (2010) "A Preliminary Report on the Efficacy of a Dynamic Jaw Opening Device (Dynasplint Trismus System) as Part of the Multimodal Treatment of Trismus in Patients with Head and Neck Cancer." *Archives of Physical Medicine and Rehabilitation* 91 (8): 1278–82.

Torres, Antoni, Santiago Ewig, Harmut Lode, and Jean Carlet. (2009) "Defining, Treating and Preventing Hospital Acquired Pneumonia: European Perspective." *Intensive Care Medicine* 35 (1): 9–29.

Tripathi, Shalini, G K Malik, Amita Jain, and Neera Kohli. (2010) "Study of Ventilator Associated Pneumonia in Neonatal Intensive Care Unit: Characteristics, Risk Factors and Outcome." *Internet Journal of Medical Update-EJOURNAL* 5 (1)

Volpato, Luiz Evaristo Ricci, Thiago Cruvinel Silva, Thaís Marchini Oliveira, Vivien Thiemy Sakai, and Maria Aparecida Andrade Moreira Machado. (2007) "Mucosite Bucal Rádio e Quimioinduzida." *Revista Brasileira de Otorrinolaringologia* 73: 562–68.

Werfalli, SG, Drangsholt M, Martin M, Leresche L (2020) Whole saliva and residual mucosal saliva in patients with burning mouth syndrome: a case-control study. *Oral Surg Oral Med Oral Pathol Oral Radiol.* 129(1):e193.

Willis, Paul J. (1965) "The Role of Dentistry in the Hospital." *Journal of the American Dental Society of Anesthesiology* 12 (2): 40.

World Health Organization (2004) *World alliance for patient safety: forward programme 2005.* WHO Library Cataloguing. www.who.int/patientsafety

Zambrano, Thaináh Bruna Santos, Caisa Batista, Ana Claudia Poletto, Nora Gavilanes, Marcos Heidy Guskuma, Marcos Antonio do Amaral, and Ricardo Sérgio Couto de Almeida. (2018) "Oral Hygiene of Patients with Cancer in the Intensive Care Unit." *Journal of Health Sciences* 20 (2): 83–86.

Zimlichman, Eyal, Daniel Henderson, Orly Tamir, Calvin Franz, Peter Song, Cyrus K Yamin, Carol Keohane, Charles R Denham, and David W Bates. (2013) "Health Care – Associated Infections: A Meta-Analysis of Costs and Financial Impact on the US Health Care System." *JAMA Internal Medicine* 173 (22): 2039–46.

Part II

Applications of Microbial Biofilms

7 The Roles of Biofilms in Corrosion

Oluwafemi Adebayo Oyewole, Japhet Gaius Yakubu and Ramat Onyeneoyiza Raji

CONTENTS

7.1   INTRODUCTION

Since before the existence of humans and other higher organisms billions of years ago, microbes have been habiting planet earth and their activities have helped in providing suitable environments for higher forms of life. Microorganisms are ubiquitous and have been found growing in extreme environmental conditions such as temperatures of 120°C in hydrothermal vents at bottom of oceans and −13°C in lakes covered with ice in Antarctica. Their presence oftentimes in such extreme environments is associated with their ability to utilize the available substrates while maintaining their cellular functions (Machuca, 2019). Aside from the beneficial roles played by microorganisms in providing a sustainable ecosystem, they have also been found to cause harmful effects to life including but not limited to diseases, morbidity and mortality,

DOI: 10.1201/9781003184942-9

as well as infrastructural deterioration known as microbial corrosion (Kannan *et al.*, 2018; Machuca, 2019).

Microbial corrosion – also known as biological corrosion or microbiologically influenced corrosion (MIC) – is a type of corrosion that takes place on ceramics, concretes, stones or metals resulting from the presence and actions of microorganisms (Li *et al.*, 2013; Kip and van Veen, 2015; Mansour *et al.*, 2016; Blackwood, 2018; Jia *et al.*, 2019; Machuca, 2019). The universal application of metals in industrial and civil structures, biomedical devices, processing plants, aviation fuel system, nuclear waste storage facilities (Jia *et al.*, 2019) and oil rigs, as well as in the transport of fluids such as water and crude oil, has raised great concerns to practitioners particularly in the petroleum industry owing to the harmful effects caused by MIC (Masali *et al.*, 2018). Although fungi and microalgae causes MIC, however, archaea and bacteria are the most significant players leading to MIC on metallic and nonmetallic surfaces, which is mostly enhanced by production of biofilm (Loto, 2017; Kannan *et al.*, 2018; Khouzani *et al.*, 2019; Machuca, 2019).

Biofilms arise from the aggregation of microbial cells, which could be of the same species or a mixture of varying microorganisms (i.e. archaea and bacteria) bonded together and onto surfaces of objects or structures. The aggregate of cells formed are usually embedded within a matrix of self-produced extracellular polymeric substances (EPS) (Loto, 2017). Biofilms oftentimes create a microenvironment for the microbial community to carry out deleterious effects on the objects to which they adhere, through the release of acidic metabolites which are corrosive to surfaces; effects of aeration resulting from cells with different aeration levels cause depletion of oxygen among various effects (Veena *et al.*, 2013; Imo *et al.*, 2016). The biofilms also help the microbial community withstand mechanical pressure caused by high velocity of fluids in pipelines, and resistance capacity against bactericidal substances applied to mitigate MIC (Berne *et al.*, 2018). The mechanisms and steps involved in biofilm formation, its structure and composition that makes it an invaluable asset for corrosion will among many others be discussed in this chapter.

7.2   MICROBIAL BIOFILM

According to the International Union of Pure and Applied Chemistry (IUPAC), biofilm is an aggregate of microbial cells embedded in a matrix of self-produced EPS, attached to one another and/or to a biotic or abiotic surface (Loto, 2017; Liaqat *et al.*, 2019). Biofilm could arise from single or mixed groups of microorganisms (i.e. bacteria, fungi, and or archaea). Biofilms arising from a single-strain microorganism are more easily detachable from surfaces to which they adhere than are those formed synergistically from mixed cultures (Jia *et al.*, 2019). Microorganisms – especially bacteria – showcase two modes of growth: either free living (planktonic) or the surface-attached sessile mode of living within a biofilm. Biofilm formation has been recognized as a common trait among bacterial species, creating a matrix that help them overcome and thrive in adverse environmental conditions (Vasudevan, 2014; Rumbaugh and Sauer, 2020). Biofilms of microorganisms have been found to be ubiquitous in environments experiencing substantial amounts of liquid both in medical and industrial settings. They've been found growing in deep-sea vents, river rocks, roots of plants, indwelling catheters in human bodies, maritime equipment, water and petroleum pipelines, among many places (Berne *et al.*, 2018).

The discovery of biofilm in 1684 by Antoine Van Leeuwenhoek made an important contribution to the understanding of microbial growth. In the calculus of his teeth he observed sessile growth of animalcules known today as dental plaque, which validated the earliest evidence of the existence of bacterial biofilms (Vasudevan, 2014). Although discovered in 1684, the word "biofilm" was not defined or used until 1978 by Costerton (Costerton *et al.*, 1978). Fifteen years later, the significance of biofilms was then recognized by the American Society of Microbiology, with lots of research carried out to understand their roles in pathogenesis of diseases, as well as in corrosion of materials, especially metals (Khatoon *et al.*, 2018). The production or formation of biofilm by microorganisms has been discovered to be an important structure for their survival and proliferation in any given environment. It is important to note that microorganisms that form biofilm thrive better than the planktonic ones; this is because they live as a community in the process and their metabolic activities are optimized, providing nutrients and protection to the vulnerable ones (Vasudevan, 2014).

7.3   COMPOSITION AND STRUCTURES OF MICROBIAL BIOFILMS

Biofilms are basically composed of water (75–90%) (Telegdi *et al.*, 2020), 10–25% of microbial cells (Liaqat *et al.*, 2019) and EPS, which accounts for 50–90% of the total organic compounds (Berhe *et al.*, 2017). Constituents of the EPS include polysaccharides, humic and nucleic acids, proteins, lipids and other polymeric compounds, as represented in Figure 7.1 (Rabin *et al.*, 2015; Berhe *et al.*, 2017; Khatoon *et al.*,

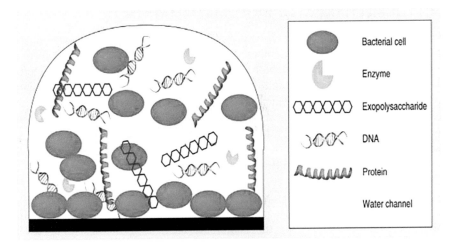

FIGURE 7.1   Basic structure of a biofilm (Rabin *et al.*, 2015).

2018; Verderosa *et al.*, 2019; Muhammad *et al.*, 2020; Telegdi *et al.*, 2020). However, the compositions and amounts of the organic compounds differs among microbial species – even within the same bacterial species – depending on the environmental conditions such as fluid velocity, pH and oxygen concentration, among other factors (Berhe *et al.*, 2017). Despite the complex and varying composition of the organic compounds, EPS are a very essential component of the biofilm matrix due to the framework they provide in housing microbial cells (Berhe *et al.*, 2017; Liaqat *et al.*, 2019). Aside from holding and housing microbial cells, the matrix of EPS imbues the biofilm with attributes including nutrient capture, antibiotic resistance, gradient formation, storage of extracellular enzymes and protection from environmental stress, among others, which in its absence cannot be achieved by single or planktonic cells. Thus, the emergent properties of biofilm are a responsibility of the matrix (Liaqat *et al.*, 2019; Verderosa *et al.*, 2019).

The structure of microbial biofilm is well organized; the cells are held together by various kind of forces such as hydrogen bonds, electrostatic interactions and van der Waals forces with a binding force of 10–30 kJ/mol, 12–29 kJ/mol and 2.5 kJ/mol, respectively (Telegdi *et al.*, 2020). Water channels and interstitial voids are distributed all over the biofilm for the transport of water, nutrients and air from where they are very much available to areas where they are needed (Vasudevan *et al.*, 2014). There is also a means of communication among the microbial cells throughout the biofilm known as quorum sensing (Verderosa *et al.*, 2019). This quorum sensing ensures the adequate distribution of necessary microbial requirements from one microcolony to another within the biofilm, as well as regulation of how cells respond to their immediate environment (Rabin *et al.*, 2015; Verderosa *et al.*, 2019). Within the biofilm, there is also horizontal gene transfer from one species to another, ensuring the exchange of some characteristics such as antibiotics resistance to cells that were once susceptible (Rabin *et al.*, 2015).

7.4   THE MECHANISM OF BIOFILM FORMATION

Generally, biofilm formation comprises cyclic yet progressive stages which are phenotypically distinct from one another. Although there are diverse biofilm-forming microorganisms with different biofilm architectures, however, numerous studies have used single species in elucidating the developmental stages in biofilm formation and they all coincide with the general features of attachment (i.e. reversible and irreversible), EPS production, maturation and dispersion of cells regardless of the type of species (Römling *et al.*, 2014; Rumbaugh and Sauer, 2020). Figure 7.2 depicts a bacterium (*Pseudomonas aeruginosa*) model used to demonstrate the different developmental stages involved in biofilm formation.

7.4.1   Initial Attachment of Bacterial Cells

Before biofilm can be formed, the planktonic bacterial cells must first be able to come in close proximity or contact to a particular surface (Rabin *et al.*, 2015;

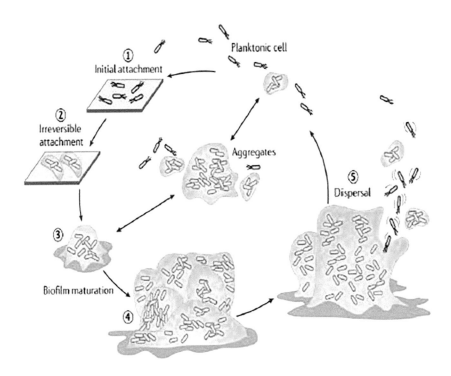

FIGURE 7.2 Schematic presentation of different developmental stages in biofilm formation by Pseudomonas aeruginosa in vitro (Rumbaugh and Sauer, 2020).

Rumbaugh and Sauer, 2020), which can be achieved through convection, sedimentation or Brownian motion (Khatoon *et al.*, 2018; Muhammad *et al.*, 2020; Yuanzhe *et al.*, 2020). Chemotaxis, a direct movement of bacterial cells toward a source of nutrient (e.g. sugars and amino acids), is a phenomenon that occurs in virtually all microorganisms which also enhances growth of bacterial cells on surfaces thereby promoting cell-surface interaction (Muhammad *et al.*, 2020). Close to a surface at a distance of 10–20 nm, several forces both for and against the bacterial cell come into play. Bacterial cells often experience repulsive forces arising from the negative charges on environmental surfaces against the negative charges on their own cell surfaces. Since like poles repel, in order to overcome this obstacle, bacterial cells often employ hydrophobic, hydrostatic and van der Waals forces to overcome the repulsive forces posed by surfaces (Rabin *et al.*, 2015; Muhammad *et al.*, 2020; Rumbaugh and Sauer, 2020; Telegdi *et al.*, 2020). Also, the use of flagella and fimbriae often provide them with mechanical support to surfaces (Rabin *et al.*, 2015). However, since these forces are not physically strong enough, they get separated and the initial attachment is reversed by shearing effects of flowing fluids or bacterial mobility (Muhammad *et al.*, 2020).

7.4.2 IRREVERSIBLE ATTACHMENT OF BACTERIAL CELLS

Following a successful initial reversible attachment, the bacterial species often initiates the irreversible attachment processes through short-range hydrophobic and dipole–dipole interactions, hydrogen and covalent and ionic bonds, and the involvement of bacterial adhesive structures such as fimbriae, pili and flagella have all been involved in process of irreversible attachment (Vasudevan, 2014; Muhammad *et al.*, 2020; Rumbaugh and Sauer, 2020). Extracellular substances such as adhesins are produced and secreted into the immediate environment of the bacterial cells (Berne *et al.*, 2018). These adhesins serve as cementing materials for the bacterial cells to one another and to surfaces (Muhammad *et al.*, 2020). Microscopic examination suggests that matrix production is initiated once the bacterial cells are irreversibly attached, preventing them from moving in the process some of the adhesive structures are lost (Rumbaugh and Sauer, 2020).

7.4.3 EPS PRODUCTION AND BIOFILM MATURATION

At this stage, the bacterial cells grow and multiply, and in the process metabolize EPS and water channels, which enhances the movement of water, oxygen and nutrients to bacterial cells, thus facilitating growth of microbial community (Flemming *et al.*, 2016; Khatoon *et al.*, 2018; Sharma *et al.*, 2019; Yuanzhe *et al.*, 2020). The EPS formed provides the biofilm with a firm architecture supporting and maintaining the microbial community. Aside from that, the EPS mediates cohesion of bacterial cells to one another and the biofilm to a particular surface through ion bridging and hydrophobic interactions, allowing maturation of biofilm to take place. Other functions of the EPS include water retention, biofilm formation and structure, cell-to-cell recognition and exchange of genetic materials, trapping nutrients, cell protection and transportation of signals. More importantly, c-di-GMP – a secondary messenger during EPS production – serves as one of the stimuli that initiate transitioning from reversible to irreversible adhesion (Muhammad *et al.*, 2020).

7.4.4 BIOFILM MATURATION

The microbial cells within the EPS matrix continue to grow and form multiple layers with more secretion of EPS (Khatoon *et al.*, 2018; Yuanzhe *et al.*, 2020) causing the biofilm to transit into a three-dimensional structure (Sharma *et al.*, 2019; Telegdi *et al.*, 2020). Chemical messengers in the EPS go on to attract diverse groups of microorganisms to the biofilm. As the microbial cells in the biofilm continue to grow, autoinducers (AIs) signals are being used in the maturation of biofilm and the formation of microcolonies (Chadha, 2014). Gene expression is caused to change following the accumulation of EPS and the formation of microcolonies to the secretion of gel-like EPS, which act as a cementing material or biological glue between bacterial cells in the embedded matrix (Telegdi *et al.*, 2020). Water channels and interstitial voids are distributed throughout the biofilm and act as a circulatory system transporting nutrients and other needed material to communities of bacterial cells, as well as removal of spent material and waste products (Patel *et al.*, 2014; Berhe *et al.*, 2017;

Muhammad *et al.*, 2020). The architectural presentation of the microcolonies in the biofilm changes to a multicellular mushroom/pyramid-shaped structure (Muhammad *et al.*, 2020). Motility within the microcolonies is restricted and bacterial surface structure production is inhibited as the biofilm matures (Rumbaugh and Sauer, 2020). Microbial gradient is formed in the biofilm based on oxygen requirement; at the top of the biofilm – close to the air/solution interface – exist aerobic bacterial, whereas the anaerobic ones are located at the bottom, close to the solid surface (Rabin *et al.*, 2015; Telegdi *et al.*, 2020).

7.4.5   CELL DISPERSAL

To end the biofilm cycle and start a new one, the bacterial cells need to detach from the biofilm and get dispersed into the environment. The detachment and dispersion is usually triggered in response to certain environmental and physiological effects or conditions. The dispersal process, although complex, is however regulated by effectors molecules, signal transduction pathways and environmental signals (Telegdi *et al.*, 2020). Although the mechanism for cell dispersal differs among bacterial species, such processes can be divided into three distinct stages: cell detachment from the microcolonies, cell movement to a fresh substrate and, finally, attachment of cells to a new surface. Cell detachment can either be active (i.e. seeding) or passive (i.e. erosion and sloughing) (Rabin *et al.*, 2015). In active detachment, cells – in response to environmental stresses such as nutrient starvation, matrix accumulation of toxic waste and degradation of enzymes and antimicrobial stress – trigger the detachment processes. Passive detachment is often caused by shear forces from the external environment. In simple terms, seeding dispersal is an active mechanism used by bacteria in cell detachment from biofilms and is associated with a rapid release of planktonic cells from the middle of the biofilm, thus leaving the cavity empty (Rabin *et al.*, 2015). A sudden detachment of large portion of biofilm is known as sloughing; erosion, on the other hand, is the release of a minute portion of bacterial cells from the biofilm (Rabin *et al.*, 2015; Telegdi *et al.*, 2020).

It is important to note that in active dispersal, genetic inducers and regulators play active roles. There is usually up-regulation of genes responsible for EPS degradation and flagella synthesis, whereas those responsible for EPS synthesis and attachment are usually down-regulated. Another means of effective dispersal of bacterial cells in biofilms is by inhibiting intracellular secondary messenger c-di-GMP, which acts as signaling pathway in promoting EPS production. In this way, biofilm development is inhibited, thereby promoting biofilm dispersal (Muhammad *et al.*, 2020). Other means of regulation of cell growth and development, production of EPS and cell-to-cell communication are achieved using quorum sensing.

7.5   ROLE OF QUORUM SENSING IN BIOFILM FORMATION

Quorum sensing (QS) is an important network system used by microorganisms to communicate among themselves using chemical messengers (Figure 7.3A) known as AIs (Rabin *et al.*, 2015) to effect a behavioral change in a density-dependent manner

(Patel *et al.*, 2014; Rodis *et al.*, 2020). Both gram-positive and gram-negative bacteria utilize QS for cell-to-cell communication in the formation of biofilm (Rabin *et al.*, 2015; Sharma *et al.*, 2019). The universal AIs used by both gram-positive and gram-negative bacteria are known as oligopeptides, also known as autoinducers-2 (AI-2), however, acyl homoserine lactones (AHLs) are primarily used by gram-negative bacteria (Rabin *et al.*, 2015; Muhammad *et al.*, 2020). Other AIs such as gamma-butyrolactones are utilized among species of *Streptomyces* whereas cis-11-methyl-2-dodecanoic acid (DSF) is used by *Xylella*, *Xanthomonas* and other related species (Patel *et al.*, 2014; Muhammad *et al.*, 2020).

Bacteria use QS to coordinate and initiate biofilm formation by first releasing the AIs into the environment (Muhammad *et al.*, 2020). Stimulated my chemotaxis or other environmental conditions, planktonic bacterial cells move toward the surface they intend to colonize. Once they approach it, the next point of action is to make the environment favorable for their activities. As such, they release signaling molecules such as protons into the liquid medium which diffuses toward the interface between the liquid medium and the substrate. These signaling molecules attract other neighboring bacterial cells toward the surface of the substrate. With time, the cells multiply and keep releasing protons, causing its concentration to increase. Once the

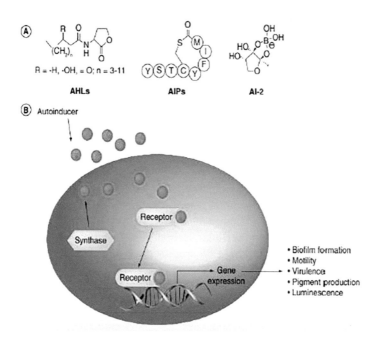

FIGURE 7.3 Various molecules involved in Quorum sensing: A – Main quorum sensing autoinducers; B – Quorum sensing circuit (Rabin *et al.*, 2015).

AHL: Acylhomoserine lactone; AIP: Autoinducing peptide; AI: Autoinducer planktonic cells sense the surface of the substrate, they begin to adhere and initiate biofilm formation processes through the synthesis and release of EPS (Rabin *et al.*, 2015; Muhammad *et al.*, 2020).

The detection of AIs is usually poor at low cell density; however, as the cells increase in density, it causes a simultaneous increase in the concentration of AIs until it reaches a threshold where the AIs/receptor protein complex acts to repress or induce the expression of targeted genes (Figure 7.3B) (Rabin *et al.*, 2015). Genes, such as the ones responsible for the synthesis of EPS, are repressed, while the ones responsible for synthesis of degradative enzymes are induced, thus promoting detachment of cells from the biofilm to kickstart another cycle on a new surface. Aside from the control of biofilm formation, QS also controls varying physiological processes including – but not limited to – dissemination of virulence factor and bioluminescence (Rabin *et al.*, 2015; Rodis *et al.*, 2020).

7.6   METAL CORROSION AND ITS TYPES

Mansour and Elshafei (2016) defined corrosion as a natural process, whereby a refined metal is converted to its oxide or hydroxide or to a more stable compound. Generally, corrosion can be defined as the gradual deterioration of materials through electrochemical, chemical or microbial interactions (Mansour and Elshafei, 2016; Ibrahim *et al.*, 2018). Classifying corrosion into different types has been a challenge to most scientists and corrosion experts around the world. This is due to types of environments, agents involved (biotic and/or abiotic), the mechanisms used, the sectors affected and the appearance of metals during and after the corrosion processes, among many other variables which have made corrosion classification ambiguous. Cicek (2017) classified corrosion into two broad categories: uniform and localized or non-uniform corrosion.

Uniform corrosion is said to occur when there is a relative reduction in the general thickness of metals due to its interactions with the environment. Metals with little or no passivation (i.e. iron) have high tendencies of this type of corrosion. It is assumed that uniform corrosion is the most common type of corrosion accountable for most loss of metal material, leaving behind residues or scales. Nonetheless, its effect is not considered dangerous, owing to its uniform distribution across the entire exposed surfaces of metal to corrosive agents resulting in a predictable reduction in the metal thickness. Various forms of corrosive agents have been discussed by Cicek (2017), which include underground or soil, atmosphere and water, which cause underground corrosion, atmospheric corrosion and water corrosion, respectively (Figure 7.4).

Localized or non-uniform corrosion, on the other hand, is said to be dangerous and causes varying degrees of environmental hazards, as well as loss of resources in the repair or maintenance of metals. There are various types of non-uniform corrosion with varying mechanisms, agents and patterns of appearance during and after the corrosive damage. Figure 7.5 represents the various types of non-uniform corrosion and the mechanisms and patterns observed by Hansson (2011), Singh (2014) and Cicek (2017).

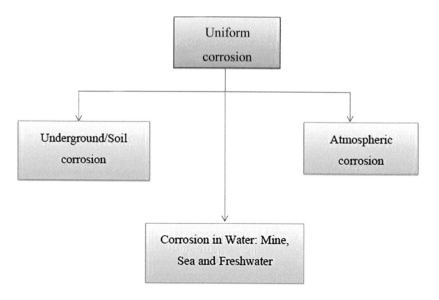

FIGURE 7.4   Types of uniform corrosion (Cicek, 2017).

7.7   IMPACTS OF CORROSION ON HUMANS AND THE ECOSYSTEM

The ecosystem comprises both living and non-living components interacting together for a sustainable environment (Ikechukwu and Pauline, 2015). The use of metal is inevitable since it is involved in daily activities of humans such as transportation, medical implants, pipelines, food processing and packaging, electronics and civil structures, just to mention a few. The longevity of metals depends on their types (i.e. iron, steel, etc.), their applications and the environmental conditions to which they are exposed (Ikechukwu and Pauline, 2015). The susceptibility of metals in the earth's atmosphere to corrosion is almost inevitable, converting them to their most stable form (i.e. carbonates, sulfides and oxides). However, their lifespan can be preserved or maintained using appropriate materials and inhibitors; otherwise, they will go on to cause major environmental problems affecting the health of humans and their environment (Hansson, 2011; Sastri, 2015).

For example, the 1984 Bhopal chemical plant explosion in India was the worst corrosion-related disaster, which resulted in the loss of both human and animal lives, as well as unsafe atmospheric conditions owing to release of carbon monoxide and other particulate compounds (Hansson, 2011). The plant was designed to enhance crop production in the country through manufacturing of pesticides and fertilizers. Instead of solving problems and helping lives, the chemical plant caused the deaths of many people thanks to management problems and design flaws which converged, resulting in a catastrophic explosion. Investigations showed that the corrosion of steel

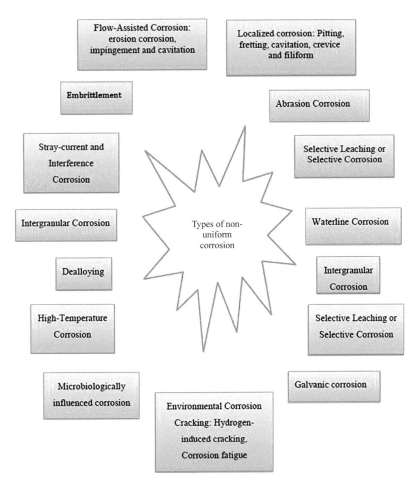

FIGURE 7.5 Various types of non-uniform corrosion (compiled from Hansson [2011], Singh [2014] and Cicek [2017]).

pipes caused water to leak into tanks holding methylisocyanate. The resulting product of iron corrosion yielded catalyst for the reaction that blew up part of the plant, releasing methylisocyanate and some other toxic gases. Over 8,000 lives were lost and an additional 15,000 died later due to the explosion, with an estimated 500,000 people who later suffered varying degrees of gas-related disorders (Hansson, 2011).

Various other deaths have been recorded in the transportation sector (aviation, maritime, rail and road) caused by corrosion of bolts, nuts and even connectors, resulting in failures of electrical wiring systems. These wiring systems, if not maintained properly, can cause accident in aircraft (Hansson, 2011). Fretting corrosion has been reported in electrical components of aircraft caused by the flaking of tin film of oxides on a mated surface of tin-containing contacts. Use of tin instead of gold (for economic reasons) has led to frequent occurrences of this problem. The crash of about six F-16 fighter aircraft have been reported to be related to shutting off of main fuel valves without the pilot's command usually caused by fretting corrosion of tin connectors (Sastri, 2015). The ferry *MV Princess Ashika* (Figure 7.6) sank in the South Pacific Ocean in 2009, killing all 78 passengers and crew members. Investigation of the incident showed that entrances and doors of the *Princess Ashika* were not able to close fully due to corrosion. Other corrosion-related incidents that led to the loss of human lives in the United States, Canada, Switzerland, Russia and Mexico have been documented by Javaherdashti (2017).

Perforation of tanks, containers and pipelines transporting petroleum products have resulted in the release of harmful contaminants into the environment. Most of these contaminants get into the environment as a result of leakages caused by MIC on the external or internal surfaces of metals (Hansson, 2011). The water bodies and wildlife are terribly affected by the restriction of mobility and even death of marine and terrestrial animals. Vegetation and microorganisms are not excluded in the havoc caused by pipeline leakages, as there is often shift in niche function, which in turn affects the survival of the ecosystem. So far, the worst leakage of petroleum pipeline

FIGURE 7.6   The sinking of the ferry *MV Princess Ashika* resulted in the deaths of all crew members and passengers (Hansson, 2011).

is the Prudhoe Bay oil spill in 2006 in Alaska (United States), which released over 1 million of liters of crude oil into the environment. Investigation showed that the Prudhoe Bay incident occurred as a result of internal corrosion of the pipeline with active involvement of microorganisms in facilitating the corrosion process (Hansson, 2011; Ibrahim *et al.*, 2018; Machuca, 2019). Sewage leaks are also detrimental to the ecosystem since they contain varying types of microorganisms, organic compounds and inorganic compounds with the ability to cause imbalance in the ecosystem (Hansson, 2011). Such is seen when there is influx of nitrate into water bodies resulting in algal blooms causing death of marine animals and making water unsafe for some particular purposes.

Aside from the impact corrosion has on environmental and human health, it also commands detrimental economic effects on the maintenance and/or repair of damaged materials. Various studies have been carried out to ascertain the impacts corrosion has on the economy of a nation. Studies conducted by the US National Bureau of Standards (NBS) and Batelle Columbus Laboratories indicated in 1975 an estimated annual cost of corrosion in the United States alone to be $70 billion, accounting for 4–5% of their gross national product (GNP) (Bennett *et al.*, 1978). In 1995, however, this cost was updated and a staggering $300 billion was reported to be lost as a result of corrosion through loss of product, structure and equipment replacement, repair and maintenance, design, redundant equipment, equipment and parts inventories, corrosion control and insurance (Sastri, 2015). Reports from countries like Finland, China, Sweden, India, Germany, Canada, Japan, Kuwait and Australia also reported a high (at least $1 billion) loss of revenue to corrosion (Sastri, 2015).

7.8   MICROBIOLOGICALLY INFLUENCED CORROSION

MIC is a type of corrosion caused by the presence and/or activities of microorganisms such as fungi, microalgae and bacteria. MIC can also be seen as impacts caused by biofilm-forming microorganisms on the kinetics of corrosion to a particular material (i.e. metals) (Ibrahim *et al.*, 2018). Although microorganisms influence/enhance corrosion rate, MIC is not unique to a particular corrosion type (Ibrahim *et al.*, 2018). Microorganisms play vital roles in corrosion processes and have the ability to alter partial reactions that take place at the cathode or anode of a reacting mixture (i.e. metal and water). The physicochemical conditions (i.e. pH level, conductivity, O_2 concentration, redox potential) are altered by the metabolism of biofilm-forming microorganisms at the interface, thereby displacing the corrosion potential toward a more positive direction. This increases metal susceptibility to localized corrosion such as hydrogen embrittlement, stress corrosion cracking, dealloying, pitting and galvanic corrosion (Khouzani *et al.*, 2019). Microbial activities on metal surfaces can result in anodic site localization, acid production and metal oxide reduction at the metal's interface (Ibrahim *et al.*, 2018).

Though discovered 100 years ago, MIC has caught attention of numerous practitioners and researchers across the globe (Kakooei *et al.*, 2012; Mansour and Elshafei, 2016). Its detrimental effects on metals cut across various sectors and industries that carry out oil and gas production, storage and pipeline transportation, wastewater and sewage treatment, and power plants generation, among many other activities (Jia *et al.*, 2019). MIC causes 20% of the total cost of corrosion globally, 20–30% of external corrosion and 40% of internal corrosion (Machuca, 2019). In the United Kingdom alone, MIC accounts for 10% of the total damage caused by corrosion, with several billions of pounds spent in the repair and maintenance of corroded materials (Ibrahim *et al.*, 2018; Inaba *et al.*, 2019). The actions of microorganisms in the speeding up of corrosion processes cannot be overemphasized. A study in Western Australia designed flow lines which are supposed to last up to 20 years but due to impacts of microbial activities, the life span of the flow lines was drastically reduced to three years (Ibrahim *et al.*, 2018). In 2006, Prudhoe Bay's largest oil spill occurred and the transit line failure was attributed to microbial corrosion with a loss of over 803,461 liters of crude oil to the environment (Ibrahim *et al.*, 2018; Machuca, 2019).

Various groups of biofilm-forming microorganisms (e.g. archaea, fungi and bacteria) have all been associated in MIC (Machuca, 2019). However, most research and studies that have been published were focused on bacteria with sulfate-reducing bacteria (SRB) and acid-producing bacteria (APB) topping the list of most destructive biological agents in MIC (Al Abbas *et al.*, 2013; Machuca, 2019; Oyewole *et al.*, 2020). In environments experiencing temperatures higher than 70°C, archaea become the major effector of MIC. However, in Southeast Asia, where the climate is humid and warm, fungi play active roles in in the corrosion of metals (Jia *et al.*, 2019). How these varying groups of microorganisms carry out this deleterious effect on metal is still a topic of hot debate among experts of corrosion across the globe (Machuca, 2019). A few aspects of MIC on metals are discussed in the following subsections.

7.8.1 MECHANISMS OF MIC

The complexity of metal interactions with microorganisms has made determining the mechanisms of MIC inconclusive knowing fully it involves operational, environmental, mechanical, electrochemical and biological factors (Li *et al.*, 2013; Machuca *et al.*, 2019). Various mechanisms have been proposed in the past such as the classical theory (also known as the cathodic depolarization theory [CDT]), biocatalytic cathodic sulfate reduction (BCSR) and anodic depolarization, among many others, to help understand the mechanism of MIC (Kakooei *et al.*, 2012; Li *et al.*, 2013; Kip and van Veen, 2015; Kannan *et al.*, 2018). However, the dynamic nature and diverse groups of microorganisms involved in MIC have made the mechanisms complex and elusive. As such, it is important to stress that evidence from a particular hypothesis can be ineluctably contradictory to another (Li *et al.*, 2013). The fact that MIC has no unique or distinctive pattern of damage on corroded material or any kind of mineralogical signature has made it unpredictable, since it can also be found taking place simultaneously with other forms of erosion (Machuca, 2019). Nevertheless, some certain metallurgical characteristics or features of corrosion such as cup-like appearances with contours or striation and deep pits that appear to be hemispherical have all been claimed to be a part of MIC, as claimed by some corrosion specialists (Machuca, 2019).

One of the most possible and general mechanism to understand MIC is that microorganisms can modify their local environment through the production of electrons, making their immediate environment corrosive in a process known as direct electron transfer (Mansour and Elshafei, 2016; Machuca, 2019). At the metal interface, the inert microbial deposits shield the area from electrolytes, giving them the leverage to initiate an oxygen gradient for diverse microbial metabolism. Areas below the microbial cell often serve as the anodes, whereas the metallic surface within this area serves as the cathode, thus promoting oxygen reduction. The metal area where the microbial deposits act on causes the metal to dissolve, resulting in pit corrosion (Mansour and Elshafei, 2016). MIC can occur under anaerobic or aerobic conditions, with the former being the most destructive (Blackwood, 2018). Under anaerobic conditions, SRB can reduce sulfate-producing oxygen and sulfide as end products. The sulfide ions produced combine with ferrous ions to give rise to iron sulfide, causing the metal surface to dissolve, whereas the oxygen molecule combines with hydrogen with a resultant production of water molecules. The overall process of MIC of metals begins with the formation of metal hydroxide (MOH) through the combination of a metal atom (M) with negative hydroxyl ions (OH$^-$), leaving free protons (H$^+$). The local area is protonated with H$^+$, causing it to be acidic; the presence of other ions such as chlorine can further accelerate the corrosion process (Mansour and Elshafei, 2016; Oyewole *et al.*, 2020).

Environmental factors can affect the rates at which MIC occurs, and the mechanism that ensures MIC varies across different environments. Conditions in freshwater environments differ with those of marine environments, which both differ from those observed in petroleum facilities. Conditions such as nutrient availability, fluid velocity, pH, temperature and light can discourage growth of a particular species while encouraging the growth of others on a particular substrate (Machuca, 2019). Types of electron acceptors such as O$_2$, CO$_2$, and sulfate also play crucial roles in shaping and promoting microbial communities that will colonize, form biofilm and carry corrosive activities on metal surfaces. A selective environment will be created when industries utilize chemical additives, chelators, biocides and other inhibitory substances to curtail fouling, scaling, MIC and other microbial activities, thus affecting the colonization and proliferation of microorganisms involved in MIC. Contrarily, stagnant water or low-velocity systems, debris and deposits can result in sediment buildup, creating an ideal environment for microbial accumulation, metabolism, growth and colonization of any metal surface (Machuca, 2019).

7.8.2 Microorganisms Involved in MIC

As stated earlier, MIC of metals is caused by a myriad of microbial species. Although SRB and APB are the most studied groups of bacteria that cause MIC, other groups of bacteria – including iron oxidizers, manganese oxidizers, sulfur-oxidizing bacteria, nitrate-reducing bacteria, iron reducers, acid producers, slime formers (Kip and van Veen, 2015) and other groups of microorganisms such as archaea and fungi – are also involved in MIC (Oyewole *et al.*, 2020).

7.8.3 Roles of Microbial Biofilms in MIC

Regardless of the type of metal, environment or a particular process or reaction governing MIC, experts in the field of MIC unanimously agreed that biofilm formation is essential and a critical step for corrosion of metals. As such, complex chemical reactions are not the only players in MIC, since active involvement of biofilms always ensure the deterioration of metals (Machuca, 2018). As earlier stated, biofilm is made up of microbial cells, water, lipids, proteins, nucleic acids, and EPS. However, it is important to note that biofilm holds an important position in MIC, since its structures, properties and functions are imperative and dynamic – both for the survival of the microbial cells and also the success of the corrosion processes. Some of the roles played by microbial biofilms in facilitating MIC include providing architectural support for microbial growth and proliferation, ensuring supply of nutrients and other necessary materials within the biofilm, differentiating aeration cells, creating diffusion barriers, localizing environmental conditions (i.e. pH and temperature) and depositing corrosive byproducts directly on metal surfaces.

Biofilm provides the architectural support needed for successful microbial colonization of metal surfaces, which is usually initiated by slime-forming bacteria (e.g. *P. aeruginosa*). Once the foundation is laid, other microbes can easily adhere and multiply without the interference of environmental factors such as fluid flow/velocity. Biofilms also provide the platform for the formation of microcolonies, which function independently of each other, thereby promoting heterogeneous communities with resultant adverse effects on the integrity of metals (Ibrahim *et al.*, 2018).

Nutrient supply is also ensured in the biofilm. As individual microcolonies carry out metabolisms, they usually deposit different kinds of organic compounds, which are usually utilized by one or more microcolonies. There is also cell-to-cell communication between the microbial communities, such that if a microcolony is short of nutrients, there is usually notification through quorum sensing and the needs of such microcolony is met, thus promoting growth and survival of microbial cells and enhancing corrosion (Machuca *et al.*, 2018).

During the formation of biofilm, various layers are formed upon one another according to the oxygen demands of the varying microbial species. These gradients of aeration of cells are very important in MIC corrosion (Telegdi *et al.*, 2020). At the top of the biofilm exist aerobic bacteria carrying out cellular respiration in the process utilizing oxygen (Ibrahim *et al.*, 2018; Telegdi *et al.*, 2020). In the middle layer are found facultative microbes whose metabolism is not affected by the function of microbes found at the top layer. However, cellular respiration of facultative microbes plays a major role in corrosion processes, since their activities help in providing an anoxic environment for SRB to carry out their own metabolisms in the process, which causes damage to metal alloys (Telegdi *et al.*, 2020).

The matrix of the biofilm is often made up of gelatinous substances that act as a diffusion barrier for the inflow and outflow of substances such as ions, biocides, O_2 and CO_2, among many others (Imo *et al.*, 2016; Meiying and Min, 2018; Inaba *et al.*, 2019). This barrier is often created at the surface of the biofilm, thereby maintaining the optimum temperature and pH of the microbial community from the bulk fluid (Machuca *et al.*, 2018; Meiying and Min, 2018). For example, APB maintain the pH of their environment through the production and release of organic and inorganic acids into the environment and becomes deposited on surfaces of metals – but absence of a diffusion barrier on biofilm would have prevented the corrosive actions of acids secreted by APB through dilution of the concentrated acid by the bulk fluid surrounding the biofilm, thus limiting its functions. Likewise, available O_2 and CO_2 concentration in the biofilm would have been hampered for microcolonies that use them as terminal electron acceptors, thereby limiting/preventing microbial metabolisms that can cause metal corrosion (Imo et $al.$, 2016; Meiying and Min, 2018).

Deleterious effects caused by the presence and actions of biofilm on metals cannot be overemphasized. Even the passivation layer that is supposed to protect metals from MIC has been destroyed by biofilm (Telegdi et $al.$, 2020). The modifications of environmental conditions in the microenvironment by heterogeneous groups of microorganisms present in a particular biofilm causes ennoblement, which goes a long way in affecting the durability and lifespan of metals. Ennoblement is simply an increase in the corrosion potential (E_{corr}) of a metal caused by activities of microbial communities within a biofilm (Ibrahim et $al.$, 2018). As the pitting potential (E_{pit}) is approached due to E_{corr}, cathodic reactions increase, thus enhancing localized corrosion (Ibrahim et $al.$, 2018). If no action is taken on metals inhabited by microbial biofilms, the cells in a matter of time will disperse and spread onto other surfaces, thereby ensuring a rapid damage to metallic structures.

7.8.4   CONTROL/MITIGATION OF MIC

There are many treatment strategies available for the restriction and mitigation of MIC. In the selection of a particular treatment strategy (i.e. chemical, physical, mechanical and/or biological), certain factors need to be considered (Machuca, 2019). They include the types of microorganisms, environment and metals, and the mechanism utilized for the MIC (Machuca, 2019). The selected treatment method should in no way affect the performance of the system. Most times, the use of one treatment method will not be enough in tackling this menace, as such; two or more treatments are used either concurrently or one after another. Some typical treatments that can be employed in the control of MIC include regular sanitation, chlorination, painting, cathodic protection, filtration, chemical biocides, velocity control, UV irradiation, mechanical cleaning (e.g. flushing and pigging) and protective coatings with antimicrobial substances and inhibitors (Soleimani, 2012; Kip and van Veen, 2015; Mansour and Elshafei, 2016; Oliveira et $al.$, 2016; Jia et $al.$, 2019; Machuca, 2019; Little et $al.$, 2020).

Environmental experts and scientists have raised concerns about the use of some of the aforementioned strategies due to their impacts on the ecosystem. It is an established fact that biofilm are ten times more resistant to antimicrobials, and for any antimicrobials to be effective against a targeted biofilm, then its concentration needs to be 1,000 times higher than the ones effective against planktonic cells (Machuca, 2019). If these high dosages are continually used, is just a matter of time, as we have known the uniqueness of microorganisms in the development of strategies to resist antimicrobial effects. This can further spread antimicrobial-resistant genes to other microorganisms when released into the environment one way or another, making treatment difficult. However, other means such as biocompetitive exclusion, biofilm formation, metabolic products of microorganisms and green technology have been proposed as safe means of mitigating MIC.

There is a gradual shift in the search for ways used in mitigation of MIC. One of the recent foci is on the use of beneficial microorganisms and their metabolites to reduce the rate of MIC a phenomenon known as microbiologically influenced corrosion inhibition (MICI) (Kip and van Veen, 2015; Deepalaxmi and Gayathri, 2018; Oyewole *et al.*, 2020). MICI can be achieved through: (a) actions of microorganisms in the neutralization or complete removal of corrosive acidic medium at the bulk solution-metal interface (an example is the removal of reactive oxygen during aerobic respiration); (b) formation of a protective matrix of EPS with metal binding potentials by non-corrosion causing bacteria; and (c) production of secondary metabolites with inhibitory properties against corrosion-causing bacteria. A combination of two or more of the aforementioned MICI can be used in biofilm containing multiple bacterial species (Kip and van Veen, 2015; Deepalaxmi and Gayathri, 2018).

The inhibition of corrosion of steel was first demonstrated in 1991 by Pedersen and Hermansson through the use of *Serratia marcescens* EF190 and *Pseudomonas sp.S9* isolated from a marine environment. Even after separating the suspension of isolates from the steel, corrosion inhibition was still detected, leading to the conclusion that aerobic respiration of the bacteria isolates is responsible for the protection against corrosion (Kip and van Veen, 2015). Ever since then, different bacterial species (i.e. *Spirulina platensi*, *Shewanella* sp. and *Bacillus* sp., among others) have been evaluated under varying conditions and material for MICI. Microbial respiration is important in mitigating MIC. Both biofilms of anaerobic *E. coli* and aerobic *Pseudomonas fragi* were demonstrated to inhibit corrosion. However, *P. fragi* was observed to decrease MIC more than *E. coli*, owing to its aerobic respiration. Rates of metal corrosion was determined between mutants of *Shewanella oneidensis* (in which their iron respiration and/or biofilm formation has been genetically deleted) and the wild type. Corrosion was observed more on metals with mutant strains compared to the wild strain, emphasizing the importance of aerobic respiration and biofilm formation in mitigating MIC (Kip and van Veen, 2015).

Mansour and Elshafei (2016) reiterate the importance of biofilm formation by non-corrosion–causing bacteria in mitigating MIC. In their report, they indicated the importance of *Bacillus sp.* as a potential tool for biotechnology in respect to mitigation of corrosion. For example, the corrosion of mild steel by iron-oxidizing *Leptothrix discophora* SP-6 and sulfate-reducing *Desulfosporosinus orientis* was inhibited by the biofilm produced by *Bacillus brevis*, as well as antimicrobial substances. However, this inhibition was not the case of non-antibiotic and biofilm-producing *Paenibacillus polymyxa* ATCC 10401 when grown on mild steel in the presence of *D. orientis* and *L. discophora* SP-6. The interaction resulted in severe corrosion of mild steel coupon. Similarly, Garcia *et al.* (2012) demonstrated MICI by bacteria isolated from surface of copper electrode. Scanning electron microscopy (SEM) analysis showed that the biofilm formed by the bacteria served as anticorrosive coating against corrosion. They hypothesized that the metal surface was prevented from interacting with the environment due to the layer of EPS formed by the bacteria isolate. Biofilm of *B. subtilis* was also found to act as a passivation layer for aluminum 2024 (Al 2024) in seawater. However, killing the bacterial cells within using antibiotics resulted in pitting corrosion with hours of the cells' exclusion. The theory, as postulated by Lin and Ballim (2012), suggested that it was due to the decrease of oxygen at the metal interface, as well as reduction of reactions at the anode, owing to the presence of biofilm. They also suggested that the negative charge of EPS and that of the bacterial cells were able to deflect corrosive ions of chloride.

The metabolism of microbes often produces secondary metabolites. Some of these secondary metabolites produced by bacteria, such as *Bacillus* sp., have inhibitory or biocidal properties against other microbes growing in the environment, thus giving them a competitive edge. The potential of *B. brevis* to be used in MICI strategy was mentioned earlier since it can produce biofilm on metal surfaces, as well as secrete metabolites (i.e. gramicidin S) that are deleterious to SRB. Genetically modified *B. subtilis* was also reported to produce bactenecin and indolicidin, which are all active against SRB. The ability to form biofilm is an added advantage to *Bacillus* sp., since the antimicrobial produced will remain within the biofilm (owing to diffusion barrier), thus maintaining its concentration by preventing dilution from the bulk fluid (Kip and van Veen, 2015). Oyewole *et al.* (2020) reported SRB inhibitory effects on corrosion causing *Desulfovibrio indonesiensis*, *D. vulgaris* and *D. alaskensis* by a low molecular weight bioproduct partially purified from a strain of *E. coli*.

Another important environmentally safe method of mitigating MIC is through biocompetitive exclusion (BE). BE is simply the supply of a particular nutrient in an environment to stimulate the growth of a desired bacterial species over another (Kip and van Veen, 2015). This is mostly used in petroleum industry, which often suffers from MIC caused by SRB. To control the presence and activities of SRB in reservoir souring, nitrate (NO_3^-) is injected into the system to spur the growth of NRB. By doing so, the NRB will grow at a faster rate while exhausting the limited nutrients which are supposed to support the growth of SRB. In other to ensure survival, the will have to switch to NO_3^- as a source of energy. This way, sulfate reduction that usually results in MIC will be restricted. The NO_2^- and NO_3^- present in the system goes on competitively bind to enzymes of SRB, thereby inhibiting reduction and sulfide production. When using BE method, one needs to be careful in the amount of NO_3^- that will be injected into the system, since S_2^- can be oxidized by NRB to SO_4^{2-} while reducing NO_3^- in the redox reaction. The NO_2^- produced is an intermediary product resulting from the incomplete reduction of NO_3^-. As such, the amount of NO_3^- to be injected should be commensurate with the right reaction leading to exclusion of SRB from the microbial community. Otherwise, as a result of excess NO_2^- in the system can cause chemical corrosion of the metal (Ibrahim *et al.*, 2018).

Another environmentally safe method of controlling MIC is through the bacterial phage treatment (BPT) strategy used as anti-biofilm agent (Akanda *et al.*, 2018). In BPT, bacteriophages are used to eliminate certain group of bacteria by lysing their cell walls using phage depolymerases. A bacteriophage isolated by Bhattacharjee *et al.* (2015) at 10^5–10^6 PFU/mL was reported to remove biofilm of *Delftia tsuruhatensis* ARB-1 from a wastewater treatment plant. Phages can either work alone or work as a group. Their synergistic effort often causes a better effect. A cocktail of three bacteriophages (LiMN17, LiMN4L and LiMN4p) were reported to reduce seven-day sessile cells of *Listeria monocytogenes* on stainless steel to undetectable levels. However, when they were used individually at 10^9 PFU/mL, they were reported to achieve a 3–4.5-log reduction of the same test isolate under the same conditions. The use of BPT in combination with other antimicrobial products has been reported in the treatment of human infections caused by a particular bacterial strain. As we mentioned earlier about the heterogeneity of bacterial biofilm involved in MIC, this has made the application of BPT in the control of MIC a challenge, due to high specificity of bacteriophages. Phages that inhibit *Desulfovibrio aespoeensis* or *D. vulgaris* may not against other species of *Desulfovibrio* (Jia *et al.*, 2019).

7.9   CONCLUSION

Microbial biofilms play a dual role in corrosion processes. They either influence it or inhibit it, with the former being prevalent, causing deterioration of metals with high economic costs. However, with the right tools and knowledge, non-corrosion–causing microorganisms (i.e. *Bacillus* sp.) with the ability to form biofilm can secrete biocidal substances, which can kill or inhibit the growth of corrosion causing microorganisms such as sulfate-reducing bacteria and acid producing bacteria, among many others.

The mechanisms used by microorganisms involved in MIC should be critically reviewed, as this will provide adequate knowledge on ways they can be stopped. Studies should be channeled more on how bacteriophages can be modified to invade microorganisms involved in MIC; this way, metals can be protected. Since biofilm formation is essential for MIC, researchers should focus more on microbial products with potential to disintegrate the biofilm created, as this will prevent surface contact with metals, as well as provide avenue for bulk fluid to dilute whatever corrosive compound they eventually produce as a result of their metabolism.

REFERENCES

Akanda ZZ, Taha M, Abdelbary H (2018) Current review-The rise of bacteriophage as a unique therapeutic platform in treating peri-prosthetic joint infections. J. Orthopaed. Res. 36:1051–1060. https://doi: 10.1002/jor.23755.

Al-Abbas FM, Williamson C, Bhola SM, Spear J, R, Olson DL, Mishra B *et al.* (2013) Influence of sulfate reducing bacterial biofilm on corrosion behaviour of low-alloy, high-strength steel (API-5L X80). *Int Biodeterior Biodegrad*. 78:34–42. http://dx.doi.org/10.1016/j.ibiod.2012.10.014

Bennett LH, Kruger J, Parker RL, Passaglia E, Reimann C, Ruff AW, Yakowitz H, Berman EB (1978) Economic effects of metallic corrosion in the United States. Part 1. Washington, DC: National Bureau of Standards; Marblehead, MA: Berman (Edward B.) Associates, Inc.

Berhe N, Tefera Y, Tintagu T (2017) Review on biofilm formation and its control options. Int. J. Advan. Res. Biolog. Sci. 4:122–133. http://dx.doi.org/10.22192/ijarbs.2017.04.08.017

Berne C, Ellison CK, Ducret A, Brun YV (2018) Bacterial adhesion at the single- cell level. Natur. Rev. Microbiol. https://doi.org/10.1038/s41579-018-0057-5

Bhattacharjee AS, Choi J, Motlagh AM, Mukherji ST, Goel R (2015) Bacteriophage therapy for membrane biofouling in membrane bioreactors and antibiotic-resistant bacterial biofilms. Biotech. Bioeng. 112:1644–1654. https://doi:10.1002/bit.25574

Blackwood DJ (2018) An Electrochemist Perspective of Microbiologically Influenced Corrosion. Corros. Mater. Degrad. 1:5976. https://doi:10.3390/cmd1010005

Chadha T (2014) Bacterial Biofilms: Survival Mechanisms and Antibiotic Resistance. J. Bacteriolog. Parasitolog. 5:3. https://doi:10.4172/2155-9597.1000190

Cicek V (2017) Types of Corrosion. In, V Cicek: editor: Corrosion Engineering and Cathodic Protection Handbook: With Extensive Question and Answer Section. New Jersey: John Wiley & Sons: 235–252. https://doi:10.1002/9781119284338

Costerton JW, Geesey GG, and Cheng KJ (1978) How bacteria stick. Scientific American. 238(1):86–95.

Deepalaxmi RK, Gayathri C (2018) Screening for the corrosion inhibition of mild steel metal using biofilm forming halophilic bacteria isolated from the saltpans of Thoothukudi district. J. Bacteriolog. Mycolog. Open Acc. 6:280–282. https://doi: 10.15406/jbmoa.2018.06.00218

Flemming HC, Wingender J, Szewzyk U, Steinberg P, Rice SA, Kjelleberg S (2016) Biofilms: an emergent form of bacterial life. Natur. Rev. Microbiol. 14:563–575. https://doi:10.1038/nrmicro.2016.94

Garcia F, Lopez ALR, Guille JC, Sandoval LH, Gonzalez CR, Castano V (2012) Corrosion inhibition in copper by isolated bacteria. Anti-Corros. Method. Material. 59:10–17. https://doi: 10.1108/00035591211190490

Hansson CM (2011) The Impact of Corrosion on Society. Metallurg. Material. Transact. 42A. https://doi: 10.1007/s11661-011-0703-2

Ibrahim A, Hawboldt K, Bottaro C, Khan F (2018) Review and analysis of microbiologically influenced corrosion: the chemical environment in oil and gas facilities. Corros. Eng. Sci. Technol. https://doi:10.1080/1478422X.2018.1511326

Ikechukwu EP, Pauline EO (2015) Environmental Impacts of Corrosion on the Physical Properties of Copper and Aluminum: A Case Study of the Surrounding Water Bodies in Port Harcourt. Open J. Social Sci. 3:143–150. http://dx.doi.org/10.4236/jss.2015.32019

Imo EO, Ihejirika CE, Orji JC, Nweke CO, Adieze IE (2016) Mechanism of Microbial Corrosion: A Review. J. Chem. Biolog Physic. Sci. 6:1173–1178.

Inaba Y, Xu S, Vardner JT, West AC, Banta S (2019) Microbially influenced corrosion of stainless steel by Acidithiobacillus ferrooxidans supplemented with pyrite: importance of thiosulfate. Appl. Environ. Microbiol. 85:e01381–19. https://doi.org/10.1128/AEM.01381-19

Javaherdashti R (2017) Microbiologically influenced corrosion (MIC) and nontechnical mitigation of corrosion: corrosion knowledge management. In Derby B (Ed.), *Microbiologically influenced corrosion an engineering insight, engineering materials and processes.* Springer International Publishing, 2nd ed., 17–96. DOI:10.1007/978-3-319-44306-5

Jia R, Unsal T, Xu D, Lekbach Y, Gu, T (2019) Microbiologically influenced corrosion and current mitigation strategies: A state of the art review. Int. Biodeter. Biodegrad. 137:4258. https://doi.org/10.1016/j.ibiod.2018.11.007

Kakooei S, Ismail MC, Ariwahjoedi B (2012) Mechanisms of Microbiologically Influenced Corrosion: A Review. World Appl. Sci. J. 17:524–531.

Kannan P, Su SS, Mannan MS, Castaneda H, Vaddiraju S (2018) A review of characterization and quantification tools for microbiologically influenced corrosion in the oil and gas industry: Current and future trends. Ind. Eng. Chem. Res. https://doi: 10.1021/acs.iecr.8b02211

Khatoon Z, McTiernan CD, Suuronen EJ, Mah T-F, Alarcon EI (2018) Bacterial biofilm formation on implantable devices and approaches to its treatment and prevention. Heliy. 4:e01067. http //doi:10.1016/j.heliyon.2018.e01067

Khouzani MK, Bahrami A, Hosseini-Abari A, Khandouzi M, Taheri P (2019) Microbiologically Influenced Corrosion of a Pipeline in a Petrochemical Plant. Metal. 9:459. https://doi:10.3390/met9040459

Kip N, van Veen JA (2015) The dual role of microbes in corrosion. Int. Societ. Microb. Ecol. 9:542551. https://doi:10.1038/ismej.2014.169

Li K, Whitfield M, Van Vliet KJ (2013) Beating the bugs: roles of microbial biofilms in corrosion. Corros. Rev. 321:3–6.2013, https://dx.doi.org/10.1515/CORRREV-2013-0019

Liaqat I, Liaqat M, Ali S, Ali NM, Haneef U, Mirza SA, Tahir HM (2019) Biofilm formation, maturation and prevention: a review. *J Bacteriolog Mycolog.* 6:1092.

Lin J, Ballim R (2012) Biocorrosion control: current strategies and promising alternatives. *Afr. J. Biotech.* 11:15736–15747. https://doi: 10.5897/AJB12.2479

Little BJ, Blackwood DJ, Hinks J, Lauroc FM Marsilic E, Okamoto A *et al.* (2020) Microbially influenced corrosion Any progress? Corros. Sci. 170:108641. https://doi.org/10.1016/j.corsci.2020.108641

Loto CA (2017) Microbiological corrosion: mechanism, control and impacta review. Int. J. Adv. Manufact. Technolog. https://doi:10.1007/s00170-017-0494-8

Machuca LL (2019) Understanding and addressing microbiologically influenced corrosion (MIC). *Corros Material.* 88:96. https://www.corrosion.com.au

Mansour R, Elshafei AM (2016) Role of Microorganisms in Corrosion Induction and Prevention. Brit. Biotech. J. 14:1–11. https://doi: 10.9734/BBJ/2016/27049

Meiying L, Min D (2018) A review: microbiologically influenced corrosion and the effect of cathodic polarization on typical bacteria. Rev. Environ. Sci. Biotech. https://doi.org/10.1007/s11157-018-9473-2

Muhammad MH, Idris AL, Fan X, Guo Y, Yu Y, Jin X, Qiu J, Guan X, Huang T (2020) Beyond Risk: Bacterial Biofilms and Their Regulating Approaches. Front. Microbiol. 11:928. https://doi: 10.3389/fmicb.2020.00928

Oliveira SH, Lima MAGA, Franc FP, Vieira MRS, Silvaa P, Filho SLU (2016) Control of microbiological corrosion on carbon steel with sodium hypochlorite and biopolymer. Int. J. Biol. Macromolec. 88:27–35. http://dx.doi.org/10.1016/j.ijbiomac.2016.03.033

Oyewole OA, Mitchell J, Thresh S, Zinkevich V (2020) The purification and functional study of new compounds produced by Escherichia coli that influence the growth of Sulphate Reducing Bacteria. Egy. J. Bas. Appl. Sci. 7:82–99. https://doi.org/10.1080/23 14808X.2020.1752033

Patel I, Patel V, Thakkar A, Kothari V (2014) Microbial Biofilms: Microbes in Social Mode. Int. J. Agricul. Food Res. 3:34–49.

Rabin N, Zheng Y, Opoku-Temeng C, Du Y, Bonsu E, Sintim HO (2015) Biofilm formation mechanisms and targets for developing antibiofilm agents. Fut. Medic. Chemistr. 7:493–512.

Rodis N, Tsapadikou VK, Potsios C, Xaplanteri P (2020) Resistance Mechanisms in Bacterial Biofilm Formations: A Review. J. Emerg. Int. Med. 4:30. https://doi: 10.36648/2576-3938.100030

Römling U, Kjelleberg S, Normark S, Nyman L, Uhlin BE, Akerlund B (2014) Microbial biofilm formation: a need to act. J. Intern. Med. 276:98110. https://doi: 10.1111/joim.12242

Rumbaugh KP, Sauer K (2020) Biofilm dispersion. Natur. Rev. Microbiol. https://doi.org/10.1038/s41579-020-0385-0

Sastri S (2015) Consequences of corrosion. In Sastri S (Ed.), *Challenges in corrosion: costs, causes, consequences, and control.* John Wiley and Sons, 317–401.

Sharma D, Misba L, Khan AU (2019) Antibiotics versus biofilm: an emerging battleground in microbial communities. Antimicrob. Resist. Infect. Contr. 8:76. https://doi.org/10.1186/s13756-019-0533-3

Singh R (2014) Corrosion principles and types of corrosion. In: R Sigh: editor: Corrosion control for Offshore Structures. Boston: Gulf Professional Publishing:7–40 https://doi: 10.1016/C2012-0-01231-8

Soleimani S (2012) *Prevention and control of microbiologically influenced concrete deterioration in wastewater concrete structures using e.coli biofilm* (Thesis). A thesis submitted to the Faculty of Graduate and Post Doctoral Affairs in partial fulfillment of the requirements for the degree of Doctor of Philosophy in Environmental Engineering Department of Civil and Environmental Engineering, Carleton University.

Telegdi J, Shaban A, Trif L (2020) Review on the microbiologically influenced corrosion and the function of biofilms. Int. J. Corros. Scale Inhib. 9:133. https://doi:10.17675/2305-6894-2020-9-1-1

Vasudevan R (2014) Biofilms: microbial cities of scientific significance. J. Microbiol. Experiment. 1:84–98. https://doi:10.15406/jmen.2014.01.00014

Veena BR, Shetty KV, Saidutta MB (2013) Characteristics of biofilms in bioreactors: a review. *Proced Eng.* https://www.researchgate.net/publication/273452184

Verderosa AD, Totsika M, Fairfull-Smith KE (2019) Bacterial Biofilm Eradication Agents: A Current Review. Front. Chemistr. 7:824. https//:doi:10.3389/fchem.2019.00824

Yuanzhe L, Xiao P, Wang Y, Hao Y (2020) Mechanisms and Control Measures of Mature Biofilm Resistance to Antimicrobial Agents in the Clinical Context. ACS Omeg. 5:22684–22690. https://dx.doi.org/10.1021/acsomega.0c02294

8 Microbial Cellulose Biofilms

Medical Applications in the Field of Urology

Andrea Sordelli, Maribel Tupa, Patricia Cerrutti and María Laura Foresti

CONTENTS

8.1 INTRODUCTION

Microorganisms were the first living beings that populated the Earth, much earlier than plants and animals. Since then, they have been naturally transforming chemical reagents into products, largely influenced by their enzymes content according to their own particular genetic "program". They can be prokaryotic cells (*Archaea* and *Bacteria* domains), or eukaryotic cells (mold and yeasts), but on the whole, they exhibit an immensurable genetic and physiological diversity that provides a wide spectrum of products and services. Also, the different species of microorganisms can grow under a wide variety of conditions, i.e. from refrigeration to teens of degrees over room temperature, and at different O_2 concentrations, yet in its complete absence. In fact, some *Archaea* live in what we call "extreme environmental conditions", similar to the primitive Earth, i.e. very high temperatures and absence of free O_2.

Although microorganisms are often associated with spoilage or diseases, most of them are innocuous or even beneficial for other living forms. Even more, many of them have established important relations with plants and animals. In the case of the human species, microorganisms play a fundamental role in the digestive system, particularly in the gut. It is also remarkable that the oldest foods – such as bread, wine, beer, cheeses, vinegar, pickles and fermented milks – have been prepared by

DOI: 10.1201/9781003184942-10

microbial fermentations from very ancient times, i.e. thousands of years BC, according to documented archaeological and historical sources. But the microbial existence was not taken into account till the 17th century, when they could be observed using microscopes. It was in 1664 when Robert Hooke reported the existence of fungi, and 20 years later, Antoni van Leeuwenhoek could observe bacteria cells. Many years later, in the mid- to late 1800s, Louis Pasteur, who defeated the theory of spontaneous generation, elucidated the mechanism of fermentation, thus giving an important boost to industrial microbiology (Madigan *et al.*, 2021; Ferrari *et al.*, 2020).

In the 20th century, large-scale fermentations were developed for obtaining a great range of high-value products like organic acids, ethanol, biomass, polymeric materials, enzymes, antibiotics, foods, pharmaceutical, fine chemicals, biomedical and cosmetics products, energy, packaging, and more. This fact was due to facility to adapt microbial growth to laboratory flasks, and then to large-scale fermenters. Another milestone occurred by the 1970s, when the studies on recombinant DNA strongly encouraged biotechnology, with multiple applications in many fields and industrial applications (Demain and Adrio, 2008; Mir, 2004).

Microbial fermentations exhibit many advantages when compared with chemical processes and even with bioproduction from plants and animals. In the first place, microorganisms are ubiquitously distributed on Earth; they can efficiently produce value-added products from renewable cheap raw materials as agro-industrial byproducts or even wastes, and their genetic modifications can be more easily performed. In addition, microbial products are usually synthesized under milder conditions, thus resulting in greener methodologies with lower energy requirements and equipment costs when compared with chemical synthetic processes. The high area-to-volume ratio of microorganisms results in high metabolic rates, and their fermentation processes are often easier to implement and render fewer undesirable byproducts. On the other hand, they have the ability to specifically produce the biological active isomers, that is D-carbohydrates and L-amino acids, unlike chemical methods that generally produce racemic mixtures. Also, kinetics of microbial growth and product formation are usually favorable (i.e. a few days for the fermentation process) over bioproduction in plants and animals. In particular, primary metabolites are obtained in shorter times than secondary ones. Finally, like other materials of natural origin, microbial products are generally biodegradable and compostable.

Nowadays, chemically or physically induced mutations contribute to improvement of microbial strains, thus enhancing yields, production and productivities, and so the economy of the overall fermentation process. In this case, genetic stability of modified strains is also required (Sivakumar *et al.*, 2010). Moreover, by means of recombinant DNA technologies, the microorganisms can be induced to produce substances typical of other microorganisms and even of plants that otherwise they would not be able to, or to improve production (i.e. g/L) by increasing or reducing a certain metabolic reaction or route.

On the other hand, it would be desirable that microbial cultures selected for industrial fermentation processes have certain features. For example, it is very important to establish the safety of the microorganisms employed. In this sense, the term GRAS (generally recognized as safe) for a microorganism or a product is a label for the security of the process for both producers and consumers. It is also desirable that the microorganism could grow on a wide variety of substrates (wide nutritional diversity), mostly renewable low-cost raw materials. The use of thermotolerant microorganisms is also frequently desirable, taking into account that the risk of contamination and the costs of fermenters refrigeration are lower than when using mesophilic ones, being also very appreciated for obtaining commercially valuable thermostable enzymes. Moreover, the use of anaerobic microorganisms that reduce energetic costs, as no aeration nor agitation are required, is also advantageous. Besides, obtaining a single valuable product with no toxic byproducts in the fermentation process would be beneficial, as well as other factors as the availability of pure strains and free from virus.

Microorganisms and their products are actually essential for human healthcare, and since a long time ago they have proved largely successful in medical treatments all around the world. In fact, although the beginning of the microbial drugs era was assigned to Alexander Fleming in 1928 with the discovery of penicillin, there are evidences that the antibiotic tetracycline had been used in humans in the Sudanese Nubia since 350–550 AD (Nelson *et al.*, 2010; Bassett *et al.*, 1980). Many largely well-known pharmaceutical products show the relevance of microorganisms in healthcare, both by means of their primary metabolites like vitamins, amino acids and nucleotides, and by more complex molecules such as antibiotics which account for secondary metabolites that are widely used for human therapies. Antibiotics are efficiently produced by *Streptomyces* and other actinomycetes, by non-filamentous bacteria and by filamentous fungi. Although many bacteria can produce interesting metabolites like the protein-digesting protease obtained from *Streptomyces griseus*, fungi are largely used for producing a wide variety of antibiotics and other drugs such as lovastatin or pravastatin with hypocholesterolemic effects, antitumoral agents such as taxol, and dietary supplements like the polyunsaturated fatty acids γ-linoleic acid and arachidonic acid. Proteins like human interferon, epidermal growth factor and hemoglobin, antigens for hepatitis-B virus, stabilizers for erythropoietin and human chorionic gonadotropin are also obtained by microbial routes, in this case by means of yeasts such as *Saccharomyces cerevisiae*, *Komagataella pastoris* or *Hansenula polymorpha*, or bacteria like *Agrobacterium tumefaciens*. Other very valuable drugs as inmunosuppressive agents like cyclosporin A are produced by the fungus *Tolypocladium nivenum* (Madigan *et al.*, 2021; Aminov, 2010; Demain and Sanchez, 2009; Demain and Adrio, 2008; Mir, 2004).

In nature, wild bacterial strains frequently synthetize extracellular polysaccharides, forming biofilms in practically all the ecosystems where essential nutrients are available. These biofilms usually impact negatively in areas like medicine, dentistry, agriculture and the environment, and in processes where piping systems are involved, but microbial polymers have also broad biotechnological applications even in the medical field. They present several advantages such as the fact that generally they do not cause toxic effects in the host (biocompatibility), and that they can be assimilated by many species (biodegradability) (Luengo *et al.*, 2003; Willians and Martin, 2002). Nevertheless, bulk production of microbial biopolymers is often limited by high production and recovery costs that restrict their application to high-value–added products. To reduce production costs, factors like fermentation efficiency and the price of the raw materials used for culture media formulation must be seriously taken into account (Corujo *et al.*, 2016). Table 8.1 mentions the microbial polymers which

TABLE 8.1

Some Important Biopolymers Produced by Microorganisms and Their Applications.

Microbial Polymer	Main Uses	Microorganisms	References
Proteins: Recombinant enzymes	Food, pharmaceutical, detergent, textile, leather, and pulp and paper industries.	Recombinant microorganisms: bacteria (e.g. *Bacillus* spp., *Corynebacterium glutamicum*), yeasts (*Kluyveromyces lactis*, *Yarrowia lipolytica*, *Saccharomyces cerevisiae*) and molds (e.g. *Aspergillus* spp., *Chrysosporium lucknowense*).	Patel *et al.* (2019); Demain and Vaishnav (2016); Najafpour (2015); Sanchez and Demain (2015).
Xanthan	Food, oil, pharmaceutical, cosmetic, paper, paint and textile industries.	*Xanthomonas* spp.	Sworn (2021); Butler (2016); Demain and Adrio (2008).
PHAs (polyhydroxyal-kanoates)	Packaging, pharmaceutical, biomedicine and agricultural industries.	*Ralstonia eutropha* and other recombinant bacteria.	Winnacker (2019); Zhao and Turng (2015).
Dextran	Food, oil, medicine, and analytical chemistry industries.	*Leuconostoc* spp. and *Streptococcus* spp.	Ullah *et al.* (2021); Munir *et al.* (2019).

currently exhibit the greatest importance, with particular focus on their most common applications. The microorganisms commonly used for the production of these biopolymers at commercial quantities have been summarized.

Among microbial polymers, and based on its outstanding properties, microbial cellulose – or, more commonly, bacterial cellulose (BC) – has received special attention during the last two decades. BC is a water-insoluble extracellular polysaccharide produced by several genera of bacteria such as *Acetobacter, Azotobacter, Agrobacterium, Pseudomonas, Achromobacter, Rhizobium, Aerobacter* and *Sarcina* by means of an aerobic fermentation. The most studied producer is the strict aerobic bacteria *Komagataeibacter xylinus* (syn. *Gluconacetobacter xylinus, Acetobacter xylinum*) (Esa *et al.*, 2014; Chawla *et al.*, 2009). The BC biofilm (hereafter referred to as membrane, mat or pellicle) is formed by *c.a.* 1–2% cellulose, and the rest is water that is mostly in a free state and can be easily removed. In contrast, bound water is more firmly attached in capillaries or by adsorption to the cellulose fibers and requires techniques such as lyophilization or drying and hot pressing to be removed (O'Neill *et al.*, 2017). Although the molecular formula of BC is the same as that of vegetable cellulose, their features are quite different, being very remarkable for its higher purity (because the bacterial polymer is free from lignin, hemicellulose and pectins), high crystallinity index, high degree of polymerization and, most importantly, its nanofibrillar structure which conditions many of its properties (e.g. huge water holding capacity) and triggers most applications.

In the current chapter, the main aspects of microbial cellulose production, as well as the most relevant properties and application areas of BC, are first described, then, in the framework of the currently hot topics of regenerative medicine and tissue engineering, the most promising uses of BC in the medical field are summarized. Finally, attention is particularly drawn to the advances made in the use of BC in the relatively less explored medicine field of urology. In the last few years, a number of contributions toward the development and efficient use of BC and BC-based materials in the treatment of different urologic pathologies and conditions have been made, and this chapter is aimed at critically revising them.

8.2 PRODUCTION OF MICROBIAL CELLULOSE BIOFILMS

BC is produced in an aerobic fermentation process whose productivity and yield mainly depend on the kind of strain used, the composition of the culture medium (nature and concentration of carbon and nitrogen sources, demand of microelements), the system implemented for the fermentation process (static or agitated and type of fermenter chosen) and the fermentation conditions (pH, agitation speed (if any), temperature, oxygen supply, culture time). Although several bacteria are capable of extracellularly generating BC (mostly gram-negative acetic acid bacteria), few species are reported to produce good yields compatible with large-scale fermentation (Corujo *et al.*, 2016). Among them, the most studied and non-pathogenic strain for producing commercially available BC is *Komagataeibacter xylinus*, due to its high-rate cellulose production and GRAS character (Wang *et al.*, 2019). *K. xylinus* efficiently produces BC from glucose and many other organic substrates, as well as from several agro-industrial byproducts and wastes. BC production takes place at mild temperatures and in a wide range of pH within neutral to slightly acidic values (Mona *et al.*, 2019; Wang *et al.*, 2019; Revin *et al.*, 2018; Lin *et al.*, 2016).

BC production implies preparation of the inoculum, production, harvesting and purification. Inoculum can be prepared under agitated or static conditions during 48–72 hours in a proper culture media. The inoculum is then added to the fermentation medium at 1–10% v/v. In the production stage, the majority of BC is produced during the exponential phase of growth. BC production can take place under agitated or static culture conditions, with the resulting characteristics of the cellulose obtained – e.g. macroscopic morphology, microstructure, crystallinity, porosity – being quite different depending on whether agitation is used or not. For example, under static fermentation, a gelatinous membrane is produced in the form of a pellicle that grows at the air–liquid interface of the fermenter; while in agitated cultivation, such as shake culture or rotating disk systems, discrete particles of different sizes and shapes (e.g. asterisk-like, sphere-like, pellet-like or irregular shapes), can be obtained within the culture medium (Figure 8.1) (Wang *et al.*, 2019; Revin *et al.*, 2018; Żywicka *et al.*, 2015). In static fermentations, BC membranes are produced in the liquid medium–air interface, entrapping the aerobic bacterial cells, allowing their contact with the atmospheric oxygen and protecting them from physical and chemical threats (Cacicedo *et al.*, 2016; Ruka *et al.*, 2014a). The production of the BC membranes in static conditions involves lower energy costs than agitated culture,

FIGURE 8.1   Bacterial cellulose obtained in static (left) and agitated (right) fermentations.

since neither agitation nor aeration are applied. Based on its simplicity, higher BC yield due to lower proportion of Cel- mutants (bacterial cells that do not produce cellulose) and the numerous potential uses of BC directly in the form of flat mats, static fermentation has always been the BC production method most widely used. In this context, the following general description of BC biosynthesis, properties and applications will be mainly focused on static systems.

BC biosynthesis involves a complex process that is regulated by a huge number of specific enzymes. During fermentation, UDP-glucose units are polymerized and assembled by synthase complexes and secreted out of the bacterial cells as subfibrils of ca. 1.5 nm in diameter that then aggregate into nanofibrils of 2–4 nm. These nanofibrils further aggregate to form nanoribbons with characteristic nanometric rectangular section and micrometric length, which produce a three-dimensional (3D) porous intricate network in the liquid medium/oxygen-rich interface (Nicolas *et al.*, 2020; Raghvendran *et al.*, 2020; Foresti *et al.*, 2017). The obligated aerobic bacterial cells get entrapped in the synthesized BC biofilm, thus allowing them to float and reach the atmospheric oxygen (Mona *et al.*, 2019; Wang *et al.*, 2019). The BC biofilms formed on the surface of the grown medium of standing cultures change their aspect with culture time (Figure 8.2). BC mats increase in thickness with fermentation time, and more compact nanoribbons networks – which result in progressively reduced porosity – are frequently observed (Cerrutti *et al.*, 2016). In this way, for a definite set of fermentation conditions, culture time is a key variable for tuning not only the BC biofilm thickness, but also BC mats properties relying on the resulting nanoribbons architecture (e.g. water holding capacity, mechanical properties, cell infiltration possibility, etc.) (Corzo Salinas *et al.*, 2021; Corzo Salinas, 2019).

On the other hand, as in static culture, the BC biofilms grow on the air-liquid surface of the culture container, adopting its shape, and can be produced in different forms depending on the design of the reactor used for the desired application. Wound dressings and face masks for the delivery of active compounds and increased skin hydration are examples of the uses of the flat BC membranes produced in vials or

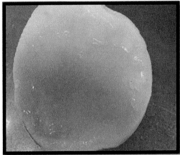

FIGURE 8.2 Bacterial cellulose membranes harvested at three days (left) and 15 days (right) of fermentation, static culture, 28°C.

trays with circular or square sections (Amorim *et al.*, 2020; Bianchet *et al.*, 2020; Portela *et al.*, 2019; Czaja *et al.*, 2007). On the other hand, hollow tubes of BC have been proposed for replacement of blood vessels and other tubular structures of the human body such as urethra, trachea or the digestive tract (Corzo Salinas *et al.*, 2021; Tang *et al.*, 2017; Huang *et al.*, 2015; Zang *et al.*, 2015; Bodin *et al.*, 2010). The production of BC tubes with the desired length, inner diameter and thickness involves the assembly of a tubular reactor with the dimensions required for the proposed use and built up with an oxygen-permeable material (a necessary condition for BC production using the aerobic bacterium *K. xylinus*). In this sense, oxygen-permeable polymers such as silicone allow the diffusion of the gas to the culture medium and the consequent formation of the BC membrane at the medium/permeable material interface (Hong *et al.*, 2015; Zang *et al.*, 2015). Figure 8.3a–c shows images of three different fermenters used for the production of BC at laboratory and pre-pilot scale including Erlenmeyers flasks with circular profile, a rectangular aluminum tray and oxygen-permeable self-standing polydimethylsiloxane (PDMS) cylindrical fermenters (Corzo Salinas, 2019). The corresponding harvested BC flat pellicles or hollow tubes are shown in Figure 8.3d–f. The brownish color of the membrane obtained in the tray is due to the fermentation medium retained in the BC film, which is later removed by washing.

In reference to culture media formulation, the traditional culture medium used for BC production is the Hestrin *et al.* (1947), which is mainly composed of glucose as carbon source, peptone and yeast extract as nitrogen sources, and other growth factors and citrate-phosphate buffer to maintain the pH of the culture (Raghvendran *et al.*, 2020; Jedrzejczak-Krzepkowska *et al.*, 2016). Regarding the carbon source, besides glucose, different commercial substrates can be used for efficient cellulose production such as sucrose, fructose, mannose, xylose, galactose and polyols as glycerol (Jedrzejczak-Krzepkowska *et al.*, 2016). *Komagataeibacter* species have a flexible metabolism and are able to assimilate a variety of monosaccharides, disaccharides, alcohols and organic acids (Fernandes *et al.*, 2020; Ullah *et al.*, 2019; Dufresne, 2017). The carbon source is considered one of the most important factors

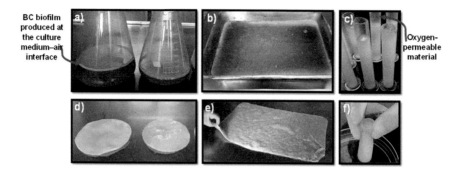

FIGURE 8.3   Images of three different fermenters used for the production of BC at laboratory scale: a) Erlenmeyers flasks; b) aluminum trays; and c) oxygen-permeable self-standing
PDMS cylindrical fermenters. Also shown are images of the corresponding harvested BC flat
pellicles (d, e) and hollow tubes (f) (Corzo Salinas, 2019).

for effective optimization of BC production, since a high concentration in the culture medium is required to reach greater productivities. Hence, this substrate may
concern up to 65% of the entire cost of BC production, determining its price and
the economic viability of the process, especially at industrial scale (Fernandes *et
al.*, 2020; Jozala *et al.*, 2015). Consequently, several researchers have evaluated the
use of cheaper carbon sources focusing on agricultural and industrial based wastes
or byproducts, frequently resulting not only in a more economical and ecofriendly
process, but also in a higher BC yield compared to the use of monosaccharides
(Wang *et al.*, 2019; Jedrzejczak-Krzepkowska *et al.*, 2016; Castro *et al.*, 2013; Chen
et al., 2013). Non-conventional raw materials which have been assayed as carbon
sources for BC production include several sugar-rich products such as fruit juices or
extracts from fruit skins and peels (e.g. apple, Japanese pear, coconut water, orange,
sugarcane, banana, pineapple, grapes, watermelon, algarroba); byproducts of other
industries (e.g. glycerol from biodiesel, beet molasses, sulfite pulping liquor, soya
bean whey, cane molasses, cheese whey); hydrolysates such as those obtained from
paper, sunflower meal, elephant grass, spruce, wheat straw, and fish; and residues or
wastes from agro-industrial processes such as saccharified food wastes, wine distill-
ery waste, rotten mangoes, rotten bananas, flour-rich wastes, dry olive mill residues
and coffee cherry husk. In most cases, these economical alternatives also provide
other essential nutrients for bacterial growth and cellulose production. On the other
hand, a pretreatment of the majority of non-conventional raw materials must be
implemented before fermentation. These pretreatments may include water extraction, acid or enzymatic hydrolysis (conversion of biomass to fermentable sugars), and
detoxification (Corujo *et al.*, 2016). Besides carbon, nitrogen supply is required in the
culture media to grow microorganisms and to produce cellulose, with peptone and
yeast extract being the most common nitrogen sources utilized in traditional growth
media. Other nitrogen sources that have been assayed include glycine, tea extracts,
corn steep liquor, fish hydrolysate, fish powder, sodium glutamate, hydrolyzed casein

and ammonium sulfate (Corujo *et al.*, 2016; Jedrzejczak-Krzepkowska *et al.*, 2016; Zhou *et al.*, 2007).

On the other hand, diverse additives have been added to BC production media for different purposes. Some examples are ethanol, which is used both as supplementary carbon source and for blocking the accumulation of bacteria that accommodates Cel-mutants (Fernandes *et al.*, 2020; Islam *et al.*, 2017; Krystynowicz *et al.*, 2002); organic nutrients, including organic acids (acetic, malic, citric and lactic acids) which mainly contribute toward energy generation during the early stages (Revin *et al.*, 2018; Jedrzejczak-Krzepkowska *et al.*, 2016; Yang *et al.*, 2014; Son *et al.*, 2003); and vitamins as ascorbic acid that possibly act as antioxidants, leading to reduced gluconic acid levels (Raghvendran *et al.*, 2020; Keshk, 2014). Fatty acids from vegetable oils such as rapeseed oil (Żywicka *et al.*, 2018); inorganic salts such as disodium phosphate (Na_2HPO_4), sulfur and magnesium or potassium salts (Fernandes *et al.*, 2020; Jozala *et al.*, 2016); and other polymers such as carboxymethylcellulose, agaragar, xanthan and sodium alginate (Dufresne, 2017; Cheng *et al.*, 2009; Zhou *et al.*, 2007; Bae and Shoda, 2005; Bae *et al.*, 2004) have also been assayed.

Culture conditions such as temperature, pH and oxygen levels have a high impact on BC production and yield. Considering each microorganism has an optimum temperature and pH for its growth, these parameters require strict control (Fernandes *et al.*, 2020; Campano *et al.*, 2016; Zeng *et al.*, 2011; Dobre *et al.*, 2008). Moderate temperature values of 28–30°C and pH values between 4 and 7 are often appropriate. High levels of gluconic acid produced during aerobic growth, when high concentrations of sugars are hydrolyzed, result in acidification of culture medium (Wang *et al.*, 2019; Vandamme *et al.*, 1998), so during fermentation, pH needs to be adjusted to the mentioned range. Oxygen, as previously mentioned, is essential for BC production. Strictly aerobic bacteria anchor in the biofilm in order to get higher levels of oxygen that are required for their metabolism (Sharma and Bhardwaj, 2019). In static systems, injection of oxygen into the fermentation vial has frequently resulted in higher BC production (Bodin *et al.*, 2007).

Once BC is produced, harvesting and purification of the BC is required. After the desired number of fermentation days (usually 5–14) have passed, the fresh BC is manually harvested by simply removing the biofilm from the fermenter surface or by filtering the culture medium, then, the BC is purified by washing with deionized water to remove residual medium components, and by a mild alkali treatment for 10–120 min (depending on the thickness of the pellicle) at 100°C to remove the entrapped microorganisms. A last step includes rinsing the membranes with plenty of water to reach a neutral pH (Raghvendran *et al.*, 2020; Mona *et al.*, 2019; Ullah *et al.*, 2019; Wang *et al.*, 2019). Finally, the purified polymer may be used in its native form, or it can also be functionalized to adapt/improve its properties for specific applications.

8.3 PROPERTIES AND APPLICATIONS OF MICROBIAL CELLULOSE

BC displays unique properties with several advantages over plant-derived cellulose. Although it consists of a linear homopolymer of glucose monomers joined by β-(1→4) glycosidic links and presents identical molecular formula as plant cellulose (*i.e.* $C_6H_{10}O_5)_n$, BC is recognized for its dense reticulated web-shaped fibril network that imparts it inherent unique properties and makes it suitable/promising for a wide variety of applications. Among the distinguishing features of microbial cellulose are its high polymerization degree (i.e. 4,000–10,000 anhydroglucose units), high crystallinity (80–90%), high stability, high chemical purity (unlike the plant-derived cellulose nanofibers, microbial cellulose is produced free of foreign compounds as lignin and hemicelluloses), light weight, extensive surface area (owing to the well-separated nanoribbons of BC); very high liquid loading capacity (i.e. 98–99% for water, much higher than plant cellulose), high degree of conformability in the wet state, high mechanical strength (owing to the hydrogen bonds between the fibrillar units that stabilize the whole structure) and high transparency. Moreover, as any other cellulose, BC is a polymer of renewable origin, highly hydrophilic (due to the three hydroxyl groups per glucose monomer), indigestible in the human intestinal tract (which promotes its use as an additive for low-calorie foods), biodegradable in soil (but not in the human body due to absence of cellulases), compostable, non-toxic, biocompatible and non-resorbable (Figure 8.4).

On the other hand, many of the mentioned properties may be greatly influenced by the culture medium composition, the strain species and the fermentation conditions used in the BC production (Cielecka *et al.*, 2020). For example, properties such as porosity, specific surface area, water holding capacity (which is positively correlated with the porosity and surface area), degree of polymerization, molecular weight, intrinsic viscosity, crystallinity index, mean crystallite size, water vapor transmission rate, oxygen gas transmission rate and mechanical properties (e.g. Young's modulus, tensile strength, stress at break, elongation at break) have been

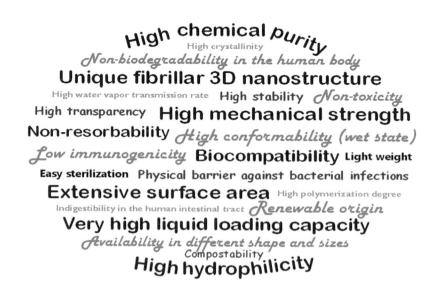

FIGURE 8.4   Main characteristics of bacterial cellulose.

reported to be significantly affected by the carbon source, the strain and/or the fermentation time chosen (Corzo Salinas *et al.*, 2021; Yim *et al.*, 2017; Cerrutti *et al.*, 2016; Kiziltas *et al.*, 2015; Yang *et al.*, 2013). In addition, the structure of BC can be altered by the introduction in the culture medium of chosen additives such as carboxymethyl cellulose, methylcellulose, hydroxypropyl methylcellulose, polyethylene oxide, polyvinyl alcohol, starch, alginate, chitosan and polyhydroxybutyrate (Castro *et al.*, 2014; Ruka *et al.*, 2014b; Osorio *et al.*, 2014; Gea *et al.*, 2010; Huang *et al.*, 2011, 2010; Cheng *et al.*, 2009; Grande *et al.*, 2009). The effect of some of these additives on the structure of BC has been associated with their interference with the aggregation of elementary fibrils within the normal ribbon assembly (Cheng *et al.*, 2009; Huang *et al.*, 2011, 2010), while others have shown their ability to integrate into de BC nanoribbons network forming in situ nanocomposites with distinguished properties (Castro *et al.*, 2014; Osorio *et al.*, 2014; Ruka *et al.*, 2014b; Gea *et al.*, 2010; Grande *et al.*, 2009).

The remarkable attributes of BC and the possibility of in situ and ex situ tailoring (i.e. during or after fermentation, respectively) of its properties make this microbial polysaccharide a material with a wide range of potential applications in different fields, some of which have already reached the market. Table 8.2 lists some of them, describing the particular function of BC addition on the mentioned products. Among its diverse uses, recent research and development trends suggest that the biomedical field seems one of the most promising fields for BC, and that will be the focus of the following sections.

TABLE 8.2
Some Applications of Bacterial Cellulose

Application	Function	Products	References
Food industry	Thickener, texturizer, stabilizing agent, calorie reducer, food packaging.	Yogurt, salads, diet foods (low-calorie ice creams, snacks, jams, and candies), BC-based films and coatings.	Raghavendran *et al.* (2020); Guo *et al.* (2018); Dufresne (2017); Okiyama *et al.* (1993).
Pulp and paper industry	Binder, absorbent.	Ultrastrength paper.	Dufresne (2017).
Cosmetic industry	Thickener, strengthener, skin moisturizer, emulsion stabilizer.	Skin cream, facial masks, astringents, base for artificial nails, fingernail polish.	Amorim *et al.* (2020); Bianchet *et al.* (2020).
Bioremediation	Adsorption/absorption of pollutants (oil or water purification).	Functionalized composites, aerogels.	Qiu *et al.* (2020); Li *et al.* (2019); Luo *et al.* (2018).
Biomedicine	Support for cell adhesion and proliferation, absorbent, gaseous exchange, physical barrier against external infection.	Wound dressing, tissue engineering, regenerative medicine, drug delivery.	Sharma and Bhardwaj (2019); Dufresne (2017).

8.4   USES OF MICROBIAL CELLULOSE IN MEDICINE

Cells are the basic structures of all living organisms, and they are the fundamental components of tissues. Usually, groups of cells form and secrete their own support structures, generating the extracellular matrix (ECM), which in humans is mainly composed of proteoglycans, hyaluronic acid, collagen, elastin, laminin and fibronectin. This matrix does much more than just existing as a cellular structural support. The ECM is – above all – a biological niche and a constant stimulus that favors cells migration and proliferation, and tissue renewal, while allowing the arrival of the necessary nutrients, the diffusion of signaling molecules and the excretion of debris and waste. The understanding that the ECM is much more than a surface that supports other components has opened an immense field of work and research, such as the development of materials indicated for use in living or culture models (Guyette *et al.*, 2014).

In this context, in the last years research in regenerative medicine and tissue engineering has resulted in huge advances toward regeneration and reconstruction of pathologically altered tissues, such as cartilage, bone, skin, heart valves, nerves and tendons (Pina *et al.*, 2019). Regenerative medicine (RM) is an emerging interdisciplinary field of research and clinical applications that develops methods and strategies to regenerate, replace or repair damaged or diseased cells, organs and/or tissues to restore the impaired function irrespectively of the reason for their damage, e.g. age, disease, congenital defects, trauma, surgery (Garcia-Gomez *et al.*, 2017; Pariente *et al.*, 2002; Scriven *et al.*, 2001). RM uses a combination of several technological approaches that exceed traditional transplantation and replacement therapies, including for example gene therapy, stem cell transplantation, tissue engineering, the reprogramming of cell and tissue types, and the patient's own healing per se or facilitated by the addition of certain molecules (Argibay, 2012; Mason and Dunnill, 2008; Greenwood *et al.*, 2006). On the other hand, tissue engineering (TE) is a biomedical engineering discipline that applies fundamental principles and methods of engineering, physics, chemistry and biology to understand the relationship between structure and function of tissues in physiological and pathological situations. Besides, TE studies and tests the development of materials, biological substitutes, devices and clinical strategies to restore, maintain or improve the fundamental function of tissues (Lanza *et al.*, 2013). Thus, TE is usually associated with the development of biomaterials (i.e. any material used for the manufacture of devices that will be used to functionally replace a certain part of the body in a safe, reliable, economical and physiologically acceptable manner (Park and Lakes, 2012), for their subsequent implementation in pursuit of RM.

In this context, TE often involves the seeding of cells on tissue scaffolds, which are porous three-dimensional matrices capable of providing an adequate environment for the regeneration of tissues and organs. Scaffolds (which may be manufactured according to the needs of the patient and adapt to the anatomical characteristics of the defect to be repaired or treated) are intended to function as transitional supports for cell colonization, migration, proliferation and differentiation; increasing the retention rate of the implanted cells in the target organ; and enabling diffusion of vital cell nutrients and expressed products, mimicking the mechanical and/

or physicochemical characteristics of the damaged tissue and providing support or containment.

To achieve this goal, scaffolds used in TE should reproduce the function of living tissues in biological systems in a safe, mechanically functional and physiologically acceptable way, regardless of being implanted temporarily or permanently in the body. The materials used should of course be biocompatible, i.e. generate an acceptable biological, chemical non-toxic and mechanical response upon specific application (Salvatierra *et al.*, 2009; Torres *et al.*, 2012). Biocompatibility is a result of the complex interactions between an implant and the surrounding tissues (Torres *et al.*, 2012). Non-biocompatible materials can cause a foreign body type inflammatory response that may result in graft rejection to necrosis of the recipient tissue, large sequelae, risk of removal surgeries, additional and more complex surgeries for the establishment of the remaining tissue, associated infections, intense drug treatment schemes and even the death of the patient. In the case of scaffold materials which are biodegradable in the human body, the degradation products should be removed from the body via metabolic pathways, causing no or at least minimal inflammation or adverse effects (Kim *et al.*, 2000). Moreover, the scaffold degradation rate should coincide as much as possible with the rate of tissue formation to guarantee that the scaffold would provide the required structural integrity until the natural matrix structure fabricated by the cells is able to take over the mechanical load.

The porosity, pore size distribution, interconnectivity of pores, elasticity, topography, wettability, surface charge, fiber diameter, presence of hydrophobic and hydrophilic domains, and density and conformation of functional groups are also important characteristics of the scaffold which would play a crucial role in the cell-material interaction conditioning the fate of the envisaged biomedical applications (Klemm *et al.*, 2018; Picheth *et al.*, 2017; Giuliak *et al.*, 2009). Besides biomaterials nature selection, a number of the mentioned characteristics of the scaffolds rely on the processing technology used for their development (e.g. electrospinning, solvent casting with particulate leaching, freeze-drying, gas foaming, 3D printing, etc.), being the processing technique chosen of significant importance to optimize their performance (Pina *et al.*, 2019). In this framework, the ability to create and develop better and cost-efficient materials to be used as scaffolds, and to attempt to control cell behavior through both physical and molecular interactions, would expand the boundaries of tissue substitutes in combination with a wide variety of cell types and modulators (de Oliveira Barud *et al.*, 2016; Giuliak *et al.*, 2009).

In recent years, a broad variety of natural and synthetic polymers have been used for the development of scaffolds for TE and RM. Decellularized ECM and also composite structures (e.g. polymers and inorganic and ceramic materials) with improved properties have also been assayed (Pina *et al.*, 2019). Among biopolymers, bacterial cellulose – which is well-recognized for its unique fibrillar 3D nanostructure that resembles that of the ECM, extraordinary mechanical properties, high chemical purity, biocompatibility, low immunogenicity, promotion of cell adhesion and growth, high water-holding capacity and the possibility of producing it in various shapes, sizes and porosity (Carvaljo *et al.*, 2019; Osorio *et al.*, 2019; de Oliveira Barud *et al.*, 2016; Jozala *et al.*, 2016) – has received great attention of the academic community.

A variety of potential applications of BC materials in medicine have been proposed, including the use of both flat membranes and 3D structures. Key properties of BC for its use in TE and RM – such as mechanical properties and porosity – have been tuned both in situ (i.e. during fermentation by adjusting culture conditions and medium formulation including additives) and ex situ (i.e. after microbial biofilm harvesting and purification, e.g. by blending or by preparing composite materials). Besides, different organic or inorganic compounds that promote cell attachment, proliferation and differentiation, or provide specific additional functions such as antimicrobial activity, have been incorporated to better suit the properties of BC for definite applications (Emre Oz *et al.*, 2021).

In this framework, BC and BC-based materials are very promising options for different health-related uses, such as – among others – wound dressings for the temporary covering of second-degree burns and different wounds such as pressure sores, skin tears, venous stasis, ischemic and diabetic wounds, skin graft donor sites, traumatic abrasions and lacerations, non-healing lower extremity ulcers and biopsy sites (Picheth *et al.*, 2017; de Oliveira Barud *et al.*, 2016; Kowalska-Ludwicka *et al.*, 2013; Portal *et al.*, 2009; Czaja *et al.*, 2007, 2006); drug delivery; cardiovascular implants as heart valve replacements (Mohammadi, 2011; Millon *et al.*, 2008; Millon and Wan, 2006); cartilage implants or replacement (Martínez *et al.*, 2012; Svensson *et al.*, 2005); bone regeneration and osseous damage healing (Tazi *et al.*, 2012; Yin *et al.*, 2011; Shi *et al.*, 2009; Wan *et al.*, 2009; Klemm *et al.*, 2006); regeneration of periodontal disease and temporary dressings for oral mucosa, which prevent dehydration of the wound and infection for (de Oliveira Barud *et al.*, 2021, 2016); dura mater substitute (Klemm *et al.*, 2018), corneal regeneration (de Oliveira Barud *et al.*, 2016; Messaddeq *et al.*, 2008) and treatment of macular degeneration (Picheth *et al.*, 2017); tympanic membrane healing (de Oliveira Barud *et al.*, 2016); and replacement of tubular structures of the human body (trachea, urethra and blood vessels) by use of BC tubes of different length, inner diameter and wall thickness (de Oliveira Barud *et al.*, 2016; Wippermann *et al.*, 2009; Schumann *et al.*, 2009; Klemm *et al.*, 2006, 2005).

8.5 USES OF MICROBIAL CELLULOSE IN UROLOGY

Urology is the branch of medicine concerned with the urinary system of both males and females, and with the genital tract of the reproductive system of males. The urinary tract is responsible for cleaning and filtering excess fluid and waste material from the blood, and it includes the kidneys (where urine is produced), the ureters (two thin tubes which carry the urine from the kidneys to the bladder), the bladder (which acts as the urine reservoir) and the urethra (a tube located at the bottom of the bladder that allows urine to exit the body) (Figure 8.5). The proper functioning of all body parts in the urinary tract is required for normal urination. Male reproductive structures under the domain of urology are testes, epididymis, ductus deferens, seminal vesicles, prostate and penis (Figure 8.5). Urologic diseases or conditions include – among others – urinary tract infections often resulting from bacteria from the digestive tract reaching the urethra; urinary incontinence; male infertility; kidney disease; renal transplantation; cancers affecting the urinary system or the

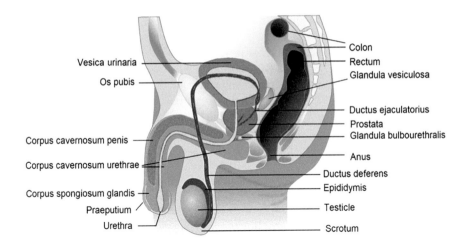

FIGURE 8.5 Some components of the male genitourinary system (from Wikipedia, the free encyclopedia).

reproductive system of men; bladder prolapse; enlarged prostate, which causes the urethra to constrict and induced problems with urination; interstitial cystitis; kidney and ureteral stones; erectile dysfunction; undescended testes; hypospadias and epispadias, two anomalous birth conditions that mainly affect males, causing the penis not to develop correctly; diseases related with soft tissues, such as phimosis (i.e. the inability to retract the skin covering the head of the penis) and Peyronie's disease (i.e. a fibrous layer of scar tissue that develops beneath the skin of the penis and leads to bending or curving during erection); and, very common, urethral stricture (i.e. narrowing of the lumen of the urethra), resulting from the scarring of the urethra which may narrow or block the path of urine flowing from the bladder.

Related to the urinary tract, defects of the ureter and the urethra are, in some cases, difficult to cure, especially if the defect exceeds a certain size. The principal problem in the urology field is the inner diameter of the conducts, so if the defect or harm is lengthy, the problem faced is the possibility of strictures, which means scarred constrictions of the uriniferous system (Klemm et al., 2018). A urinary tract obstruction can cause damage to the urinary tract and kidneys because urine backs up and pools in various areas along the tract, which may lead to infection, scarring and long-term kidney damage.

Repair of the urethral strictures is still a challenging problem in urology. As well as with other congenital and acquired pathologies, strictures require surgical urethral reconstruction, being the graft sources most commonly used penile skin and buccal, colonic and lingual mucosa (Bodea et al., 2020; Huang et al., 2015). However, usual shortness of those tissues, morbidity of the donor site and/or surgical complications, such as prolapse and stricture recurrence, have triggered the development of different biomaterials to be used in urethral reconstruction (Zhang and Xiao, 2016). Among them, bacterial cellulose has been proposed as a particularly promising scaffold for urethral substitution, although tuning of the scaffold porosity is frequently required.

In this direction, Huang *et al.* (2015) assayed the reconstruction of urethral defects in rabbit models using 3D porous BC scaffolds seeded with lingual keratinocytes. Aiming to promote cell adhesion, growth and penetration, the scaffold's porosity was increased by addition of a medical gelatin sponge to the fermentation medium, which interfered with cellulose assembly during static culture. Authors concluded that 3D porous BC seeded with lingual keratinocytes enhanced urethral tissue regeneration without inducing inflammatory reactions. Mechanical strength (which is essential for urethral reconstruction since the scaffold would have to bear voiding urine pressures and prevent the occurrence of fistulas or diverticula), and suture retention strength were both considered adequate for implantation (Huang *et al.*, 2015). Lv *et al.* (2016a) also used BC and gelatin to develop in this case a bilayer 3D microporous composite scaffold composed of an upper smooth layer and a porous or sponge-like sublayer, and evaluated it for long-segment urethral regeneration in a dog model. The results not only showed that the composites were proper for keratincocytes and muscle cells adhesion and growth in vitro, but also that the bilayer scaffold with lingual keratinocytes and muscle cells enhanced the repair in dog urethral defect models. Besides, the combination of both materials, BC and gelatin, resisted a higher mechanical load than the gelatin alone (Lv *et al.*, 2016a). Later on, the same group developed another bilayer scaffold comprising a microporous network of silk fibroin and a nanoporous surface of BC, and evaluated its potential for long-segment urethral regeneration in the dog model. Cultures established with lingual keratinocytes and lingual muscle cells also confirmed the suitability of these scaffolds to support cell adhesion and proliferation; and the urethra reconstructed with the seeded scaffold showed better structure compared to the non-seeded biomaterial (Lv *et al.*, 2018). Similar results were obtained by this research upon development of a bilayer nanofibrous material using potato starch to interrupt BC assembly during static culture and control the scaffold´s porosity (Lv *et al.*, 2016b). In all cases, authors claimed that the produced bi-layered scaffolds could be used for other hollow organs of the urinary tract, such as bladder and ureter.

In a later contribution, these authors pointed out the importance in urethral TE of designing and fabricating urethral scaffolds that possess biomimetic multi-scale structures, and indirect 3D printing technology and template biosynthesis were used to fabricate a 3D gelatin/silk-BC urethral scaffold, with defined macro, micro and nanostructure which better mimicked the native urethra ECM (Li *et al.*, 2017). More recently, this group combined BC and bladder acellular matrix to design a 3D biomimetic scaffold that could accelerate urethral regeneration by enhancing angiogenesis (i.e. growth of blood vessels from the existing vasculature) and epithelialization, since limited angiogenesis and epithelialization using conventional tissue-engineered grafts are known to hinder urethral regeneration. In vitro studies revealed that the BC + bladder acellular matrix scaffold promoted angiogenesis, and its implantation in a rabbit urethral defect model demonstrated that the improved blood vessels formation in the scaffold significantly promoted epithelialization and accelerated urethral regeneration (Wang *et al.*, 2020).

Urinary incontinence is another very common urological problem which strongly affects the quality of life and self-esteem of the patient. It may be caused by bladder dysfunction and by weakness of the urethral rhabdosphincter resulting from surgical trauma, vaginal delivery, neurological conditions and congenital anomalies (Lima *et al.*, 2017). Pubovaginal sling surgery is a procedure used to manage stress urinary incontinence (i.e. leakage of urine associated with activities like coughing, sneezing, laughing, lifting or even walking or standing). A sling of tissue or mesh is placed like a hammock under the bladder and urethra to help support them in place, and keep the urethra closed to prevent urine leakage. Lucena *et al.* (2015) evaluated the interaction of BC with urethral tissue in comparison with a polypropylene mesh when used as a pubovaginal sling by use of a rat animal model. The BC implant showed good integration to the host tissue, preserving its architecture, whereas stereological analysis at the suburethral area showed a significant difference in collagen presence in favor of BC (Lucena *et al.*, 2015).

Treatment of urinary incontinence by implantation of anti-incontinence devices such as artificial sphincters devoted to producing urethral occlusive force often results in the structural weakening of the chronically occluded urethral wall, leading to atrophy and/or erosion (Kim *et al.*, 2008). In this context, the application of BC membrane as urethral strengthening wrap was assayed in an animal model, aiming to test the suitability of BC as a protecting barrier to the urethra following the implantation of anti-incontinence devices. BC strips were applied around the urethra just below the bladder neck of female rats with urethral dissection to mimic a urinary incontinence model. Based on the morphometric and histological aspects of BC membrane integration, authors concluded that the cellulosic grafts showed satisfactory integration to the surrounding tissue and promoted its remodeling and strengthening (Lima *et al.*, 2017). The same research group later assayed BC as urethral reinforcement for urethrovesical anastomosis. Prostatic cancers are often treated by radical prostatectomy aimed at curing cancer and also maintaining urinary continence and erectile function. However, radical prostatectomy may result in urinary fistulas, bladder neck stenosis and urinary incontinence (da Silva Maia *et al.*, 2018). Urethrovesical anastomosis is a complex task which depends on the surgeon's skill, the type of material and technique used. The results obtained in a rabbit animal model showed that no extrusion, stenosis or urinary fistula occurred, demonstrating the biocompatibility and biointegration of BC, but no definite conclusion on the ability of BC to promote urethral reinforcement could be reached, since a decrease in the urethral wall thickness was observed after 14 weeks of implantation (da Silva Maia *et al.*, 2018).

After prostate cancer, bladder cancer is the second most common urologic malignancy in the United States (Bodin *et al.*, 2010). Treatment of bladder cancer may require bladder removal (cystectomy), after which urine needs to be redirected with a urinary diversion (i.e. a surgery that makes a new way for urine to leave the body) and reconstruction procedure. Urinary diversion is also required in other situations as when the normal flow of urine is blocked, some birth defects or when the bladder is damaged upon radiation therapy or other cancer treatments. Usually, in urinary diversions, a part of the intestine is surgically converted to either a passage tube for urine to exit the body, or a reservoir to store urine (like a normal bladder). However, urinary diversion may cause urinary tract infections, stone formation, skin break-down around the stoma (i.e. opening in the abdominal wall covered with a bag that gathers the urine), stenosis of the stoma or impaired renal function, among other complications. On the other hand, TE technology may provide an alternative approach for building a functional urinary conduit to store urine for patients with bladder cancer who require total cystectomy. With this aim, porous BC tubes were biosynthesized by adding sterile paraffin particles of chosen micrometric size to a tubular fermentation vessel. The BC porous scaffolds were seeded with induced human urine-derived stem cells and assayed for their capacity to form a tissue-engineered conduit for use in urinary diversion. Urothelial cells and smooth muscle cells derived from urine-derived stem cells proved to form multilayers on the BC scaffold surface, and some of them infiltrated into the scaffold. The scaffolds were implanted into athymic mice, where cells appeared to differentiate and express urothelial and smooth muscle cells markers. Authors highlighted the potentiality of these cell-seeded BC tissue-engineered urinary conduits for patients with end-stage bladder diseases who may need bladder reconstruction (Bodin *et al.*, 2010).

BC scaffolds loaded with vascular endothelial growth factor (VEGF, a potent angiogenic factor with a fundamental role in the process of blood vessel formation) were also assessed for their feasibility for bladder reconstruction in comparison with unmodified BC scaffolds. In this way, keratinocytes and lingual muscle cells were seeded onto the scaffolds. Results suggested that the bladder capacity and compliance were better in the VEGF-loaded BC scaffolds which could release VEGF for at least four weeks, and they evidenced thicker epithelial layers, better-organized muscle bundles and denser microvasculature than the control group. Authors claimed that the VEGF-nanoparticle loaded BC scaffolds could be an alternative biomaterial for bladder reconstruction (Feng *et al.*, 2016). Further research of this group assayed the effect of adding rabbit lingual epithelial cells and tongue muscle cells onto the VEGF-loaded BC and the unmodified BC scaffolds (Zhang *et al.*, 2016).

Besides diseases that affect the urinary tract, urology also cares about conditions of the male reproductive organs, including erectile dysfunction. Erectile dysfunction is very common condition defined as the consistent or recurrent inability to attain and/or maintain penile erection sufficient for sexual satisfaction, including satisfactory sexual performance (Burnett *et al.*, 2018). In mild cases of erectile dysfunction, its symptoms may often be significantly improved by simple behavior changes such as quitting smoking, reducing weight and doing exercise regularly. However, treatment of severe erectile dysfunction requires medical therapies, including pharmacotherapy with oral phosphodiesterase-5 (PDE-5) inhibitors that enable increased blood to flow and result in erection, use of vacuum erection devices which pull blood into the penis, self-injection of vasodilator medications into the side of the penis, intraurethral prostaglandin suppositories and, if necessary, surgical placement of a penile prosthesis, which results in irreversible change in penile body tissue after implantation with destruction of cavernous tissue (Mobley *et al.*, 2017). Considering the associated surgery risks and derived consequences, a model of penile prosthesis involving an injectable intracavernous BC gel was proposed and evaluated in a rabbit model. Results indicated that the implant of BC increased both the length and thickness of the penis three and six months after the last injection, with a consequent increase in the diameter of the corpus cavernosum, and evidenced the biocompatibility and biointegration of BC to the host tissue. Results suggested that the applicability of BC gel as penile filling material, with local application and being minimally invasive. Authors claimed that injectable BC prostheses may be applied in the future in cases of penile surgery, penile reconstruction and erectile dysfunction (Lima *et al.*, 2021).

One possible cause of erectile dysfunction is Peyronie's disease, which is another pathologic model in the field of urology. Peyronie's disease is characterized by the presence of an induration in the tunica albuginea of the penis resulting from fibrous scar tissue development, which generates curvature or deformity during erection, pain and – in some cases – erectile dysfunction (Ostrowski *et al.*, 2016). In this context, Sordelli *et al.* (2021) designed a protocol for the treatment of this disease using a scaffold of BC and tested it in a rat animal model. Authors demonstrated that the suture of a BC mat, as well as the instillation of platelet-rich plasma, favored the recovery and remodeling of the extracellular matrix and tissue repair. Similar results were obtained with the addition of adipose tissue–derived stem cells contained in stromal vascular fraction in conjunction with platelet-rich plasma dispensed after BC scaffold graft surgery, but in this case, a certain inflammatory response associated with infection was found (Sordelli, 2021). The obtained results suggest that BC membranes, together with autologous platelet-rich plasma and stromal vascular fraction with the patient's own adipose tissue–derived stem cells, may be a promising option for the innovative, less invasive, less traumatic and effective treatment of Peyronie's disease (Sordelli *et al.*, 2021).

In the field of urology, BC has also been assayed as dressing for surgical male wound healing at the urogenital area. Despite surgical correction of genitalia anomalies showing significant evolution during the last few years, proper selection of a wound dressing for the best postoperative wound healing is a very important aspect, since healing of the urogenital area is a complex phenomenon that usually takes longer than dermatological healing. In fact, and based on many of its distinguishing properties (e.g. high water holding capacity and high water vapor transmission rate, which contribute to maintaining a moist environment at the wound/dressing surface), as a nanoporous structure which prevents penetration of external bacteria into the wound bed, the possibility to sterilize by steam radiation, mechanical strength, biocompatibility, ability to be molded in situ, high elasticity, conformability, etc. (Czaja *et al.*, 2007), BC has been long recognized as an ideal wound dressing material for dermatological healing. With respect to urological injuries, Phase I studies showed BC effectiveness as a mechanical barrier and as a safe adjuvant in the treatment of a surgical wound remaining from hypospadias correction (Martins *et al.*, 2013). In a later contribution, in an open Phase II non-controlled clinical trial with 141 patients including children, adolescents and adults with hypospadias, epispadias, phimosis and Peyronie's disease, patients were applied a BC dressing over the operated area after the corresponding surgery. Based on the absence of adverse events, complete healing time (i.e. 8–10 days after surgery), adhesiveness in the wound area, moisture retention, transudation capacity, and pain and discomfort relief, authors concluded that BC dressings are a satisfactory alternative for postoperative wound healing in the urogenital area (Oliveira *et al.*, 2016).

Finally, BC has also been used in the treatment of human urinary tract infections which can be caused by bacteria, fungi and viruses. Urinary tract infections mainly affect the lower tract (urethra and bladder) and cause pain and burning with increased frequency of urination (Zmejkoski *et al.*, 2021). Therapies for urinary tract infections (including those associated with the use of catheters that might be complicated with encrustation by ureases) imply the use of antibiotics, which are in turn challenged by development of bacteria resistance. Alternatively, BC-chitosan composite hydrogels were assayed. The produced composites showed strong antibacterial activity against the strains typically found in patients with bacterial urinary tract infections, suggesting their promising use in their treatment as well as in urinary catheter coating (Zmejkoski *et al.*, 2021).

8.6 CONCLUSIONS

Based on several distinguishing characteristics of bacterial cellulose already described, this microbial polymer is recognized worldwide for being particularly well suited for biomedical applications. In this framework, during recent years, a number of biomedical uses of BC pellicles and 3D hollow structures with very promising results have been proposed, and several review articles on the topic have been published. However, the particular use of BC in the treatment of urological diseases and conditions had not yet been reviewed, this being the aim of this chapter.

The literature review performed evidenced that in the last five years, the use of BC in the treatment of several urologic pathologies and conditions has shown significant advances, and different promising alternatives for urethral and bladder reconstruction, as well as for the treatment of urinary incontinence, erectile dysfunction, urinary tract infections and postoperative wounds, have been proposed. Most developments were made by a still relatively low number of (very active) research groups.

Researchers' results have shown the importance of a proper architecture of the BC scaffolds and dressings, which may be tuned in situ by use of porogens and templates that interfere with cellulose assembly during static culture. As for any other scaffold, mechanical properties and usage of compounds that promote cell attachment, proliferation and differentiation have also proved to be of huge importance. Overall, BC appears as a promising biomaterial for further innovation in the treatment of different urological disorders, especially those requiring tissue regeneration and healing.

REFERENCES

Aminov, R. I. (2010) A brief history of the antibiotic era: lessons learned and challenges for the future. *Front Microbiol.* 1(134):1–7. https://doi.org/10.3389/fmicb.2010.00134.

Amorim, J. D. P., Galdino, C. J. S., Costa, A. F. S., Almeida, H., Vinhas, G. M., Sarubbo, L. A. (2020) BioMask, a polymer blend for treatment and healing of skin prone to acne. *Chemical Engineering Transactions.* 79:205–210. https://doi.org/10.3303/CET2079035.

Argibay P (2012) *Medicina regenerativa: hechos y fantasías en relación con los potenciales terapéuticos de las células madre (2012).* Medicina Regenerativa, Ediciones del Hospital. ISBN:978-987-1639-14-4.

Bae, S., Shoda, M. (2005) Statistical optimization of culture conditions for bacterial cellulose production using Box-Behnken design. *Biotechnology and Bioengineering.* 90(1):20–28. https://doi.org/10.1002/bit.20325.

Bae, S., Sugano, Y., Shoda, M. (2004) Improvement of bacterial cellulose production by addition of agar in a jar fermentor. *Journal of Bioscience and Bioengineering.* 97(1):33–38. https://doi.org/10.1016/S1389-1723(04)70162-0.

Bassett, E. J., Keith, M. S., Armelagos, G. J., Martin, D. L., Villanueva, A. R. (1980) Tetracycline-labeled human bone from ancient Sudanese Nubia (AD 350) *Science.* 209(4464):1532–1534. https://doi.org/10.1126/science.7001623.

Bianchet, R. T., Cubas, A. L. V., Machado, M. M., Moecke, E. H. S. (2020) Applicability of Bacterial Cellulose in Cosmetics – Bibliometric Review. *Biotechnology Reports.* e00502. https://doi.org/10.1016/j.btre.2020.e00502.

Bodea, I. M., Cătunescu, G. M., Stroe, T. F., Dîrlea, S. A., Beteg, F. I. (2020) Applications of bacterial-synthesized cellulose in veterinary medicine – a review. *Acta Veterinaria Brno.* 88(4):451–471. https://doi.org/10.2754/avb201988040451.

Bodin, A., Bäckdahl, H., Fink, H., Gustafsson, L., Risberg, B., Gatenholm, P. (2007) Influence of cultivation conditions on mechanical and morphological properties of bacterial cellulose tubes. *Biotechnology and Bioengineering.* 97(2):425–434. https://doi.org/10.1002/bit.21314.

Bodin, A., Bharadwaj, S., Wu, S., Gatenholm, P., Atala, A., Zhang, Y. (2010) Tissue-engineered conduit using urine-derived stem cells seeded bacterial cellulose polymer in urinary reconstruction and diversion. *Biomaterials.* 31(34), 8889–901. https://doi.org/10.1016/j.biomaterials.2010.07.108.

Burnett, A. L., Nehra, A., Breau, R. H., Culkin, D. J., Faraday, M. M., Hakim, L. S., . . ., Shindel, A. W. (2018) Erectile dysfunction: AUA guideline. *The Journal of Urology.* 200(3):633–641. https://doi.org/10.1016/j.juro.2018.05.004.

Butler M (Ed.) (2016) *Xanthan gum: applications and research studies.* Nova Publishers.

Cacicedo, M. L., Castro, M. C., Servetas, I., Bosnea, L., Boura, K., Tsafrakidou, P., Castro, G. R. *et al.* (2016) Progress in bacterial cellulose matrices for biotechnological applications. *Bioresource Technology.* 213:172–180. https://doi.org/10.1016/j.biortech.2016.02.071.

Campano, C., Balea, A., Blanco, A., Negro, C. (2016) Enhancement of the fermentation process and properties of bacterial cellulose: a review. *Cellulose.* 23(1):57–91. https://doi.org/10.1007/s10570-015-0802-0.

Carvalho, T., Guedes, G., Sousa, F. L., Freire, C. S., Santos, H. A. (2019) Latest Advances on Bacterial Cellulose-Based Materials for Wound Healing, Delivery Systems, and Tissue Engineering. *Biotechnology Journal.* 14(12):1900059. https://doi.org/10.1002/biot.201900059.

Castro, C., Cleenwerck, I., Trček, J., Zuluaga, R., De Vos, P., Caro, G., Aguirre, R., Putaux, J.-L., Gañán, P. (2013) *Gluconacetobacter medellinensis* sp. nov., cellulose-and non-cellulose-producing acetic acid bacteria isolated from vinegar. *International Journal of Systematic and Evolutionary Microbiol.* 63(3):1119–1125. https://doi.org/10.1099/ijs.0.043414-0.

Castro, C., Vesterinen, A., Zuluaga, R., Caro, G., Filpponen, I., Rojas, O. J., *et al.* (2014) In situ production of nanocomposites of poly(vinyl alcohol) and cellulose nanofibrils from *Gluconacetobacter* bacteria: Effect of chemical crosslinking. *Cellulose.* 21(3):1745–1756. https://doi.org/10.1007/s10570-014-0170-1.

Cerrutti, P., Roldán, P., García, R. M., Galvagno, M. A., Vázquez, A., Foresti, M. L. (2016) Production of bacterial nanocellulose from wine industry residues: Importance of fermentation time on pellicle characteristics. *Journal of Applied Polymer Science.* 133(14):43109. https://doi.org/10.1002/app.43109.

Chawla, P. R., Bajaj, I. B., Survase, S. A., Singhal, R. S. (2009) Microbial cellulose: fermentative production and applications. *Food Technology and Biotechnology.* 47(2):107–124.

Cielecka, I., Ryngajłło, M., Bielecki, S. (2020) BNC biosynthesis with increased productivity in a newly designed surface air-flow bioreactor. *Applied Sciences.* 10(11):3850. https://doi.org/10.3390/app10113850.

Chen, L., Hong, F., Yang, X. X., Han, S. F. (2013) Biotransformation of wheat straw to bacterial cellulose and its mechanism. *Bioresource Technology.* 135:464–468. https://doi.org/10.1016/j.biortech.2012.10.029.

Cheng, K.-C., Catchmark, J. M., Demirci, A. (2009) Effect of different additives on bacterial cellulose production by *Acetobacter xylinum* and analysis of material property. *Cellulose.* 16(6):1033–1045. https://doi.org/10.1007/s10570-009-9346-5.

Corujo, V. F., Cerrutti, P., Foresti, M. L., Vázquez, A. (2016) Production of bacterial nanocellulose from non-conventional fermentation media. In D. Puglia, E. Fortunati, J. Kenny (Eds.), *Multifunctional polymeric nanocomposites based on cellulosic reinforcements* (pp. 39–59) William Andrew Publishing. https://doi.org/10.1016/B978-0-323-44248-0.00002-X.

Corzo Salinas DR (2019) *Producción de formas tubulares de nanocelulosa bacteriana para aplicaciones biomédicas* (Bachelor's Thesis). Universidad de Buenos Aires.

Corzo Salinas, D. R., Sordelli, A., Martinez, L. A., Villoldo, G., Bernal, C., Perez, M. S., Cerrutti, P., Foresti, M. L. (2021) Production of bacterial cellulose tubes for biomedical applications: analysis of the effect of fermentation time on selected properties. International Journal of Biological Macromolecules. – (Accepted contribution, july 2021)

Czają, W.K., Krystynowicz, A., Bielecki, S., Brown, R.M. (2006) Microbial cellulose: the natural power to heal wounds. *Biomaterials.* 27(2):145–151. https://doi.org/10.1016/j.biomaterials.2005.07.035.

Czaja, W.K., Young, D.J., Kawecki, M., Brown, R.M. (2007) The future prospects of microbial cellulose in biomedical applications. *Biomacromolecule.* 8(1):1–12. https://doi.org/10.1021/bm060620d.

da Silva Maia, G. T., de Albuquerque, A. V., Martins, E. D., de Souza, V. S. B., da Silva, A. A., de Melo Lira, M. M., Lima, S. V. C. (2018) Bacterial cellulose to reinforce urethrovesical anastomosis. A translational study1. *Acta Cirúrgica Brasileira.* 33(8):673–683. https://doi.org/10.1590/s0102-865020180080000003.

Demain, A. L., Adrio, J. L. (2008) Contributions of microorganisms to industrial biology. *Molecular Biotechnology.* 38(1):41. https://doi.org/10.1007/s12033-007-0035-z.

Demain, A. L., Sanchez, S. (2009) Microbial drug discovery: 80 years of progress. *The Journal of Antibiotics.* 62(1):5–16. https://doi.org/10.1038/ja.2008.16.

Demain, A. L., Vaishnav, P. (2016) Production of recombinant enzymes. In *Reference module in food science.* Elsevier, 1–12. https://doi.org/10.1016/B978-0-08-100596-5.03023-7.

de Oliveira Barud, H. G., da Silva, R. R., Borges, M. A. C., Castro, G. R., Ribeiro, S. J. L., da Silva Barud, H. (2021) Bacterial nanocellulose in dentistry: perspectives and challenges. *Molecules.* 26(1):49. https://doi.org/10.1016/j.carbpol.2016.07.059.

de Oliveira Barud, H. G., da Silva, R. R., da Silva Barud, H., Tercjak, A., Gutierrez, J., Lustri, W. R., . . ., Ribeiro, S. J. (2016) A multipurpose natural and renewable polymer in medical applications: bacterial cellulose. *Carbohydrate Polymers.* 153: 406–420. https://doi.org/10.1016/j.carbpol.2016.07.059.

Dobre, T., Stoica, A., Parvulescu, O. C., Stroescu, M., Iavorschi, G. U. S. T. A. V. (2008) Factors influence on bacterial cellulose growth in static reactors. *Revista De Chimie-Bucharest-Original Edition.* 59(5):591. https://doi.org/10.37358/RC.08.5.1835.

Dufresne, A. (2017) *Nanocellulose: from nature to high performance tailored materials.* Walter de Gruyter GmbH, Co KG. https://doi.org/10.1515/9783110480412.

Emre Oz, Y., Keskin-Erdogan, Z., Safa, N., Esin Hames Tuna, E. (2021) A review of functionalised bacterial cellulose for targeted biomedical fields. *Journal of Biomaterials Applications.* 1–34. https://doi.org/10.1177/0885328221998033.

Esa, F., Tasirin, S. M., Abd Rahman, N. (2014) Overview of bacterial cellulose production and application. *Agriculture and Agricultural Science Procedia.* 2:113–119. https://doi.org/10.1016/j.aaspro.2014.11.017.

Feng, C., Xiang-Guo, L. V., Zhu, W. D., Lu, W. L., Peng, X. F., Zhang, X. R. (2016) VEGF-Loaded Nanoparticle-Modified Bacterial Cellulose Combined with Multiple Cells in Tissue-Engineered Bladder. *Journal of Biomaterials and Tissue Engineering.* 6(3):216–223. https://doi.org/10.1166/jbt.2016.1434.

Fernandes, I. D. A. A., Pedro, A. C., Ribeiro, V. R., Bortolini, D. G., Ozaki, M. S. C., Maciel, G. M., Haminiuk, C. W. I. (2020) Bacterial cellulose: From production optimization to new applications. *International Journal of Biological Macromolecules.* 164:2598–2611. https://doi.org/10.1016/j.ijbiomac.2020.07.255.

Ferrari A, Vinderola G, Weill R (2020) *Alimentos fermentados: microbiología, nutrición, salud y cultura.* Instituto Danone del Cono Sur. ISBN:978-987-25312-2-5.

Foresti, M. L., Vázquez, A., Boury, B. (2017) Applications of bacterial cellulose as precursor of carbon and composites with metal oxide, metal sulfide and metal nanoparticles: A review of recent advances. *Carbohydrate Polymers.* 157:447–467. https://doi.org/10.1016/j.carbpol.2016.09.008.

Garcia-Gomes, B., Ralph, D., Levine, L., Moncada Iribarren, I., Djinovic, R., Albersen, M., Garcia-Cruz, E., Romero-Otero, J. (2017) Grafts for Peyronie's disease: a comprehensive review. *Andrology.* 6:117–126. https//.doi.org/10.1111/andr.12421.

Gea, S., Bilotti, E., Reynolds, C. T., Soykeabkeaw, N., Peijs, T. (2010) Bacterial cellulose – poly (vinyl alcohol) nanocomposites prepared by an in-situ process. *Materials Letters.* 64(8): 901–904. https://doi.org/10.1016/j.matlet.2010.01.042.

Giuliak, F., Cohen, D.M., Estes, B.T., Gimble, J.M., Liedtke, W., Chen, C.S. (2009) Control of Stem Cell Fate by Physical Interactions with the Extracellular Matrix. *Cell Stem Cell.* 5(1):17–26. https://doi.org/10.1016/j.stem.2009.06.016.

Grande, C. J., Torres, F. G., Gomez, C. M., Troncoso, O. P., Canet-Ferrer, J., Martínez-Pastor, J. (2009) Development of self-assembled bacterial cellulose – starch nanocomposites. *Materials Science and Engineering: C.* 29(4):1098–1104. https://doi.org/10.1016/j.msec.2008.09.024.

Greenwood, H. L., Thorsteinsdóttir, H., Perry, G., Renihan, J., Singer, P., Daar, A. (2006) Regenerative medicine: new opportunities for developing countries. *International Journal of Biotechnology.* 8(1–2): 60–77. https://doi.org/10.1504/IJBT.2006.008964.

Guo, Y., Zhang, X., Hao, W., Xie, Y., Chen, L., Li, Z., *et al.* (2018) Nano-bacterial cellulose/soy protein isolate complex gel as fat substitutes in ice cream model. *Carbohydrate Polymers.* 198:620–630. https://doi.org/10.1016/j.carbpol.2018.06.078.

Guyette, J.P., Gilpin, S.E., Charest, J.M., Tapias, L.F., Ren, X., Ott, H.C. (2014) Perfusion decellularization of whole organs. *Nature Protocols.* 9:1451–1468. https://doi.org/10.1038/nprot.2014.097.

Hestrin, S., Aschner, M., Mager, J. (1947) Synthesis of cellulose by resting cells of *Acetobacter xylinum. Nature.* 159(4028):64–65. https://doi.org/10.1038/159064a0.

Hong, F., Wei, B., Chen, L. (2015) Preliminary study on biosynthesis of bacterial nanocellulose tubes in a novel double-silicone-tube bioreactor for potential vascular prosthesis. *BioMed Research International.* 560365. https://doi.org/10.1155/2015/560365.

Huang, H.-C., Chen, L.-C., Lin, S.-B., Chen, H.-H. (2011) Nano-biomaterials application: In situ modification of bacterial cellulose structure by adding HPMC during fermentation. *Carbohydrate Polymers.* 83(2):979–987. https://doi.org/10.1016/j.carbpol.2010.09.011.

Huang, H.-C., Chen, L.-C., Lin, S.-B., Hsu, C.-P., Chen, H.-H. (2010) In situ modification of bacterial cellulose network structure by adding interfering substances during fermentation. *Bioresource Technology.* 101(15):6084–6091. https://doi.org/10.1016/j.biortech.2010.03.031.

Huang, J. W., Lv, X. G., Li, Z., Song, L. J., Feng, C., Xie, M. K., Li, C., Li, H. B., Wang, J. H., Zhu, W. D., Chen, S. Y., Wang, H. P., Xu, Y. M. (2015) Urethral reconstruction with a 3D porous bacterial cellulose scaffold seeded with lingual keratinocytes in a rabbit model. *Biomedical Materials.* 10(5):055005. https://doi.org/10.1088/1748-6041/10/5/055005.

Islam, M. U., Ullah, M. W., Khan, S., Shah, N., Park, J. K. (2017) Strategies for cost-effective and enhanced production of bacterial cellulose. *International Journal of Biological Macromolecules.* 102:1166–1173. https://doi.org/10.1016/j.ijbiomac.2017.04.110.

Jedrzejczak-Krzepkowska, M., Kubiak, K., Ludwicka, K., Bielecki, S. (2016) Bacterial nanocellulose synthesis, recent findings. In *Bacterial Nanocellulose* (pp. 19–46). Elsevier. https://doi.org/10.1016/B978-0-444-63458-0.00002-0.

Jozala, A. F., de Lencastre-Novaes, L. C., Lopes, A. M., de Carvalho Santos- Ebinuma, V., Mazzola, P.G., Pessoa-Jr, A., Grotto, D., Gerenutti, M., Chaud, M.V. (2016) Bacterial nanocellulose production and application: a 10-year overview, *Applied Microbiol and Biotechnology.* 100(5):2063–2072. https://doi.org/10.1007/s00253-015-7243-4.

Jozala, A. F., Pértile, R. A. N., dos Santos, C. A., de Carvalho Santos-Ebinuma, V., Seckler, M. M., Gama, F. M., Pessoa, A. (2015) Bacterial cellulose production by *Gluconacetobacter xylinus* by employing alternative culture media. *Applied Microbiol and Biotechnology.* 99(3):1181–1190. https://doi.org/10.1007/s00253-014-6232-3.

Keshk, S. M. (2014) Vitamin C enhances bacterial cellulose production in Gluconacetobacter xylinus. *Carbohydrate Polymers.* 99:98–100. https://doi.org/10.1016/j.carbpol.2013.08.060.

Kim, B.S., Baez, C.E., Atala, A. (2000) Biomaterials for tissue engineering. *World Journal of Urology.* 18:2–9. https://doi.org/10.1007/s003450050002.

Kim, S. P., Sarmast, Z., Daignault, S., Faerber, G. J., McGuire, E. J., Latini, J. M. (2008) Long-term durability and functional outcomes among patients with artificial urinary sphincters: a 10-year retrospective review from the University of Michigan. *The Journal of Urology.* 179(5):1912–1916. https://doi.org/10.1016/j.juro.2008.01.048.

Kiziltas, E.E., Kiziltas, A., Gardner, D.J. (2015) Synthesis of bacterial cellulose using hot water extracted wood sugars. *Carbohydrate Polymers.* 124(9):131–138. https://doi.org/10.1016/j.carbpol.2015.01.036.

Klemm, D., Heublein, B., Fink, H.P., Bohn, A. (2005) Cellulose: fascinating biopolymer and sustainable raw material. *Angewandte Chemie.* 44(22):3358–3393. http://doi.org/10.1002/anie.200460587.

Klemm, D., Cranston, E.D., Fischer, D., Gama, M., Kedzior, S.A., Kralich, D., Kramer, F., Lindström, T., Nietzsche, S., Petzold-Welcke, K., Rauchfu, F. (2018) Nanocellulose as a natural source for groundbreaking applications in materials science: Today's state. *Materials Today.* 21(7):720–748. http://doi.org/10.1016/j.mattod.2018.02.001.

Klemm, D., Schumann, D., Kramer, F., Heßler, N., Hornung, M., Schmauder, H.P., Marsh, S. (2006) Nanocelluloses as Innovative Polymers in Research and Application. *Advances in Polymer Science.* 205:49–96. http://doi.org/10.1007/12_097.

Kowalska Ludwicka, K., Cala, J., Grobelski, B., Sygut, D., Jesionek-Kupnicka, D., Kołodziejczyk, M., Bielecki, S., Pasieka, Z. (2013) Modified bacterial cellulose tubes for regeneration of damaged peripheral nerves. *Archives of Medical Science.* 9(3):527–534. http://doi.org/10.5114/aoms.2013.33433.

Krystynowicz, A., Czaja, W., Wiktorowska-Jezierska, A., Gonçalves-Miśkiewicz, M., Turkiewicz, M., Bielecki, S. (2002) Factors affecting the yield and properties of bacterial cellulose. *Journal of Industrial Microbiol and Biotechnology.* 29(4):189–195. https://doi.org/10.1038/sj.jim.7000303.

Lanza R, Langer R, Vacanti J (2013) *Principles of tissue engineering.* Academic Press, 4th ed.

Li, Z., Lv, X., Yan, Z., Yao, Y., Feng, C., Wang, H., Chen, S. (2017, July) Fabrication of Urethral Tissue Engineering Scaffolds Based on Multi-scale Structure of Bacterial Cellulose Matrix Materials: A Preliminary Study. In *Chinese Materials Conference* (pp. 437–445) Springer, Singapore. https://doi.org/10.1007/978-981-13-0110-0_49.

Li, Z., Zhong, L., Zhang, T., Qiu, F., Yue, X., Yang, D. (2019) Sustainable, flexible, and superhydrophobic functionalized cellulose aerogel for selective and versatile oil/water separation. *ACS Sustainable Chemistry & Engineering.* 7(11):9984–9994. https://doi.org/10.1021/acssuschemeng.9b01122.

Lima, S. V. C., Chagas, H. M., Monteiro, C. C. P., Ferraz-Carvalho, R. S., Albuquerque, A. V., Silva, A. A., . . ., Vilar, F. O. (2021) Injectable semi rigid penile prosthesis: study in rabbits and future perspectives. *Translational Andrology and Urology*. 10(2):841. https://doi.org/ 10.21037/tau-20-1128.

Lima, S. V. C., Machado, M. R., Pinto, F. C. M., Lira, M. M. D. M., Albuquerque, A. V. D., Lustosa, E. S., da Silva, G. M., Campos, O. (2017) A new material to prevent urethral damage after implantation of artificial devices: an experimental study. *International Brazilian Journal of Urology*. 43:335–344. https://doi.org/10.1590/S1677-5538.IBJU.2016.0271.

Lin, S. P., Huang, Y. H., Hsu, K. D., Lai, Y. J., Chen, Y. K., Cheng, K. C. (2016) Isolation and identification of cellulose-producing strain *Komagataeibacter* intermedius from fermented fruit juice. *Carbohydrate Polymers*. 151:827–833. https://doi.org/10.1016/j.carbpol.2016.06.032.

Lucena, R. G., Lima, S. V., Aguiar, J. L. D. A., Andrade, R. T., Pinto, F., Vilar, F. O. (2015) Experimental use of a cellulosic biopolymer as a new material for suburethral sling in the treatment of stress urinary incontinence. *International Brazilian Journal of Urology*. 41:1148–1153. https://doi.org/10.1590/S1677-5538.IBJU.2014.0155.

Luengo, J. M., García, B., Sandoval, A., Naharro, G., Olivera, E. R. (2003) Bioplastics from microorganisms. *Current Opinion in Microbiol*. 6(3):251–260. https://doi.org/10.1016/s1369-5274(03)00040-7.

Luo, H., Xie, J., Wang, J., Yao, F., Yang, Z., Wan, Y. (2018) Step-by-step self-assembly of 2D few-layer reduced graphene oxide into 3D architecture of bacterial cellulose for a robust, ultralight, and recyclable all-carbon absorbent. *Carbon*. 129:824–832. https://doi.org/10.1016/j.carbon.2018.07.048.

Lv, X., Feng, C., Liu, Y., Peng, X., Chen, S., Xiao, D., . . ., Lu, M. (2018) A smart bilayered scaffold supporting keratinocytes and muscle cells in micro/nano-scale for urethral reconstruction. *Theranostics*. 8(11):3153. https://doi.org/ 10.7150/thno.22080.

Lv, X., Feng, C., Xu, Y., Li, Z., Wang, H.P. (2016a) Three-dimensional microporous gelatin sponge/nanofibrous bacterial cellulose bilayer scaffold for urethral reconstruction. *Journal of Urology*. 195(4S):e347. https://doi.org/10.1016/j.juro.2016.03.108.

Lv, X., Feng, C., Yang, J., Li, Z., Chen, S., Xie, M., Xu, Y. *et al.* (2016b) Bacterial cellulose-based biomimetic nanofibrous scaffold with muscle cells for hollow organ tissue engineering. *ACS Biomaterials Science & Engineering*. 2(1):19–29. https://doi.org/10.1021/acsbiomaterials.5b00259.

Madigan MT, Aiyer J, Buckley DH, Sattley WM, Stahl DA (2021) *Brock biology of microorganisms*. eBook. Pearson Higher Ed.

Martínez, H., Brackmann, C., Enejder, A., Gatenholm, P. (2012) Mechanical stimulation of fibroblasts in micro-channeled bacterial cellulose scaffolds enhances production of oriented collagen fibers. *Journal of Biomedical Materials Research*. 100A(4):948–957. https://doi.org/10.1002/jbm.a.34035.

Martins, A. G. S., Lima, S. V. C., Araujo, L. A. P. D., Vilar, F. D. O., Cavalcante, N. T. P. (2013) A wet dressing for hypospadias surgery. *International Brazilian Journal of Urology*. 39:408–413. https://doi.org/10.1590/S1677-5538.IBJU.2013.03.15.

Mason, C., Dunnill, P. (2008) A brief definition of regenerative medicine. *Regenerative Medicine*. 3(1):1–5. https://doi.org/10.2217/17460751.3.1.1.

Messaddeq Y, Ribeiro SJL, Thomazini W (2008) Contact lens for therapy, method and apparatus for their production and use. *Brazil Patent BR*. PI0603704-6.

Millon, L.E., Guhados, G., Wan, W. (2008) Anisotropic polyvinyl alcohol – Bacterial cellulose nanocomposite for biomedical applications. *Journal of Biomedical Materials Research*. 86B(2):444–452. https://doi.org/10.1002/jbm.b.31040.

Millon, L.E., Wan, W.K. (2006) The polyvinyl alcohol – bacterial cellulose system as a new nanocomposite for biomedical applications. *Journal of Biomedical Materials Research*. 79B(2):245–253. https://doi.org/10.1002/jbm.b.30535.

Mir, J. (2004) Industrial microbiology: a new challenge. *International Microbiol.* 7(2):81–82. ISSN:1139-6709.

Mobley, D. F., Khera, M., Baum, N. (2017) Recent advances in the treatment of erectile dysfunction. *Postgraduate Medical Journal.* 93(1105):679–685. http://dx.doi.org/10.1136/postgradmedj-2016-134073.

Mohammadi, H. (2011) Nanocomposite biomaterial mimicking aortic heart valve leaflet mechanical behaviour. *Proceedings of the Institution of Mechanical Engineers.* 225(7). 718–722. https://doi.org/10.1177/0954411911399826.

Mona, S., Bajar, S., Deepak, B., Kiran, B., Kaushik, A. (2019) Microbial cellulose: production and application. In *Materials for Biomedical Engineering* (pp. 309–322) Elsevier. https://doi.org/10.1016/B978-0-12-818415-8.00011-5.

Munir, F., Saleem, Y., Munir, N., Iqbal, M. S., Sattar, S. (2019) Production and characterization of dextran from Leuconostoc mesenteroides NRRL B-512 (f) Fermentation. *Life Science Journal of Pakistan.* 1(1):19–23.

Najafpour GD (2015) Industrial microbiology. In Najafpour GD (Ed.), *Biochemical engineering and biotechnology.* Elsevier. https://doi.org/10.1016/B978-0-444-63357-6.00001-8.

Nelson, M. L., Dinardo, A., Hochberg, J., Armelagos, G. J. (2010) Brief communication: mass spectroscopic characterization of tetracycline in the skeletal remains of an ancient population from Sudanese Nubia 350–550 CE. *American Journal of Physical Anthropology.* 143(1):151–154. https://doi.org/10.1002/ajpa.21340.

Nicolas, W. J., Ghosal, D., Tocheva, E. I., Meyerowitz, E. M., Jensen, G. J. (2020) Structure of the bacterial cellulose ribbon and its assembly-guiding cytoskeleton by electron cryotomography. *BioRxiv.* 203(3):e00371. https://doi.org/10.1101/2020.04.16.045534.

Okiyama, A., Motoki, M., Yamanaka, S. (1993) Bacterial cellulose IV. Application to processed foods. *Food Hydrocolloids.* 6(6):503–511. https://doi.org/10.1016/s0268-005x(09)80074-x.

Oliveira Vilar, F., Moreno Pinto, F.C., Vasconcelos Albuquerque, A., Santos Martins, A.G., Pereira de Araújo, L.A., Lamartine de Andrade Aguiar, J., Vilar Correia Lima, S. (2016) A wet dressing for male genital surgery: A phase II clinical trial. *International Brazilian Journal of Urology.* 42:1220–122. https://doi.org/10.1590/S1677-5538.IBJU.2016.0109.

O'Neill, H., Pingali, S. V., Petridis, L., He, J., Mamontov, E., Hong, L., . . ., Davison, B. H. (2017) Dynamics of water bound to crystalline cellulose. *Scientific Reports.* 7(1):1–13. https://doi.org/10.1038/s41598-017-12035-w

Osorio, M., Cañas, A., Puerta, J., Diaz, L., Naranjo, T., Ortiz, I., Castro, C. (2019) Ex Vivo and In Vivo Biocompatibility Assessment (Blood and tissue) of three-Dimensional Bacterial Nanocellulose Biomaterials for soft tissue Implants. *Scientific Reports.* 9(1):10553. https://doi.org/10.1038/s41598-019-46918-x.

Osorio, M. A., Restrepo, D., Velásquez-Cock, J. A., Zuluaga, R. O., Montoya, U., Rojas, O., . . ., Castro, C. I. (2014) Synthesis of thermoplastic starch-bacterial cellulose nanocomposites via in situ fermentation. *Journal of the Brazilian Chemical Society.* 25:1607–1613. http://dx.doi.org/10.5935/0103-5053.20140146.

Ostrowski, K.A., Gannon, J.R., Walsh, T.J. (2016) A review of the epidemiology and treatment of Peyronie's disease. *Research and Reports in Urology.* 8:61–70. https://doi.org/10.2147/RRU.S65620.

Pariente, J.L., Kim, B.S., Atala, A. (2002) In vitro biocompatibility evaluation of naturally derived and synthetic biomaterials using normal human bladder smooth muscle cells. *The Journal of Urology.* 167:1867–1871. https://doi.org/10.1016/S0022-5347(05)65251-2.

Park J, Lakes RS (2012) *Biomaterials: an introduction.* Springer.

Patel, N., Rai, D., Shahane, S., Mishra, U. (2019) Lipases: sources, production, purification, and applications. *Recent Patents on Biotechnology.* 13(1):45–56. http://doi.org/10.2174/1872208312666181029093333.

Pina, S., Ribeiro, V. P., Marques, C. F., Maia, F. R., Silva, T. H., Reis, R. L., Oliveira, J. M. (2019) Scaffolding strategies for tissue engineering and regenerative medicine applications. *Materials.* 12(11):1824. http://doi.org/10.3390/ma12111824.

Picheth, G. F., Pirich, C. L., Sierakowski, M. R., Woehl, M. A., Novak Sakakibara, C., Fernandes de Souza, C., Amado Martin, A., da Silva, R., Alves de Freitas, R. (2017) Bacterial cellulose in biomedical applications: a review. *International Journal of Biological Macromolecules.* 104:97–106. https://doi.org/10.1016/j.ijbiomac.2017.05.171.

Portal, O., Clark, W.A., Levinson, D.J. (2009) Microbial cellulose wound dressing in the treatment of nonhealing lower extremity ulcers. *Wounds.* 21(1):1–3.

Portela, R., Leal, C. R., Almeida, P. L., Sobral, R. G. (2019) Bacterial cellulose: A versatile biopolymer for wound dressing applications. *Microbial Biotechnology.* 12(4):586–610. https://doi.org/10.1111/1751-7915.13392.

Qiu, Z., Wang, M., Zhang, T., Yang, D., Qiu, F. (2020) In-situ fabrication of dynamic and recyclable TiO2 coated bacterial cellulose membranes as an efficient hybrid absorbent for tellurium extraction. *Cellulose.* 27:4591–4608. https://doi.org/10.1007/s10570-020-03096-8.

Raghavendran, V., Asare, E., Roy, I. (2020) Bacterial cellulose: Biosynthesis, production, and applications. In R. K. Poole (Ed.) *Advances in Microbial Physiology,* (pp. 89–138) Academic Press. https://doi.org/10.1016/bs.ampbs.2020.07.002.

Revin, V., Liyaskina, E., Nazarkina, M., Bogatyreva, A., Shchankin, M. (2018) Cost-effective production of bacterial cellulose using acidic food industry by-products. *Brazilian Journal of Microbiol.* 49:151–159. https://doi.org/10.1016/j.bjm.2017.12.012.

Ruka, D. R., Simon, G. P., Dean, K. M. (2014 a) Bacterial cellulose and its use in renewable composites. In Thakur VK (Ed.), *Nanocellulose polymer nanocomposites: fundamentals and applications,* 89–130. https://doi.org/10.1002/9781118872246.ch4.

Ruka, D. R., Simon, G. P., Dean, K. (2014 b) Harvesting fibrils from bacterial cellulose pellicles and subsequent formation of biodegradable poly-3-hydroxybutyrate nanocomposites. *Cellulose.* 21(6):4299–4308. https://doi.org/10.1007/s10570-014-0415-z.

Salvatierra, N.A., Oldani, C.R., Reyna, L., Taborda, R.A.M. (2009) ¿Qué es la biocompatibilidad?. *Revista Argentina de Bioingeniería.* 15:28–32.

Sanchez, S., Demain, A. L. (2015) *Antibiotics: current innovations and future trends.* Caister Academic Press. https://doi.org/10.21775/9781908230546.

Schumann, D.A., Wippermann, J., Klemm, D.O., Kramer, F., Koth, D., Kosmehl, H., Wahlers, T., Salehi-Gelani, S. (2009) Artificial vascular implants from bacterial cellulose: preliminary results of small arterial substitutes. *Cellulose.* 16:877–885. https://doi.org/10.1007/s10570-008-9264-y.

Scriven, S.D., Trejdosiewicz, L.K., Thomas, D.F.M., Southgate, J. (2001) Urothelial cell transplantation using biodegradable synthetic scaffolds. *Journal of Materials Science: Materials in Medicine.* 12:991–996. https://doi.org/10.1023/a:1012869318205.

Sharma, C., Bhardwaj, N. K. (2019) Bacterial nanocellulose: Present status, biomedical applications and future perspectives. *Materials Science and Engineering: C.* 104:109963. https://doi.org/10.1016/j.msec.2019.109963.

Shi, S., Chen, S., Zhang, X., Shen, W., Li, X., Hu, W., Wang, H. (2009) Biomimetic mineralization synthesis of calcium-deficient carbonate-containing hydroxyapatite in a three-dimensional network of bacterial cellulose. *Journal of Chemical Technology and Biotechnology.* 84(2):285–290. https://doi.org/10.1002/jctb.2037.

Sivakumar PK, Joe MM, Sukesh K (2010) *An introduction to industrial microbiology.* S. Chand and Company Ltd., 1st ed. ISBN:8121935199.

Son, H. J., Kim, H. G., Kim, K. K., Kim, H. S., Kim, Y. G., Lee, S. J. (2003) Increased production of bacterial cellulose by *Acetobacter* sp. V6 in synthetic media under shaking culture conditions. *Bioresource Technology.* 86(3):215–221. https://doi.org/10.1016/s0960-8524(02)00176-1.

Sordelli A (2021) *Replacement of tunica albuginea from rat corpora cavernosa using bacterial nanocellulose scaffolds seeded with adipose tissue-derived stem cells and platelet-rich plasma* (Doctoral Thesis). Universidad de Buenos Aires.

Sordelli A, Cerrutti P, Foresti ML, Gueglio G (2021, March 25–27) *Reemplazo de túnica albugínea de cuerpos cavernosos de ratas utilizando andamios de nanocelulosa bacteriana.* IV Jornadas de Investigadores en Formación CyT – UNQ, Quilmes, Argentina. https://www.conicet.gov.ar/new_scp/detalle.php?keywords=&id=24232&congresos=ye s&detalles=yes&congr_id=9781964

Svensson, A., Nicklasson, E., Harrah, T., Panilaitis, B., Kaplan, D.L., Brittberg, M., Gatenholm, P. (2005) Bacterial cellulose as a potential scaffold for tissue engineering of cartilage. *Biomaterials.* 26(4):419–431. https://doi.org/10.1016/j.biomaterials.2004.02.049.

Sworn G (2021) Xanthan gum. In Phillips GO, Williams PA (Eds.), *Handbook of hydrocolloids.* Woodhead Publishing, 833–853. https://doi.org/10.1016/B978-0-12-820104-6.00004-8.

Tang, J., Li, X., Bao, L., Chen, L., Hong, F. F. (2017) Comparison of two types of bioreactors for synthesis of bacterial nanocellulose tubes as potential medical prostheses including artificial blood vessels. *Journal of Chemical Technology & Biotechnology.* 92(6):1218–1228. https://doi.org/10.1002/jctb.5111.

Tazi, N., Zhang, Z., Messaddeq, Y., Almeida-Lopes, L., Zanardi, L.M., Levinson, D., Rouabhia, M. (2012) Hydroxyapatite bioactivated bacterial cellulose promotes osteoblast growth and the formation of bone nodules. *AMB Express.* 2(1):1–10. https://doi.org/10.1186/2191-0855-2-61.

Torres, F.G., Commeaux, S., Troncoso, O.P. (2012) Biocompatibility of Bacterial Cellulose Based Biomaterials. *Journal of Functional Biomaterials.* 3(4):864–878. https://doi.org/864-878, doi:10.3390/jfb3040864.

Ullah MW, Manan S, Kiprono SJ, Ul-Islam M, Yang G (2019) Synthesis, structure, and properties of bacterial cellulose. In Huang J, Dufresne A, Lin N (Eds.), *Nanocellulose: from fundamentals to advanced materials.* John Wiley and Sons, 81–113.

Ullah, M. W., Ul-Islam, M., Khan, T., Park, J. K. (2021) Recent developments in the synthesis, properties, and applications of various microbial polysaccharides. In G. O. Phillips, P. A. Williams (Eds.), *Handbook of Hydrocolloids* (pp. 975–1015) Woodhead Publishing. https://doi.org/10.1016/B978-0-12-820104-6.00032-2.

Vandamme, E.-J., Baets, S.-D., Vanbaelen, A., Joris, K., Wulf, P.-D. (1998) Improved production of bacterial cellulose and its application potential. *Polymer Degradation and Stability.* 59:93–99. https://doi.org/10.1016/S0141-3910(97)00185-7.

Wan, Y.Z., Luo, H., He, F., Liang, H., Huang, Y., Li, X.L. (2009) Mechanical, moisture absorption, and biodegradation behaviours of bacterial cellulose fibre-reinforced starch biocomposites. *Composites Science and Technology.* 69(7–8):1212–1217. https://doi.org/10.1016/j.compscitech.2009.02.024.

Wang, B., Lv, X., Li, Z., Zhang, M., Yao, J., Sheng, N., . . ., Chen, S. (2020) Urethra-inspired biomimetic scaffold: a therapeutic strategy to promote angiogenesis for urethral regeneration in a rabbit model. *Acta Biomaterialia.* 102:247–258. https://doi.org/10.1016/j.actbio.2019.11.026.

Wang, J., Tavakoli, J., Tang, Y. (2019) Bacterial cellulose production, properties and applications with different culture methods – A review. *Carbohydrate Polymers.* 219:63–76. https://doi.org/10.1016/j.carbpol.2019.05.008.

Williams, S. F., Martin, D. P. (2002) Applications of PHAs in medicine and pharmacy. *Biopolymers.* 4:91–127. https://doi.org/10.1002/3527600035.bpol4004

Winnacker, M. (2019) Polyhydroxyalkanoates: recent advances in their synthesis and applications. *European Journal of Lipid Science and Technology.* 121(11):1900101. https://doi.org/10.1002/ejlt.201900101.

Wippermann, J., Schumann, D., Klemm, D., Kosmehl, H., Salehi-Gelani, S., Wahlers, T. (2009) Preliminary results of small arterial substitute performed with a new cylindrical biomaterial composed of bacterial cellulose. *European Journal of Vascular and Endovascular Surgery.* 37(5):592–596. https://doi.org/10.1016/j.ejvs.2009.01.007.

Yang, X. Y., Huang, C., Guo, H. J., Xiong, L., Li, Y. Y., Zhang, H. R., Chen, X. D. (2013) Bioconversion of elephant grass (Pennisetum purpureum) acid hydrolysate to bacterial cellulose by *Gluconacetobacter xylinus. Journal of Applied Microbiol.* 115:995–1002. https://doi.org/10.1111/jam.12255.

Yang, X. Y., Huang, C., Guo, H. J., Xiong, L., Luo, J., Wang, B., Chen, X. F., Lin, X. Q., Chen, X.D. (2014) Beneficial effect of acetic acid on the xylose utilization and bacterial cellulose production by *Gluconacetobacter xylinus. Indian Journal of Microbiol.* 54:268–273. https://doi.org/10.1007/s12088-014-0450-3.

Yim, S.M., Song, J.E., Kim, H.R. (2017) Production and characterization of bacterial cellulose fabrics by nitrogen sources of tea and carbon sources of sugar. *Process Biochemistry* 59(8): 26–36. https://doi.org/10.1016/j.procbio.2016.07.001.

Yin, X., Yu, C., Zhang, X., Yang, J., Lin, Q., Wang, J., Zhu, Q. (2011) Comparison of succinylation methods for bacterial cellulose and adsorption capacities of bacterial cellulose derivatives for Cu^{2+} ion. *Polymer Bulletin.* 67:401–412. https://doi.org/10.1007/s00289-010-0388-5.

Zang, S., Zhang, R., Chen, H., Lu, Y., Zhou, J., Chang, X., . . ., Yang, G. (2015) Investigation on artificial blood vessels prepared from bacterial cellulose. *Materials Science and Engineering: C.* 46:111–117. https://doi.org/10.1016/j.msec.2014.10.023.

Zeng, X., Small, D. P., Wan, W. (2011) Statistical optimization of culture conditions for bacterial cellulose production by *Acetobacter xylinum* BPR 2001 from maple syrup. *Carbohydrate Polymers.* 85(3):506–513. https://doi.org/10.1016/j.carbpol.2011.02.034.

Zhang, X., Wenlong, L. U., Feng, C., Xiangguo, L. V., Zhu, W. (2016) Constructing tissue-engineered bladder by vascular endothelial growth factor nanoparticle-bacterial cellulose composite scaffold with various kinds of cells. *Chinese Journal of Tissue Engineering Research.* 20(21):3088–3096. https://doi.org/10.3969/j.issn.2095-4344.2016.21.007.

Zhang, L., Xiao, Y. (2016) Biomatrices in Urethral Reconstruction. *Journal of Biotechnology & Biomaterials.* 6(231):2. https://doi.org/ 10.4172/2155–952X.1000231.

Zhao, H., Turng, L. S. (2015) Mechanical performance of microcellular injection molded biocomposites from green plastics: PLA and PHBV. In M. Misra, J. K. Pandey, A. K. Mohanty (Eds.), *Biocomposites: Design and Mechanical Performance* (pp. 141–160) Woodhead Publishing Series in Composites Science and Engineering. https://doi.org/10.1016/B978-1-78242-373-7.00015-9.

Zhou, L. L., Sun, D. P., Hu, L. Y., Li, Y. W., Yang, J. Z. (2007) Effect of addition of sodium alginate on bacterial cellulose production by *Acetobacter xylinum. Journal of industrial Microbiol and Biotechnology.* 34(7):483. https://doi.org/10.1007/s10295-007-0218-4.

Zmejkoski, D. Z., Marković, Z. M., Zdravković, N. M., Trišić, D. D., Budimir, M. D., Kuzman, S. B., . . ., Marković, B. M. T. (2021) Bactericidal and antioxidant bacterial cellulose hydrogels doped with chitosan as potential urinary tract infection biomedical agent. *RSC Advances.* 11(15):8559–8568. https://doi.org/10.1039/D0RA10782D.

Żywicka, A., Junka, A. F., Szymczyk, P., Chodaczek, G., Grzesiak, J., Sedghizadeh, P. P., Fijałkowski, K. (2018) Bacterial cellulose yield increased over 500% by supplementation of medium with vegetable oil. *Carbohydrate Polymers.* 199:294–303. https://doi.org/10.1016/j.carbpol.2018.06.126.

Żywicka, A., Peitler, D., Rakoczy, R., Konopacki, M., Kordas, M., Fijalkowski, K. (2015) The effect of different agitation modes on bacterial cellulose synthesis by *Gluconacetobacter xylinus* strains. *Acta Scientiarum Polonorum Zootechnica.* 14(1):137–150.

9 Algal Biofilms and Their Role in Bioremediation

Rahul R. Patil and Umar Faruk J. Meeranayak

CONTENTS

9.1   INTRODUCTION

Bioremediation is the ecological process of detoxifying detrimental soil, water, and air contaminants with the aid of microorganisms. The principal mechanism involved in bioremediation is translating the harmful and highly toxic contaminants into innocuous, nontoxic components. The major advantages of employing bioremediation in environment cleaning process are its smooth operation and avoidance of other concomitant effects (Vidali, 2001). Bioremediation is a relatively cost-effective method in comparison to traditional methods. Microbes with the ability to grow in nearly all habitats can sustain wide range of artificial conditions provided under experimental laboratory setups (Prince, 2000; Das and Dash, 2014). This helps in the ex situ studies on bioremediation of environmental samples.

In different environmental settings, the micro biome exists in the form of colonies of diverse microbial association. Consortia of microbial populations lead to the formation of mats. These biological mats formed by the association of microbial colonies adhering to the biotic or abiotic material are called biofilm (O'Toole *et al.*, 2000). The mechanism of biofilm formation is also a naturally designed technique of the microbial world which can be employed for the degradation of environmental pollutants effectively. The products of metabolic activities of these microorganisms play a crucial role in biofilm formation, as well as bioremediation (Gieg *et al.*, 2014).

Biofilms are a more advantageous and beneficial system in the bioremediation process. This is because the counterpart of the biofilm, such as free-floating phytoplankton, are more susceptible to changing environmental patterns (Davies *et al.*, 1998). Slight variations in the ecosystem manipulates and destroys the survivability of the freely existing organisms. In comparison to this, the biofilm consists of a

DOI: 10.1201/9781003184942-11

177

mixture of microbial communities which have gained the ability to withstand slight to modest changes in the environment (Edwards *et al.*, 2013). This adeptness is due to the coexistence and tolerance of the deprived nutrition and anxious conditions, along with the counter-protection of species taking part in biofilm (Von Canstein *et al.*, 2002). The added advantage of the utilization of biofilm rather than the free plankton in the bioremediation process is clear separation of treated water and the biofilm. However, in the case of the free planktonic cells employed for this purpose, they cannot end up with clear separation of treated water; hence, an additional harvesting and filtration of cells is required in conventional methods of water treatment (Roeselers *et al.*, 2008).

Extensive research in the field of bacteria and fungi sheds light on the applications of these microbes in ecological studies. However, in addition, algae can also be utilized for effective bioremediation.

9.2 BIOLOGICAL GLUE (EPS) EXOPOLYSACCHARIDES

Exopolysaccharides or extracellular polymeric substances (EPS) are the matrix of secondary metabolites produced by the microbial cells in a biofilm complex which acts as a cementing material. Primarily, the EPS is composed of polysaccharides, along with lipids, proteins and nucleic acids (Flemming and Wingender, 2001). EPS helps in protecting the biofilm from desiccation, maintains integrity, helps in motility and supports the microalgae to form consortia with bacteria providing an added advantage in bioremediating natural waters (Bruckner *et al.*, 2011; Lind *et al.*, 1997; Stal, 2003). Initial adhesion of cells is carried out by EPS and is influenced by multiple factors like the age of a biofilm, availability of nutrients, composition of species, stress response, light and temperature thorough photosynthesis carried out by algae (Berner *et al.*, 2015; Kesaano and Sims, 2014). Exopolysaccharide has a sorption property by which it can remove the toxic components from environment; the presence of exopolysaccharide makes the biofilm resistant to biocides and antibiotics (Morton *et al.*, 1998). Deposition of toxic metals like copper, cadmium, uranium, zinc and lead on exopolysaccharides was noticed by Ford and Ryan (1995). In industrial scale-up processes, the EPS became advantageous in providing viscosity to the materials also acts as biological glue in pharmacological applications as the glue-like property of polysaccharides helps in attachment of various particles. The exclusive hydrophobicity of the polymeric substances extensively supports the substratum attachment in water treatment applications. At organism level, these cascades of polymers protect the cells and organisms from freezing during extremely low temperatures (Kim *et al.*, 2016) and resist dehydrating environmental conditions (Hill *et al.*, 1997). The exopolysaccharides provide sustenance to the organisms at variable degrees of salinity in marine environments (Krembs *et al.*, 2011). Along with these properties, the exopolysaccharides provide protection against the shifts in the ionic nature, osmotic condition and change in the pH level of the surroundings (Deming and Young, 2017).

9.3 BIOREMEDIATION BY ALGAL BIOFILM

With increased industrial activities, the addition of pollutants into the water resources increased the threat of reduced water quality and loss of aquatic life forms, along with degradation of natural habitats. Recently, increased interest in bioremediation has led to the use of microalgal cells, which has shown considerable interest.

Algal biofilms offer a challenging technology for bioremediation of wastewater, as they do not require artificial encapsulation (Figure 9.1). Microalgae form important constituents of wastewater, as their presence is noticed in effluents of industrial origin, as well as in domestic wastewater (Renuka *et al.*, 2015). Hence, these bioforms

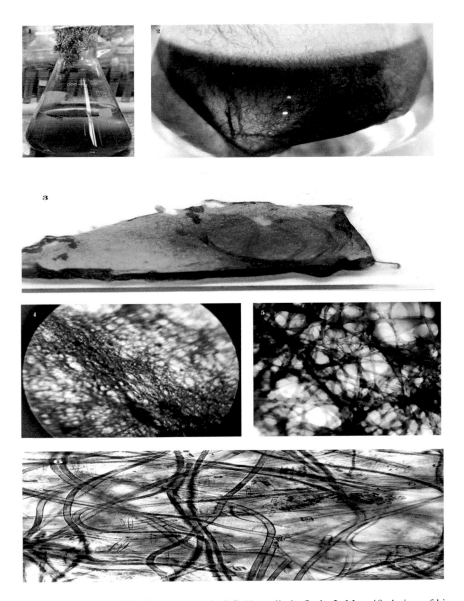

FIGURE 9.1   1. Algal Biofilm growing in BG-11 media in flask; 2. Magnified view of biofilm; 3. Texture of biofilm on glass slide; 4–5. Biofilm matrix under microscopic view; 6. Blue-green algae in biofilm networking.

can be creatively utilized as scavengers to remove metal pollutants (De-Bashan and Bashan, 2010; Marchello *et al.*, 2015; Wollmann *et al.*, 2019), and also used as bioindicators of water quality (Patil *et al.*, 2021; Meeranayak *et al.*, 2020). In general, the practice of employing algae in wastewater treatment is termed as phycoremediation (Emparan *et al.*, 2019).

The affinity of the microalgae toward the inorganic nitrogen and phosphorus provides a greater advantage for water treatment plants. De Godos *et al.* (2009) reported the reduction of 90% of the ammonium in presence of the microalgal biofilm. Similarly, Shi *et al.* (2007) noticed 80% phosphate removal efficiency from biofilm. The microalgae *Chlorella* and *Nitzschia* were successfully employed for the treatment of the settled domestic sewage by McGriff and McKinney (1972), and in diluted pig slurry, 54–98% reduction in nitrogen and 42–89% reduction of phosphorus was observed by Fallowfield and Garrett (1985) using *Chlorella vulgaris* in a raceway reactor. Pacheco *et al.* (2020) used *Chlamydomonas oblonga* to degrade a pharmacological drug carbamazepine up to 35%. A microalgal consortium was used to achieve 98% and 99% of nitrate and phosphate removal, respectively, in raw dairy industry wastewater (Mulbry and Wilkie, 2001). Table 9.1 lists several algae types reported so far with proven potential to degrade various organic compounds.

9.3.1 Bioremediation of Phosphate and Nitrate

Increased concentration of phosphates in the aquatic environments threatens the ecosystem by resulting in eutrophication of water bodies. Removal of phosphorus from wastewater, use of microalgal biofilm at tertiary treatment stage was suggested by Sukacova *et al.* (2020) as it displayed a high phosphorus recovery potential, amounting to about 1132 kg of phosphorus per year. Hence, this technology can be used to deal with the problems of eutrophication in aquatic environments, and the biomass can be utilized as a source of phosphorus to overcome the shortage of it in agriculture. Biofilms of *Scenedesmus vacuolatus* and *Chlorella vulgaris* are capable of removing 90% of nutrients in semi batch processes in photobioreactors (PBR) when they are attached rather than suspended freely, whereas in natural conditions, the major limiting factor during the remediation was found to be depletion of one of the nutrients (Moreno Osorio *et al.*, 2019).

9.3.2 Removal of Total Dissolved Solids

Anthropogenic activities – including modern agricultural practices with usage of large quantities of fertilizers which leach out and reach the natural waters, increased urbanization, mining activities, industrial development etc. – raise the levels of total dissolved solids (TDS) in natural waters (Canedo-Arguelles *et al.*, 2013; Steele and Aitkenhead-Peterson, 2011). The chief components added to the water are calcium, magnesium, sodium, potassium, chlorides and sulfates. The surplus TDS can be noxious to diverse organisms living in aquatic habitats. Revolving algal biofilm (RAB) reactors provide an eco-friendly solution for minimizing the

TABLE 9.1
List of Microalgae Involved in Removal of Nutrients

Sl. No.	Algae	Degrading Compound	Reference
1	*Scenedesmus obliquus*	P, N	Martınez *et al.*, 2000
2	*Tetradesmus obliquus*	Phenol	Ellis, 1977
3	*Spirulina maxima*	Phenol	Scragg, 2006
4	*Auxenochlorella pyrenoidosa*	Fe, Cu	Mahan *et al.*, 1989
5	*Chlamydomonas reinhardtii*	Fe, Cu	Mahan *et al.*, 1989
6	*Stichococcus bacillaris*	Fe, Cu	Mahan *et al.*, 1989
7	*Anabaena doliolum*	Fe, Cu	Rai and Mallick, 1992
	Microcysis aeruginosa, Aphanothece halophytica	Cu, Pb, Zn	Parker *et al.*, 2000
8	*Tetraselmis marina*	Metal pollutants	De Schryver *et al.*, 2008
9	*Stigeoclonium tenue*	Metal pollutants	De Schryver *et al.*, 2008
10	*Tetradesmus obliquus*	Salicylic acid	Shi *et al.*, 2007
11	*Scenedesmus* sp.	Cr, nitrates, phosphate	Ballen-Segura *et al.*, 2016
12	*Synechococcus elongatus, Phormidium tenue, Nostoc linckia, Nostoc muscorum*	Organophosphorus, organochlorine insecticides	Megharaj *et al.*, 1987; Ibrahim *et al.*, 2014

levels of TDS in municipal wastewater treatment facilities and industrial effluents. A 27% decrease in TDS was noticed in industrial effluents by the use of RAB reactors as compared to only 3% removal achieved by suspended algal culture system (Peng *et al.*, 2020). For removal of TDS from industrial effluents and municipal sewage, RAB reactors provide sustainable and eco-friendly technology. Cellular absorption by algae, precipitation and absorption by exopolysaccharides present on biofilm contributed to minimize the TDS effectively in RAB reactor, whereas adsorption and absorption are the only ways of decreasing TDS in suspended culture systems.

9.4   CONCLUSION

Algal biofilms are an eco-friendly way of remediating wastewater from varied sources; the biomass developed can also be used for production of biodiesel and as manure in agricultural field. The dynamic approach of using algal biofilm is effective over the suspended culture techniques used for bioremediation, and as it greatly reduces the cost of harvesting algal biomass, this technique is extensively used at the commercial level. Further, work is needed to optimize the conditions at which maximum bioremediation of nutrients can be achieved using algal biofilms in natural conditions.

REFERENCES

Ballen-Segura M, Hernandez Rodriguez L, Parra Ospina D, Vega Bolanos A, Perez K (2016) Using *Scenedesmus* sp. for the phycoremediation of tannery wastewater. Tecciencia 11(21):69–75. http://dx.doi.org/10.18180/tecciencia.2016.21.11.

Berner F, Heimann K, Sheehan M (2015) Microalgal biofilms for biomass production. Journal of Applied Phycology 27:1793–1804. https://doi.org/10.1007/s10811-014-0489-x

Bruckner CG, Rehm C, Grossart HP, Kroth PG (2011) Growth and release of extracellular organic compounds by benthic diatoms depend on interactions with bacteria. Environmental Microbiology 13(4):1052–1063. https://doi.org/10.1111/j.1462-2920.2010.02411.x

Canedo-Arguelles M, Kefford BJ, Piscart C, Prat N, Schafer RB, Schulz C (2013) Salinization of rivers: An urgent ecological issue. Environmental Pollution 173:157–167. https://doi.org/10.1016/j.envpol.2012.10.011

Das S, Dash HR (2014) Microbial Bioremediation: A Potential Tool for Restoration of Contaminated Areas. In Microbial Biodegradation and Bioremediation. Oxford: Elsevier pp. 1–21. https://doi.org/10.1016/B978-0-12-800021-2.00001-7

Davies DG, Parsek MR, Pearson JP, Iglewski BH, Costerton JW, Greenberg EP (1998) The involvement of cell-to-cell signals in the development of a bacterial biofilm. Science 280(5361):295–298 https://doi.org/10.1126/science.280.5361.295

De-Bashan LE, Bashan Y (2010) Immobilized microalgae for removing pollutants: review of practical aspects. Bioresource Technology 101(6):1611–1627. https://doi.org/10.1016/j.biortech.2009.09.043

De Godos I, Gonzalez C, Becares E, Garcia-Encina PA, Munoz R (2009) Simultaneous nutrients and carbon removal during pretreated swine slurry degradation in a tubular biofilm photo bioreactor. Applied Microbiology and Biotechnology 82(1):187–194. https://doi.org/10.1007/s00253-008-1825-3

Deming JW, Young, JN (2017) The role of exopolysaccharides in microbial adaptation to cold habitats. In Psychrophiles: from biodiversity to biotechnology pp. 259–284. Springer, Cham. https://doi.org/10.1007/978-3-319-57057-0_12

De Schryver P, Crab R, Defoirdt T, Boon N, Verstraete W (2008) The basics of bio-flocs technology: the added value for aquaculture. Aquaculture 277(3–4):125–137. https://doi.org/10.1016/j.aquaculture.2008.02.019

Edwards SJ, Kjellerup BV (2013) Applications of biofilms in bioremediation and biotransformation of persistent organic pollutants, pharmaceuticals/personal care products, and heavy metals. Applied microbiology and biotechnology 97(23):9909–9921. https://doi.org/10.1007/s00253-013-5216-z

Ellis BE (1977) Degradation of phenolic compounds by fresh-water algae. Plant Science Letters 8(3):213–216. https://doi.org/10.1016/0304-4211(77)90183-3

Emparan Q, Harun R, Danquah MK (2019) Role of phycoremediation for nutrient removal from wastewaters: A review. Applied Ecology and Environmental Research 17(1):889–915. http://dx.doi.org/10.15666/aeer/1701_889915

Fallowfield HJ, Garrett MK (1985) The photosynthetic treatment of pig slurry in temperate climatic conditions: a pilot-plant study. Agricultural wastes 12(2):111–136. https://doi.org/10.1016/0141-4607(85)90003-4

Flemming HC, Wingender J (2001) Relevance of microbial extracellular polymeric substances (EPSs)-Part II: Technical aspects. Water Science and Technology 43(6):9–16. https://doi.org/10.2166/wst.2001.0328

Ford T, Ryan D (1995) Toxic metals in aquatic ecosystems: a microbiological perspective. Environmental health perspectives 103(suppl1):25–28. https://doi.org/10.1289/ehp.95103s125

Gieg LM, Fowler SJ, Berdugo-Clavijo C (2014) Syntrophic biodegradation of hydrocarbon contaminants. Current opinion in biotechnology 27:21–29. https://doi.org/10.1016/j.copbio.2013.09.002

Hill DR, Keenan TW, Helm RF, Potts M, Crowe LM, Crowe JH (1997) Extracellular polysaccharide of Nostoc commune (Cyanobacteria) inhibits fusion of membrane vesicles during desiccation. Journal of Applied Phycology 9(3):237–248. https://doi.org/10.1023/A:1007965229567

Ibrahim WM, Karam MA, El-Shahat RM, Adway AA (2014) Biodegradation and utilization of organophosphorus pesticide malathion by cyanobacteria. BioMed Research International http://dx.doi.org/10.1155/2014/392682

Kesaano M, Sims RC (2014) Algal biofilm based technology for wastewater treatment. Algal Res 5:231–240. https://doi.org/10.1016/j.algal.2014.02.003

Kim SJ, Kim BG, Park HJ, Yim JH (2016) Cryoprotective properties and preliminary characterization of exopolysaccharide (P-Arcpo 15) produced by the Arctic bacterium *Pseudoalteromonas elyakovii* Arcpo 15. Preparative Biochemistry and Biotechnology 46(3):261–266. https://doi.org/10.1080/10826068.2015.1015568

Krembs C, Eicken H, Deming JW (2011) Exopolymer alteration of physical properties of sea ice and implications for ice habitability and biogeochemistry in a warmer Arctic. Proceedings of the National Academy of Sciences 108(9):3653–3658. https://doi.org/10.1073/pnas.1100701108

Lind JL, Heimann K, Miller EA, Van Vliet C, Hoogenraad NJ, Wetherbee R (1997) Substratum adhesion and gliding in a diatom are mediated by extracellular proteoglycans. Planta 203(2):213–221. https://doi.org/10.1007/s004250050184

Mahan CA, Majidi V, Holcombe JA (1989) Evaluation of the metal uptake of several algae strains in a multicomponent matrix utilizing inductively coupled plasma emission spectrometry. Analytical chemistry 61(6):624–627. https://doi.org/10.1021/ac00181a026

Marchello AE, Lombardi AT, Dellamano-Oliveira MJ, de Souza CW (2015) Microalgae population dynamics in photobioreactors with secondary sewage effluent as culture medium. Brazilian Journal of Microbiology 46(1):75–84. https://doi.org/10.1590/S1517-838246120131225

Martınez ME, Sanchez S, Jimenez JM, El Yousfi F, Munoz L (2000) Nitrogen and phosphorus removal from urban wastewater by the microalga *Scenedesmus obliquus*. Bioresource technology 73(3):263–272. https://doi.org/10.1016/S0960-8524(99)00121-2

McGriff Jr EC, McKinney RE (1972) The removal of nutrients and organics by activated algae. Water Research 6(10):1155–1164. https://doi.org/10.1016/0043-1354(72)90015-2

Meeranayak UFJ, Nadaf RD, Toragall MM, Nadaf U, Shivasharana CT (2020) The Role of Algae in Sustainable Environment: A Review. Journal of Algal Biomass Utilization 11(2):28–34.

Megharaj M, Venkateswarlu K, Rao AS (1987) Metabolism of monocrotophos and quinalphos by algae isolated from soil. Bulletin of Environmental Contamination and Toxicology 39:251–256. https://doi.org/10.1007/BF01689414

Moreno Osorio JH, Gabriele P, Antoninio P, Luigi F, Piet Nicolaas LL, Giovanni E (2019) Start-up of a nutrient removal system using *Scenedesmus vacuolatus* and *Chlorella vulgaris* biofilms. Bioresour Bioprocess. 6:2. https://doi.org/10.1186/s40643-019-0259-3

Morton, L. H. G., Greenway, D. L. A, Gaylarde, C. C, Surman, S. B. (1998) Consideration of some implications of the resistance of biofilms to biocides. *International Biodeterioration, Biodegradation.* 41(3–4):247–259. https://doi.org/10.1016/S0964-8305(98)00026-2

Mulbry WW, Wilkie AC (2001) Growth of benthic freshwater algae on dairy manures. Journal of Applied Phycology 13(4):301–306. https://doi.org/10.1023/A:1017545116317

O'Toole G, Kaplan HB, Kolter R (2000) Biofilm formation as microbial development. Annual Reviews in Microbiology 54(1):49–79. https://doi.org/10.1146/annurev.micro.54.1.49

Pacheco D, Rocha AC, Pereira L, Verdelhos T (2020) Microalgae Water Bioremediation: Trends and Hot Topics. Applied Sciences 10(5):1886. https://doi.org/10.3390/app1005 1886

Parker DL, Mihalick JE, Plude JL, Plude MJ, Clark TP, Egan L, Flom JJ, Rai LC, Kumar HD (2000) Sorption of metals by extracellular polymers from the cyanobacterium *Microcystis aeruginosa* f. flos-aquae strain C3–40. Journal of Applied Phycology 12:219–224. https://doi.org/10.1023/A:1008195312218

Patil RR, Kambhar SV, Giriyappanavar BS, Chakraborty S (2021) Algae as Environmental Biotechnological Tool for Monitoring Health of Aquatic Ecosystem. In: Maddela NR, Garcia Cruzatty LC, Chakraborty S (eds) Advances in the Domain of Environmental Biotechnology. Environmental and Microbial Biotechnology. Springer, Singapore. https://doi.org/10.1007/978-981-15-8999-7_20

Peng J, Kuldip K, Martin G, Thomas K, Zhiyou W (2020) Removal of total dissolved solids from wastewater using a revolving algal biofilm reactor. Water Environment Research 92: 766–778. https://doi.org/10.1002/wer.1273

Prince RC (2000) Bioremediation. Kirk-Othmer Encyclopedia of Chemical Technology: John Wiley Sons, Inc https://doi.org/10.1002/0471238961.0209151816180914.a01

Rai LC, Mallick N (1992) Removal and assessment of toxicity of Cu and Fe to *Anabaena doliolum* and *Chlorella vulgaris* using free and immobilized cells. World J. Microbiol. Biotechnol. 8:110–14. https://doi.org/10.1007/BF01195827

Renuka N, Sood A, Prasanna R, Ahluwalia AS (2015) Phycoremediation of wastewaters: a synergistic approach using microalgae for bioremediation and biomass generation. International Journal of Environmental Science and Technology 12(4): 1443–1460. https://doi.org/10.1007/s13762-014-0700-2

Roeselers G, Van Loosdrecht MC, Muyzer G. (2008) Phototrophic biofilms and their potential applications. Journal of applied phycology 20(3):227–235. https://doi.org/10.1007/s10811-007-9223-2

Scragg AH (2006) The effect of phenol on the growth of *Chlorella vulgaris* and *Chlorella* VT-1. Enzyme and microbial technology 39(4):796–799. https://doi.org/10.1016/j.enzmictec.2005.12.018

Shi J, Podola B, Melkonian M (2007) Removal of nitrogen and phosphorus from wastewater using microalgae immobilized on twin layers: an experimental study. Journal of Applied Phycology 19(5):417–423. https://doi.org/10.1007/s10811-006-9148-1

Stal LJ (2003) Microphytobenthos, their extracellular polymeric substances, and the morphogenesis of intertidal sediments. Geomicrobiology Journal 20(5):463–478. https://doi.org/10.1080/713851126

Steele MK, Aitkenhead-Peterson JA (2011) Long-term sodium and chloride surface water exports from the Dallas/Fort Worth region. Science of the Total Environment 409:3021–3032. https://doi.org/10.1016/j.scitotenv.2011.04.015

Sukacova K, Daniel V, Jiri D (2020) Perspectives on Microalgal Biofilm Systems with Respect to Integration into Wastewater Treatment Technologies and Phosphorus Scarcity. Water 12:2245. https://doi:10.3390/w12082245

Vidali M (2001) Bioremediation. An overview. Pure and applied chemistry 73(7):1163–1172. https://doi.org/10.1351/pac200173071163

Von Canstein H, Kelly S, Li Y, Wagner-Dobler I (2002) Species diversity improves the efficiency of mercury-reducing biofilms under changing environmental conditions. *Appl Environ Microb.* 68(6):2829–2837. https://doi.org/10.1128/AEM.68.6.2829-2837.2002

Wollmann F, Dietze S, Ackermann JU, Bley T, Walther T, Steingroewer J, Krujatz F (2019) Microalgae wastewater treatment: Biological and technological approaches. Engineering in Life Sciences 19(12):860–871. https://doi.org/10.1002/elsc.201900071

10 Impact and Application of Microbial Biofilms in Water Sanitation

Abioye O.P., Babaniyi B.R., Victor-Ekwebelem M.O., Ijah U.J.J., and Aransiola S.A.

CONTENTS

10.1 INTRODUCTION

Water is a substance made up of elements and exists in gaseous, liquid, and solid states. It's one of the foremost important and essential compounds. A tasteless and odorless liquid at room temperature, it has the potential to dissolve several solutes. Hence, the flexibility of water as a solvent is important to living organisms. In small quantities, water seems colorless; however, water really has an intrinsic blue color because of penetration of sunshine at red wavelengths Zumdahl (2021). There has been much reduction in water availability for human use albeit, it is an important

DOI: 10.1201/9781003184942-12

necessity as water resources are becoming more and more scarce and contaminated by anthropogenic activities corresponding to industrial effluents and agricultural and domestic waste, so biological effluent treatment systems play an important role in water quality and human health. These world water crises are because of speedy increases in population, climatic variation, environmental pollution, urbanization, industry and contamination of existing water reservoirs. Therefore, the treatment of effluent remains crucial before its discharge (Sehar and Naz, 2016). Water sanitation is the method of improvement and purifying water so that it becomes safe for use. Adequate sanitation, along with smart hygiene and safe water treatment, are elementary to healthiness and to social and economic development. Sanitation refers to public health conditions involving clean drink and adequate treatment and disposal of human excrement and sewage. Reprocessing activities inside the sanitation network could concentrate on the nutrients, water, energy or organic matter contained in excreta and effluent (Sehar and Naz, 2016). In line with Seow *et al.* (2016), water quality is vulnerable because of the presence of an outsized variety of pathogens and phylogenic chemicals that come into urban and rural water bodies. Discharges of wastewater from municipal and industrial treatment plants are recognized as of the key factors of aquatic pollution around the world. In several developing countries, the majority of domestic and industrial wastewater is directly discharged into water streams, with less or no treatment. Untreated wastewater that contains an outsized quantity of organic matter will consume the dissolved oxygen for satisfying the organic chemistry oxygen demand (BOD) of wastewater, and thus exhaust the dissolved oxygen of the water stream needed by the aquatic lives. Untreated effluent typically contains an outsized amount of pathogens, or disease-inflicting microorganisms and poisonous compounds, that may dwell within the human viscus tract, so threatening the human health. Effluent may contain specific amounts of nutrients, which might stimulate the expansion of aquatic plants and Protoctista blooms, thus resulting in eutrophication of the lakes and streams. The decomposition of organic compounds in wastewater can cause the assembly of huge quantities of high gases (Topare *et al.*, 2011). Moreover, microbic reduction of sulfate to sulfide in healthful sewer systems could be a universal sewer maintenance drawback thanks to pestilent odors, health hazards, and corrosion of in the main concrete sewers. Although the development is most widespread in countries with hot climates, the issues also are occurring in cold climate conditions.

The microbial reduction of sulfate takes place mainly within the biofilm at the sewer pipe inner surface (Bengtsson *et al.*, 2018). Impacts on public health, safe water, sanitation and hygiene – also called "WASH" – are essential for human health and well-being. Notably, large number of people globally lack ample WASH services, and thereby suffer many avertable infections. Improper WASH negatively affects standard of life and hinders elementary human rights. Inadequate water sanitation services cause frailty in health systems and creates a serious threat to economies (WHO, 2019). WASH-associated diseases and risks are numerous, including infections contracted through fecal or oral routes, health risks because of chemical exposure and alternative contaminants in drinking water on well-being. WASH-related unwellness is often exacerbated by numerous agents corresponding to climate change, population growth, speedy urbanization, or – within the case of antimicrobial resistance – antibiotic use. Globally, in 2016, 1.9 million deaths and 123 million disability-adjusted life-years (DALYs) were recorded that might have been prevented with correct sanitation. The WASH-attributable disease burden amounts to about 4.6% of worldwide DALYs and 3.3% of global deaths; the burden of deaths among kids underneath five years is 13.1% WHO (2019). Adequate water sanitation management practices will curb mosquitoes' breeding. Mosquitoes are vectors for dengue fever and alternative arboviruses, mainly in urban centers. Half the world's population is in danger of dengue; therefore, access to economical water facilities and safe food handling routines that stop foodborne infections ought to be encouraged. Clean water is significant for the management of neglected tropical diseases (NTD) morbidity, as well as surgical procedures for trachomatous trichiasis and self-care for bodily fluid filariasis, yaws, kala azar, and Buruli ulcers, in line with WHO (2019). Improper sanitation could lead to diarrhea, viscous parasite infections, and probably through environmental enteric dysfunction. In 2018, 149 million (21.9%) children aged younger than 5 years had stunted growth and 49.5 million (7%) globally were at risk of wasting. Inefficient disposal of effluent adds to the proliferation of antimicrobial-resistant microorganisms, and genes within the surroundings also boost individuals' exposure to resistant bacteria in native communities (Bouzid *et al.*, 2018). Antimicrobial residues in wastewater also can promote resistance inside environmental bacteria and invariably resistant diseases. Distinguishably to the acute and prompt nature of waterborne microbic infections, most chemical contaminants solely have a control when a protracted amount of exposure. Chemicals with public health impact to which people are exposed through drinking water are arsenic and cadmium, lead (from family plumbing materials), and nitrate (from waste product contamination or agricultural runoff). Regarding 140 million people in numerous countries worldwide are drinking water containing excess arsenic; long-term results of arsenic in drinking water and food intake include skin lesions, cancer, and polygenic disorders; additionally, childhood exposure has been linked to impaired psychological feature development, and young adults are at risk of death. Safe WASH contributes to social and economic well-being. Safe sanitation ensures dignity and safety, particularly for girls and adolescent girls. In healthcare networks, improper WASH has undesirable influence on worker morale, behavior of patients seeking healthcare, and their overall healthcare services.

10.2   MICROBIAL BIOFILM FORMATION IN WATER PIPELINES

A biofilm is a layer of microorganisms contained in a lattice (ooze layer), which structures on surfaces in touch with water. The surface on which the biofilm structures (the line divider) is known as the foundation Van Vuuren and Van Dijk (2012). Microorganisms can enter pipelines either by enduring the treatment interactions or by reintroduction of pollution (Mains, 2008). The essential colonizers attach to the line surface. Since it pastes to the line divider, the piece of the biofilm is adjusted by the base and furthermore by the inorganic atoms (Unais, 2015). Their essence in drinking water pipe organization can explain a wide scope of water quality and operational issues. They are bacterial totals connected to different biotic and abiotic surfaces which associate with one another to adjust to natural stressors contrasted with planktonic presence (Ashoka and Dash, 2015).

When there are structural changes to a pipe, optional colonizers set in because of adsorption of organisms to the molded surface (Kerr *et al.*, 2003). The adsorption is in two stages: reversible bond and irreversible grip (Marshall *et al.*, 1971). At the point when nearby conditions are conducive to biofilm development, irreversible bonds happen, and if the neighborhood states are not good, the miniature living beings relocate until great states are achieved (Kerr *et al.*, 2003). The bond between cells is one of the vital components for bacterial progression. There are two kinds of grip: co-collection and co-attachment. Co-collection happens between suspended cells, while co-attachment happens on surfaces (Bos *et al.*, 1996).

The irreversible connection causes cells to duplicate in this manner, yield miniature settlements, and create a lot of extracellular polymeric substances (EPS), framing a network that overwhelm the cells present (Kerr *et al.*, 2003). The EPS gives the biofilm its foul nature (Mains, 2008). The EPS is supporting design of the biofilm. It serves two primary capacities: it keeps up connection onto the substrate, and it holds cells together inside the state (Kerr *et al.*, 2003). The EPS is a network that comprises natural polymers that are created and discharged by the miniature organic entities present in the biofilm (Momba *et al.*, 2000). It contributes 70–95% of the natural matter of biofilms and, by correlation, miniature creatures just address a minor part (by volume and by weight) of the biofilm (Van Vuuren and Van Dijk, 2012). The substance structure of the EPS is not something similar among the different kinds of life forms, and is likewise depend on the general climate (Momba *et al.*, 2000).

Regularly, microorganisms structure the bigger segment of the biofilm populace; microbes that need natural mixtures as wellsprings of energy and carbon are known as heterotrophic microscopic organisms, and this is the most widely recognized microscopic organism found in biofilms (Butler and Boltz, 2014). Additionally, there are other miniature creatures that make up the biofilm, which incorporates astute microbes, protozoa, green growth, parasites, helminths, and different spineless creatures (Kambam, 2006). After development of the ooze layer, the ooze layer helps trap natural particles that microscopic organisms can use as a wellspring of energy and food (Mains, 2008). Any microorganism, including essential and artful microbes present in water, may append or get enmeshed in the biofilm. Notwithstanding, the endurance time for some microorganisms in biofilms is questionable and likely fluctuates, depending on the organic entity. Amphibian microorganisms are all around adjusted to the low supplement level and cool water temperature of the conveyance framework. The interspecific or vague blends between gram-positive and gram-negative, as seen in past investigations, is more extensive for biofilm development than for intraspecific bacterial populace (Ashoka and Dash, 2015).

10.3   IMPACTS OF MICROBIAL BIOFILM ON WATER SANITATION

Biofilms in drinking water account for numerous unpleasant outcomes in the quality of water distributed. Among the major setbacks of biofilms is their ability to serve as a cover niche for waterborne pathogens that are capable for different occurrences of disease due to contaminated water distributed for consumption (Simoes and Simoes, 2013).

The presence of biofilm has an effect on the hydraulic capacity of the pipeline, but it is difficult to quantify this effect because biofilm growth is always swinging due to changes in the water quality, flow conditions, temperature, and pH levels; more so, normal pipe wall roughness change due to presence of biofilms (Van Vuuren and Van Dijk, 2012). The severity of the biofilm impact was highlighted by Van Vuuren and Van (2012); the tests carried out revealed increase in the friction of the pipeline was not due to pipeline degradation but due to biofilm growth in the pipeline. Their experiments on a ten-year-old pipe found that the roughness after ten years of the pipeline was 1.76 mm, while the designed roughness was 0.5 mm. This leads to an increase in friction, which leads to an increase in energy input costs, since biofilms are viscoelastic in nature (Van Vuuren and Van Dijk, 2012). Ascertainment of the frictional head loss is prime since it also controls the operating costs of the pipeline. Biofilm growth due to the increase in hydraulic roughness affects a given flow and pressure head due to increase in the friction per unit length, which means a higher energy input and increased costs in water sanitation process.

10.3.1 Impact on Public Health

Biofilms have the potential to harbor opportunistic pathogens and are seen as the prevalent cause of water quality deterioration (Kerr *et al.*, 2003). Opportunistic pathogens exposure cause diseases in people (Table 10.1) with compromised

TABLE 10.1

Some Reported Outbreaks of Diseases due to Microbial-Contaminated Drinking Water (Simoes and Simoes, 2013)

Disease	Source
Cholera	Water from the Broad Street pump
Typhoid fever and dysentery	Municipal water supply
Cryptosporidiosis	Public water supply, of which the filtration system was contaminated
Hepatitis A infection	Consumption of water from a feces-contaminated well
Viral gastroenteritis	Drinking water highly contaminated with enteric viruses
Norovirus gastroenteritis	Contaminated drinking water obtained from private wells
Giardiasis	Community water supply
Enterovirus	Pollution of the water supply by enteroviruses
E. coli infection	Contaminated drinking water
Campylobacteriosis	Municipal water supply

TABLE 10.2

Pathogens Associated with Waterborne Diseases and Their Resistance to Disinfection by Chlorine (Simoes and Simoes, 2013)

Bacteria	Resistance to Chlorine	Protozoa	Resistance to Chlorine	Viruses	Resistance to Chlorine
Aeromonas spp.	Low	*Blastocystis hominis*	High	*Astrovirus*	Moderate
Escherichia coli	Low	*Acanthamoeba castellani*	High	*Noroviru*	Moderate
E. coli enterohaemorrhagic	Low	*Balantidium coli*	High	*Hepatitis A virus*	Moderate
Francisella tularensis	Moderate	*Giardia intestinali*	High	*Poliovirus*	Moderate
Mycobacterium spp.	High	*Sarcocytis spp.*	High	*Sapovirus*	Moderate
Salmonella typhi	Low	*Microsporidia*	Moderate	*Rotavirus*	Moderate

immune systems, such as AIDS patients, diabetic patients, cancer patients, and other vulnerable groups such as children and older people (Mains, 2008). Therefore, the pathogenic microorganisms must be removed by disinfection, thereby preventing the microbes to reach the end users and outbursts of disease within a community. Notably, some of the opportunist pathogens (Table 10.2) are recalcitrant and manifest high aversion to chlorination, while many others show mild aversion (Simoes and Simoes, 2013). Contamination of a water sanitation network occurs in two ways: either due to microorganisms which are not totally eliminated at the disinfection plant, or microorganisms which detach from the biofilm present on the pipe walls (Unais, 2015). Opportunistic bacteria that can survive free chlorine residuals of 0.5–1.0 mg/L include species of *Pseudomonas aeruginosa*, *Klebsiella* spp., *Mycobacterium*, *Legionella* spp., and *Flavobacterium* spp. (Unais, 2015). The most identified disease associated with waterborne outbreaks in developed countries is gastroenteritis (Simoes and Simoes, 2013). The health effects differ in severity and can range from moderate gastroenteritis to severe diarrhea, dysentery, hepatitis, and typhoid fever (World Health Organisation, 2011). Biofilm posed difficulty in water sanitation; to restrict biofilm growth, chlorine is most frequently applied to drinking water. However, due to the risk of elevating the levels of harmful disinfection byproducts of chlorine in drinking water, researchers are in dilemma finding alternative means to water sanitation, factors such as the nutrient content of water, the concentration of residual disinfectant, the hydrodynamic conditions of the network, the pipe materials, and their conservation conditions, the diversity of microorganisms present in drinking water distributions, and environmental factors like pH and the temperature of water, should be considered in any new approach – some of which is tasking, time consuming, and capital intensive.

10.3.2 MICROBIOLOGICALLY INDUCED CORROSION

Microbiologically induced corrosion refers to the disintegration of metal by chemical and electrochemical reactions, together with its environment or by the physical wearing away of the metal with its impact on the water quality (Unais, 2015). It is explained as an electrochemical process in which the presence of microorganisms is able to initiate, facilitate, or accelerate the corrosion reaction without altering its electrochemical nature (Videla, 2001). This form of corrosion is found extensively in pipes carrying treated water, raw water, and wastewater. Microbiologically induced corrosion is caused by different types of microorganisms, normally bacteria, yeasts, and algae (Ringas, 2007); different species of microorganisms affect the corrosion processes in different forms. Sulfate-induced bacteria are responsible for most of microbiologically induced corrosion. They are found at the interface between the biofilm and metallic substrate, and since they are anaerobic, their survival rate is high in that environment because they are protected from the bulk of fluid. Sulfate-induced bacteria derive energy from conversion of sulfates and phosphates into sulfides, which react to emit either hydrogen sulfides or iron sulfides. Hydrogen sulfides are extremely aggressive and attack metal surfaces (Ringas, 2007). Microbial biofilms persist and corrode water pipelines, making sanitation inefficient; there is a lot of capital and labor involved in replacing water pipeline damage due to corrosion.

10.4 SIGNIFICANCE OF MICROBIAL BIOFILM IN WATER TREATMENT

Biofilm formation, also referred to as biofouling, could be a well-developed technology during which solid media are added to suspended growth reactors to supply attachment surfaces for biofilms proliferation (Pandit *et al.*, 2020). Biofilms, in total, refers to the unpleasant aggregation of biotic matter on a surface. it's been shown to be of considerable hygienic, operational, and economical significance, not only in treated drinking water but also in other purified water systems, like dental unit waterlines, dialysis units, reverse osmosis systems, laboratories, pharmaceutics, the semiconductor industry, and even the international artificial satellite water recovery and management system. Different problems encountered in installation are related to microbes, biofilm development, nitrification, microbial-mediated corrosion, and also the occurrence and persistence of pathogens (Wingender and Flemming, 2011).

Biofoulings are said to be the prime source of microorganisms in beverage supplies which are treated and have not any pipeline linkages (Deepike *et al.*, 2018). Flemming *et al.* (2002) showed that about 95% of the whole biomass in water is estimated to be connected to pipe walls, while only 5% is within the water phase. For that reason, microbial development in biofilms is very significant for water quality as long as they directly act on cell density within the bulk phase. Extracellular polymers are the key substances keeping biofilm organisms together, bonding them to the surface and ensuring protection against force or tension. Any inorganic particle moving close, like corrosion products, clays, and sand, might also combine with the formation of biofilms, thereby increasing its "mechanical strength" (Hassard *et al.*, 2015).

Bacteria, in general, form the greater part of biofilms because of their high development levels, their adaptation capacities, and also the ability to provide extracellular polymers (Deepike *et al.*, 2018); hence, viruses, protozoa, fungi, and algae might also be present in treated water biofilms (Simoes and Simoes, 2013). Emulating the sessile type of life, microorganisms in biofilm enjoy some benefits over their planktonic equivalents; among these merits are the ability of the extracellular polymeric matrix, which they discharge, to apprehend and target nutrients within the environment like carbon, nitrogen, and phosphate (Maes *et al.*, 2019). More so, biofilms form of development account for resistance to a form of strategies for removal, most significantly, of antimicrobial and mechanical tensions (Simoes and Simoes, 2013). Chlorination, as an example, may eliminate or bring down bacteria load in plankton, but have little to no effect on the load of bacteria in biofilm. A study showed that effects of chlorine at 10 mg/L1 on the viability of multispecies biofilms in water revealed the presence of viable cells after treatment and significant recovery after 24 hours (Simoes and Simoes, 2013).

The antimicrobial resistance of biofilms can be controlled through very low metabolic levels and drastically reduced levels of biological processes of the deeply embedded microorganisms. Additionally, slowing assimilation of antimicrobial agents is said to s act as a diffusion barrier in biofilms; antimicrobial agents react with extracellular polymers compounds, reducing the agent's concentration and its effectiveness (Deepike *et al.*, 2018). Survival of microorganisms in the biofilm formation depends on the interchanges between bacteria with different physiological requirements (Stewart *et al.*, 2008); this can promote the event of spatial niches in a biofilm in response to environmental conditions and therefore the activity of their neighbors so as to maximize the nutritive resources, and also, bacterial correspondence through excreted signaling molecules is another significant activity of living organisms in biofilm (Simoes *et al.*, 2007). A major merit of the biofilm mothed of development is the potential for spreading through detachment (Maes *et al.*, 2019). Under the direction of fluid flow, detached microorganisms migrate to other regions to fuse and promote biofilm growth on clean surfaces. Consequently, this advantage allows a persistent bacterial source community, often opposing antimicrobial agents, at the identical time promoting the continual braking to enable bacterial dispersion (Simoes *et al.*, 2007).

The co-existence of pathogens with other microorganisms in biofilm has been a serious drawback in beverage supply. Biofilms' growth within potable-water networks may contain bacterial pathogens, like *Legionella pneumophila* and coliforms of intestinal and non-intestinal origin; commonly found within portable installation are protozoa embedded in biofilms and are related to pathogen persistence and invasiveness (Thomas and Ashbolt, 2011). Organisms transmitting disease, like *Mycobacterium* spp., *L. pneumophila, P. aeruginosa, Klebsiella* spp., *Burkholderia* spp., *Giardia, Cryptosporidium*, and many more transmit disease through contaminated water, and biofilms are a decent convener as they act as a protective niche for their survival; this means a significant role of an efficient treatment arrangement is to control microorganisms within the bulk phase and their biofilms so as to supply high portable water. It is important to notice that biological treatment is usually just a step-in wastewater treatment procedures.

Several other methods also are employed to treat the recycled water that our world is so desperately in need of; however, it's been concluded through extensive experimentation and analysis that biofilms do effectively clean wastewater (Sehar and Naz, 2016).

The microbial communities within the biofilm break down different nutrients, like phosphorus and nitrogen-containing compounds, and carbonaceous materials further as trapped pathogens from the wastewater. Once pollutants are removed, treated water of a biofilter is either released to the environment or used for agriculture and other recreational purposes. Wastewater treatment with biofilm systems has several advantages, including operational flexibility, low space requirements, reduced hydraulic retention time, resilience to changes within the environment, increased biomass duration, high active biomass concentration, and enhanced ability to degrade recalcitrant compounds also as a slower microbial rate, leading to lower sludge production (Sehar and Naz, 2016).

10.5   CONTROL OF MICROBIAL BIOFILMS IN DRINKING WATER

The formation of biofilms is influenced by many factors, such as temperature and water quality. The methods used to control biofilm formation are presented in the following sections

10.5.1   Disinfectant

Any disinfectant chosen for water treatment should be able to penetrate the biofilm and de-activating attached microorganisms, and must be suitable, stable, and persistent in the water network. Chlorine is the most cost-effective disinfectant (Kerr *et al.*, 2003). Treatment of water at the plant before it enters the pipe networks is not efficient enough; there are some microorganisms present in the pipelines that may enter through open reservoirs or cracks, joints, valves, cross-connections, and backflow valves in the pipeline (Mains, 2008), or through incomplete disinfection at the treatment plant. Therefore, there is need for excess chlorination. (Kerr *et al.*, 2003). Meanwhile, chorine residual in the water is beneficial, but excess chlorine residual can also lead to problems in the water system.

10.5.2   Material Selection

Identifying materials that do not contribute to or that can suppress biofilm growth is essential when preventing biofouling (Simoes and Simoes, 2013). Different materials, such as ethylene-propylene, natural latex, stainless steel, mild steel, polypropylene, polyethylene (PE), chlorinated, and polyvinyl chloride (PVC), were ranked according to their biofilm growth propensity, which unfortunately led to the conclusion that there is hardly any material that does not allow biofilm formation. Hence, the type and stability of the material used in water supply is an important factor that can influence biofilm proliferation. There is a distinct development rate of microbial population structure of biofilms in different types of pipe. Bacteria are capable of leaching nutrients from the materials. A study showed that iron pipes can support 10–45 times more growth than plastic pipes (Niquette *et al.*, 2000). Also, iron pipes are more reactive with disinfectants and quench their antimicrobial effects. Therefore, the type of material can also affect the disinfectant efficiency of biofilms. Biofilms grown on copper, PE, PVC, and cement-lined ductile iron were inactivated with a much lower amount of free chlorine than those grown on unlined iron surfaces. This was explained by the interaction of chlorine with iron. In cement-lined ductile iron, the cement provides a layer of protection for the iron against attack by chlorine. The pipe service age is key in response to limiting biofilms growth.

10.5.3 Hydrodynamics

To prevent zones of high water retention time in pipes and sediment accumulation, the distribution system must be stationed. Pipes with long water residence times and dead ends are found with zones of high organic material precipitates, leading to formation of biofilm. High bacterial numbers are linked with periods of non-flow or the storage of water in household pipes or tanks. Therefore, operational measures should be taken into account in order to alleviate sediment build-up in pipes, particularly the optimization of pre-treatment, to minimize particles in water supply network, and the application of sufficiently high flow velocities that may result in a self-cleaning process and regular flushing of the water network.

10.5.4 Targeting Key Microbes

Another preventative measure aims to target the prime microorganisms in biofilm formation. It is conceivable that interaction with prime bacteria in the biofilm may cause dispersion. This could be done by interfering with the biofilm command language referred to as quorum sensing (QS). The method by which cells communicate and are arranged in a social community is controlled by the secretion of signaling molecules and this process is called quorum sensing. Bacteria have the potential to signal and sense the state of population density in order to change physiological needs under different developmental states. The discovery that many bacteria use QS molecules to form biofilms makes it an attractive source for their prevention.

10.5.5 Application of Microbial Biofilms in Wastewater Treatment

One of the most widely used tactics for wastewater treatment is activated sludge; recently, biofilm technologies utilization has been gaining momentum for the purpose of water treatment. Biofilms offer efficient advantages over suspended single cells, by promoting cell–liquid separation through filtration in downstream networking. Though design of trickling filters, is very well established for wastewater treatment that the application of particulate biofilm, moving on some carriers, is trending and highly accepted method in many industries (Deepika *et al.*, 2018). Removal of the pollutants from wastewater by biofilm on the filter media is schematically represented in Figure 10.1 and Figure 10.2.

FIGURE 10.1   Life cycle of biofilms showing the attachment phase, growth phase, and dispersal phase (Cunningham *et al.*, 2011).

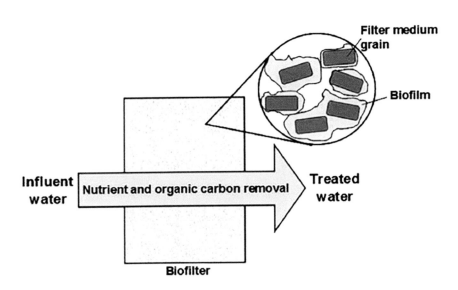

FIGURE 10.2   Biofilm in wastewater treatment.

10.5.6 FLUIDIZED-BED REACTOR

This is another method for utilizing biofilms to treat fouled water; in this strategy, water is gradually siphoned vertically and keeps biofilm globules suspended during treatment of the defiled water (Arindam and Suman, 2016). Solids are raised by stream of fluid or gas at a specific speed. Fluidization grants biofilms to develop on a spread surface region and produce high biomass. Air circulation is done either by means of an oxygenator or provided from the lower part of the reactor. Fluidized-bed reactor has been utilized for the treatment of streams polluted with natural and inorganic mixtures (Kumar and Saravanan, 2009).

10.5.7 BIOREMEDIATION

Microbial biofilm procedures are productive to intercede different sorts of toxins like tireless natural poisons, oil slicks, substantial metals pesticides, and xenobiotics. Biofilm remediation has been especially valuable in treatment of substantial metal-debased examples from groundwater and soil for oftentimes experienced weighty metals like cadmium, copper, uranium, and chromium (Valls and de Lorenzo, 2002). Phosphatase chemicals in the presence of biofilm networks support metal sedimentation in oxygen-consuming microorganisms and anaerobic conditions (Shukla *et al.*, 2014). In some cases, biofilm development is hindered by augmentation of carbon sources in dirtied ground water to blockade or restrict the progression of poisons from the place of defilement in this way, limiting its spread.

Persistent organic pollutants (POP) that include polycyclic aromatic hydrocarbons (PAH) and polychlorinated biphenyls. With their hydrophobicity nature facilitates longer half-lives and can be found in air, soil, and water. Many bacterial-forming biofilms have been identified from surroundings that POP are degraded, which may be further modified and used in bioremediation (Chen *et al.*, 2015). Bacterial biofilms have been engineered for detoxification of POP. Studies have shown that biofilm formation and co-metabolism in biofilms are primary factors in remediating polycyclic aromatic hydrocarbons (Adetunji and Anthony, 2021).

10.6 MEDICAL APPLICATIONS

Antimicrobial drug resistance is one of the traits of microbial biofilms. Full grown biofilms resist antimicrobial drug therapy better than their planktonic cell equivalents due to the extracellular matrix of the biofilm. Bacteria transfer genetic material from cell to cell by horizontal gene transfer (HGT), either transformation or conjugation or transduction. Within the course of natural transformation, bacterial DNA from lysed cells is obsessed by actively developing competent cells of the identical species or a distinct species and incorporates into their own genome by genetic recombination. Any genetic traits like antimicrobial drug resistance are retained by natural action. HGT markedly promote interspecies and intraspecies transmission of antibiotic resistance determinants among bacterial cells in biofilms (Vazquez and Manavathu, 2020). Chronic infections may prolong for periods of time, and thus, the infected body sites of the host provide perfect conditions for protracted interaction of the infective microbes to ascertain synergy, leading to biofilm formation and possible gene transfer. Microbial biofilm is additionally capable of modulating host reaction toward invading pathogens. Actually, the inflammatory response projected by the host is directed against pathogens and is meant to guard the host cells and destroy the invading pathogen, but biofilms act as barriers. This system might be employed to forestall the spread of ailments or targeted toward a site reaching to another site.

10.7   AGRICULTURAL APPLICATIONS

Sustainable agriculture practices are aided by biofilm-forming bacteria by promoting and biocontrolling plant growth. The relationship between plant growth–promoting rhizobacteria is noted for its potential to push yield and growth of plants through direct or indirect mechanisms. Formation of biofilm improves health and growth of plants by stimulating production of growth hormones like indole carboxylic acid (Bandara *et al.*, 2006). Plant growth–promoting microbes (PGPMs) are a live aggregate of beneficial microbiomes answerable for solubilizing of phosphorus (Yadav *et al.*, 2016), potassium (Verma *et al.*, 2017a), and zinc biological organic processes (Suman *et al.*, 2016); and production of siderophores, hydrolytic enzymes, compound (HCN), and ammonia (Yadav *et al.*, 2018). These microbes are capable of performing various tasks like dissolving various macronutrients, making atmospheric nitrogen available, and reducing different phytopathogens (Verma *et al.*, 2017b). Thus, these plant growth–promoting microbes have potential biotechnological application as biofertilizers and biocontrol agents for sustainable agriculture.

The microbiome of rhizosphere supports plant growth and hence are very important factor. The genome of microbial community of rhizosphere is greater than plant genome, which is why it is considered to be plant's second genome (Berendsen *et al.*, 2012). The microbiome of a root can affect disease-oppressive soil and provide the host system with valuable microbes within the rhizosphere (Berendsen *et al.*, 2012). The rhizosphere of wheat has been found with various antifungal compounds, with reports of phenazine and its role in biofilm formation. the method of biofilm formation and root colonization was enhanced by the assembly of surfactin. Soil biofilm formation may be a complex process which needs correct study. The health of soil relies on its structure, availability of nutrients, the assorted transformation processes of carbon, and also the regulation of pests and diseases. Microbes, either naturally or inoculated into soil, influence the wholeness of the soil. Autotrophic organisms are known to be a source of carbon, and chemoautotrophic bacteria and eubacteria are noted to help this (Divjot *et al.*, 2021). The biofilm formation by these microbes, together with improved soil, considerably supports the existence of those microbes (Berg and Smalla, 2009). Consequently, inoculation of probiotic biofilms on soil and plants may develop better conditions within the field (Malusa *et al.*, 2012).

10.8   CONCLUSION

This chapter is a review of water sanitation and its impact on public and environmental health, as well as microbial biofilms in wastewater treatment and biofilm technology applications. Without a doubt, having a thorough understanding of the nature of wastewater and the types of pollutants present is essential for designing an effective treatment system that ensures the safety, efficacy, and quality of treated wastewater. As a result, as the use of biofilm processes in water and wastewater treatment grows, a new analytical approach for fixed biomass characterization is needed. The use of biofilm to increase crop yields and the synergy between plant roots and biofilm bacteria are, however, critical. Nonetheless, more research is needed to take advantage of the biofilm process in water and wastewater treatment, including developing simple and precise analytical methods for estimating bacterial biomass and activity, and gaining a better understanding of the composition and functions of biofilm exopolymers.

REFERENCES

Adetunji CO, Anthony A (2021) Utilization of microbial biofilm for the biotransformation and bioremediation of heavily polluted environment. *Microb Rej Poll Environ* 227–245. doi:10.1007/978-981-15-7447-4_9.

Arindam Mitra, and Suman Mukhopadhyay (2016) Biofilm mediated decontamination of pollutants from the environment. *AIMS Bioengineering*. 3(1): 44–59. doi:10.3934/bioeng.2016.1.44.

Ashoka M, Dash D (2015) Study of biofilm in bacteria from water pipelines. *Clin Diagn Res.* 9(3): 9–11. doi:10.7860/JCDR/2015/12415.571.

Bandara, W., Seneviratne, G., Kulasooriya, S.A. (2006) Interactions among endophytic bacteria and fungi: effects and potentials. *J. Biosci.* 31:645–650.

Bengtsson, S., de Blois, M., Wilén, B.-M. and Gustavsson, D. (2018) Treatment of municipal wastewater with aerobic granular sludge. *Critical Reviews in Environmental Science and Technology*. 48(2):119–166.

Berendsen, R.L., Pieterse, C.M., Bakker, P.A., (2012) The rhizosphere microbiome and plant health. *Trends Plant Sci.* 17:478–486.

Berg, G., Smalla, K., (2009) Plant species and soil type cooperatively shape the structure and function of microbial communities in the rhizosphere. *FEMS Microbiol Ecol.* 68:1–13.

Bos R, Me, H, van de Busscher, H (1996) co-adhesion of oral microbial pairs under flow in the presence of saliva and lactose. *Journal of Dental Research*. 75(2):809–815.

Bouzid M., Cumming O., Hunter PR. (2018) What is the impact of water sanitation and hygiene in healthcare facilities on care seeking behaviour and patient satisfaction? A systematic review of the evidence from low-income and middle-income countries. *BMJ Glob Health*.3:e000648. doi:10.1136/bmjgh-2017–000648.

Butler CS, Boltz JP (2014) Biofilm Processes and Control in Water and Wastewater Treatment. In: Ahuja S. (ed.) Comprehensive Water Quality and Purification, vol. 3, 90–107. United States of America: Elsevier.

Chen M, Xu P, Zeng G, (2015) Bioremediation of soils contaminated with polycyclic aromatic hydrocarbons, petroleum, pesticides, chlorophenols and heavy metals by composting: Applications, microbes and future research needs. *Biotechnol Adv* 33: 745–755.

Cunningham John, Francesco Locatelli, Mariano Rodriguez. (2011) Secondary hyperparathyroidism: pathogenesis, disease progression, and therapeutic options. Rev. *Am Soc Nephrol.* 6(4):913–21. doi: 10.2215/CJN.06040710

Deepika Rajwar. Mamta Bisht and J. P. N. Rai (2018) Wastewater Treatment: Role of Microbial Biofilm and Their Biotechnological Advances. Applied Microbiology 162–174. DOI: 10.4018/978-1-5225-3126-5.ch010.

Divjot Koura, Kusam Lata Ranaa, Tanvir Kaura, Neelam Yadavb, Ajar Nath Yadava, Ali A. Rastegaric and Anil Kumar Saxenad (2021) Microbial biofilms: Functional annotation and potential applications in agriculture and allied sectors. In book: New and future developments in microbial biotechnology and bioengineering. 18, 283–301.

Flemming H.C., Percival S. and Walker J., (2002,) Contamination potential of biofilms in water distribution systems. Water Sci. Technol.: Water Supply, 2, 271–280.

Hassard, F., Biddle, J., Carmell, E., Jefferson, B., Tyrell, S., Stephenson, T. (2015) Rotating biological contactors for wastewater treatment- A review. Process Safety and Environmental Protection, 94, 285–306. doi: 10.1016/j.psep.2014.07.003.

Kambam Y (2006) *Potable water treatment: biofilms in water distribution.* Iowa State University.

Kerr C, Osborn K, Robson G, Handley P (2003) Biofilms in water distribution systems. In *Handbook of water and wastewater microbiology.* Elsevier, Academic Press, 757–775.

Kumar TA, Saravanan S (2009) Treatability studies of textile wastewater on an aerobic fluidized bed biofilm reactor (FABR): a case study. Water Sci Technol 59: 1817–1821.

Maes Sharon, Thijs Vackier, Son Nguyen Huu, Marc Heyndrickx, Hans Steenackers5, Imca Sampers, Katleen Raes, Alex Verplaetse, Koen De Reu (2019) Occurrence and characterisation of biofilms in drinking water systems of broiler houses. *BMC Microbiolog.* 9: 77. doi: 10.1186/s12866-019-1451-5.

Mains C (2008) *Biofilm control in distribution systems.* National Environmental Services Center.

Malusa, E., Sas-Paszt, L., Ciesielska, J., 2012. Technologies for beneficial microorganisms inocula used as biofertilizers. Sci. World J. https://doi.org/ 10.1100/2012/491206.

Marshall F, James, Steve Senia E., Samuel Rosen (1971) The solvent action of sodium hypochlorite on pulp tissue of extracted teeth. Oral Surgery, Oral Medicine, Oral Pathology. 3, 96–103.

Momba, M., Kfir, R., Venter, S. CLoete, T. (2000) An overview of biofilm formation in distribution systems and its impact on the deterioration of water quality. Water SA, 26(1), pp. 59–66.

Niquette P., Servais P. and Savoir R., (2000) Impacts of pipe materials on densities of fixed bacterial biomass in a drinking water distribution system. Water Res., 34, 1952–1956.

Pandit A., Adholeya A., Cahill D., Brau L. Kochar M. (2020) Microbial biofilms in nature: unlocking their potential for agricultural applications. REV. Applied Microbiology. 129, 199–211. doi:10.1111/jam.1460.

Ringas, C., (2007) Internal corrosion of slurry pipelines caused by microbial corrosion: causes and remedies. *The Journal of The Southern African Institute of Mining and Metallurgy,* 107, 381–384.

Sehar S, Naz I (2016) Role of the biofilms in wastewater treatment. *Microbial Biofilms-Impor Appli.* 121–144.

Seow, T. W., Lim, C. K., Nor, M. H. M., Mubarak, M. F. M., Lam, C. Y., Yahya, A., Ibrahim, Z. (2016) Review on wastewater treatment technologies. *Int J Appl Environ Sci, 11*(1), 111–126.

Shukla SK, Mangwani N, Rao TS, (2014) Biofilm-Mediated Bioremediation of Polycyclic Aromatic Hydrocarbons. In: Das S, editor. *Microbial Biodegradation and Bioremediation.* Oxford: Elsevier. 8: 203–232.

Simoes LC, Simoes M (2013) Biofilms in drinking water: problems and solutions. *RSC Advances* 3: 2520. doi: 10.1039/c2ra22243d.

Simoes M., Pereira M. O., S. Sillankorva, Azeredo J. and Vieira M. J., (2007) The effect of hydrodynamic conditions on the phenotype of Pseudomonas fluorescens biofilms, 23, 249–258.

Stewart P. S. and Franklin M. J., Nat. Rev. (2008) Physiological heterogeneity in biofilms Microbiol. 6, 199–210

Suman A, Yadav AN, Verma P (2016) Endophytic microbes in crops: diversity and beneficial impact for sustainable agriculture. In Singh D, Abhilash P, Prabha R (Eds.), *Microbial inoculants in sustainable agricultural productivity, research perspectives.* Springer-Verlag, 117–143.

Thomas J. M. and Ashbolt N. J., (2011) Do Free-Living Amoebae in Treated Drinking Water Systems Present an Emerging Health Risk? Environ. Sci. Technol., 45, 860–869.

Topare, N. S., Attar, S. J., and Manfe, M. M. (2011) "Sewage/wastewater treatment technologies: a review," *Sci Revs Chem Commun.* 1, 18–24

Unais Kadwa (2015) *A study on biofilm formation in pipelines* (Dissertation). University of KwaZulu-Natal.

Valls M, de Lorenzo VC (2002) Exploiting the genetic and biochemical capacities of bacteria for the remediation of heavy metal pollution. FEMS *Microbiol Rev* 26: 327–338.

Van Vuuren SJ, Van Dijk M (2012) *Determination of the change in hydraulic capacity in pipelines.* Water Research Commission.

Vazquez MD, Manavathu EK (2020) *Medical importance of microbial biofilms in the management of infectious diseases.* https://www.contagionlive.com

Verma, P., Yadav, A.N., Khannam, K.S., Saxena, A.K., Suman, A. (2017a) Potassium-solubilizing microbes: diversity, distribution, and role in plant growth promotion. In: Panpatte, D.G., Jhala, Y.K., Vyas, R.V., Shelat, H.N. (Eds.), Microorganisms for Green Revolution. In: Microbes for Sustainable Crop Production, vol. 1. Springer, Singapore. 125–149. https://doi.org/10.1007/978-981-10-6241-4_7

Verma, P., Yadav, A.N., Kumar, V., Singh, D.P., Saxena, A.K. (2017b) Beneficial plant-microbes interactions: biodiversity of microbes from divers extreme environments and its impact for crops improvement. In: Singh, D.P., Singh, H.B., Prabha, R. (Eds.), Plant-Microbe Interactions i n AgroEcological Perspectives. Springer Nature, Singapore. 543–580. https://doi.org/10.1007/978-981-10-6593-4_22.

Videla HA (2001) Microbially induced corrosion: an updated overview. *Int Biodeterior Biodegradation.* 48(1–4):176–201.

Wingender J. and Flemming H.C., (2011) Biofilms in drinking water and their role as reservoir for pathogens.Int. J. Hyg. Environ. Health, 214, 417–423.

World Health Organisation (2011) *Guidelines for drinking-water quality.* WHO Press, 4th ed.

World Health Organization (2019) *Water, sanitation, hygiene and health: a primer for health professionals.* WHO Press. WHO/CED/PHE/WSH/19.149. Licence:CC BY-NC-SA 3.0 IGO.

Yadav AN, Kumar V, Prasad R, Saxena AK, Dhaliwal HS (2018) Microbiome in crops: diversity, distribution and potential role in crops improvements. In Prasad R, Gill SS, Tuteja N (Eds.), *Crop improvement through microbial biotechnology.* Elsevier, 305–332.

Yadav, A.N., Rana, K.L., Kumar, V., Dhaliwal, H.S., (2016) Phosphorus solubilizing endophytic microbes: potential application for sustainable agriculture.EU Voice 2, 21–22.

Zumdahl SS (2021) Water. *Ency Brit.* https://www.britannica.com/science/water. Accessed 9 June 2021.

11 Versatility of Bacterial Cellulose and Its Promising Application as Support for Catalyst Used in the Treatment of Effluents

Marina Gomes Silva, Rayany Magali da Rocha Santana, Karina Carvalho de Souza, Helenise Almeida do Nascimento, Alex Leandro Andrade de Lucena, Joan Manuel Rodriguez-Diaz, Daniella Carla Napoleão and Glória Maria Vinhas

CONTENTS

DOI: 10.1201/9781003184942-13

11.1   INTRODUCTION

Among the existing polymers, cellulose has stood out due to its structural properties and biodegradability, in addition to being non-toxic and easy to handle (Xie *et
al.*, 2020). It is considered the polysaccharide with the greatest natural abundance,
being the main constituent in the composition of plant cell walls, and therefore, it is
traditionally extracted from various vegetables. But this polysaccharide can also be
synthesized from microorganisms, such as fungi and bacteria with specific physicochemical properties (Parte *et al.*, 2020).

In 1886, Adrian J. Brown documented a study on the production of cellulose by
bacteria, after observing that some acetic acid-producing microorganisms (*Bacterium
xylinum*) were able to synthesize a gelatinous-looking membrane on the surface of
the vinegar fermentation broth, where glucose and oxygen were present. Through
analysis of the film, a chemical structure similar to that of vegetable cellulose (VC)
was noted, and then, this type of material came to be called bacterial cellulose (BC)
(Brown, 1886).

Over the years, for the synthesis of the biopolymer, several genera of microorganisms could be used, including *Aerobacter*, *Achromobacter*, and *Agrobacterium*
(Fontana *et al.*, 2017). However, bacterial included in the genus *Komagataeibacter* –
such as the species *K. hansenii*, *K. europaeus*, *K. oboediens*, *K. intermedius*, and *K.
xylinus* – have been the most studied, due to the significantly high cellulose yield.
Furthermore, by using such strains of bacteria, cellulose can be produced from different sources of carbon and nitrogen, making it an alternative of great economic
interest. The bacteria of the genus *Komagataeibacter* are gram-negative, aerobic and
non-pathogenic, being commonly found in fruits, vegetables, and fermented products
such as vinegar and alcoholic beverages (Cazón and Vázquez, 2021).

Cellulose formation is part of an important physiological mechanism of bacteria.
This is responsible for transporting nutrients through the diffusion process, thus generating an increased availability of oxygen for growth. During biosynthesis, there is
the formation of a film that protects the microorganism's cells from drying processes,
in addition to acting as a protective barrier against ultraviolet (UV) radiation (Silva *et
al.*, 2021; Gromovykh *et al.*, 2017).

Bacterial cellulose has a higher degree of purity than VC, as it does not consist
of lignin, hemicellulose, and other biogenic products. CB has a high mechanical
strength, porosity, non-toxicity, and crystallinity, and greater water retention capacity. Due to these characteristics and its nanofibrillar structure since VC have micro-
fibrillar structure (Gallegos *et al.*, 2016; Inoue *et al.*, 2020), and the relevant research
in obtaining, charactering, and applying this biopolymer, it is important to understand certain aspects related to its structure and production, in order to understand
its important application in different areas of knowledge.

11.2   STRUCTURE OF BACTERIAL CELLULOSE

CB, with the molecular formula of $(C_6H_{10}O_5)n$, is a non-branched straight-chain
homopolysaccharide, composed of D-glucopyranose units $(C_6H_{12}O_6)$ oriented in par-
allel and joined together through stable glycosidic bonds of the $\beta(1-4)$ type. In it, free

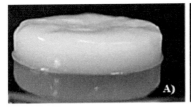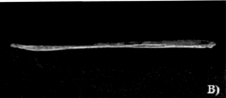

FIGURE 11.1   Bacterial cellulose obtained after ten days of fermentation: A) before drying; B) after 24 hours of drying at 30°C (adapted from Amorim *et al.* [2020]).

hydroxyl groups form hydrogen bonds with each other (intramolecular) and with other hydroxyl groups from adjacent glucose chains (intermolecular). The union of two glucose molecules forms a repeating unit known as cellobiose (Azeredo *et al.*, 2019).

A chain formed by glycans can contain from 2,000–25,000 glucose monomers, resulting in a high molecular weight polymer. In this regard, the degree of polymerization is responsible for the rigidity of the polymer chain and for the formation of straight and stable nanofibers, which give cellulose a high mechanical strength, thus enabling the insertion of materials for the production of composites (Wahid *et al.*, 2019). Furthermore, CB can be considered a biopolymer with high crystallinity due to its structural repeating unit being relatively simple and linear. This occurs since the degree of crystallization is influenced by aspects such as the chemical composition of the molecules and the configuration of the chain (Callister, 2008; Liu *et al.*, 2019; Yue *et al.*, 2017).

As mentioned, another characteristic linked to bacterial cellulose is its high water retention capacity. This is possible due to the organization of its microfibrils, which form a three-dimensional nanofibrillar network capable of producing a highly porous and hydrophilic structure (Amorim *et al.*, 2020). According to Albuquerque *et al.* (2020), depending on the culture medium, about 97% of the total mass of BC can be composed of water, as can be seen in Figure 11.1.

As can be seen in Figure 11.1, after forced drying, CB loses a large part of its mass due to water evaporation. It is important to emphasize that this elimination can also be done in a natural way, requiring more time. Despite this, the porous structure of bacterial cellulose means that it can be used in the industrial biotechnology and biomedical sectors. Currently, its use has been investigated in the development of wound dressings (Bundjaja *et al.*, 2021), devices for controlled drug release (Meneguin *et al.*, 2020), cosmetics (Fonseca *et al.*, 2021), bioactive packaging (Gedarawatte *et al.*, 2021), and bone tissue regeneration (Klinthoopthamrong *et al.*, 2020).

Given these applications, it is important to know that the BC production process takes place in different ways, depending on the type of application/use intended.

11.3   PRODUCTION OF BACTERIAL CELLULOSE

As mentioned earlier, cellulose is a polysaccharide that, in addition to being abundantly produced by vegetables, can be biosynthesized from different types of organisms, such as algae (*Vallonia*, *Micrasterias rotate*, and *Micrasterias denticulata*), fungi (*Saprolegnia*), and bacteria (*Acetobacter* [*Gluconacetobacter*], *Achromabacter*, *Aerobacter*, and *Agrobacterium*) (Garba *et al.*, 2019; Yang *et al.*, 2019).

Regarding the production of CB, although there is a wide variety of producing organisms, the basic mechanisms involve several biochemical steps, in which different individual enzymes and complexes of catalytic and regulatory proteins act. In general, the process is summarized in the steps of polymerization of monomers and formation of β(1–4) chain from the carbon source and extracellular excretion of linear chains, as well as though the organization and crystallization of glucan chains. The latter occurs from the hydroxyl groups available in the cellulose chains. This is how, though hydrogen bonds, a network of nanofibers is formed (Ross *et al.*, 1991).

Bacteria of the *Komagataeibacter* genus are capable of using a variety of carbon substrates for the synthesis of cellulose, and – depending on the carbon source used — the microorganisms follow different metabolic pathways. For example, when the glucose monosaccharide is used as a substrate, the carbohydrate is initially transported into the bacterial cytoplasm where the following enzymatic steps occur: (1) glucose phosphorylation though the hexokinase enzyme, giving rise to glucose-6-phospate; (2) isomerization of glucose-6-phosphate to glucose-1-phosphate by phosphoglycomutase; (3) synthesis of uridine diphosphoglucose (UDPGlc) though UDPGlc-pyrophosphorylase; and finally, (4) production of cellulose from its precursor UDPGlc by the enzyme cellulose synthase. In cases when disaccharide sources are used, the metabolic pathway of polymer synthesis starts with the hydrolysis of the compound to monosaccharides such as glucose, and then the final steps follow the aforementioned pathway (Ross *et al.*, 1991).

Although CB has a molecular formula and chemical structure similar to VC, it has superior physical properties, which are mainly attributed to the size of the fibers formed and the higher degree of polymerization (Gao *et al.*, 2019). Regarding the production of cellulose by bacterial fermentation in contrast to vegetable cellulose, some benefits should be highlighted. Among them are the fact that it does not depend on regional and climatic conditions, associated with the ability to control the growth of microorganisms, in order to produce the necessary amount in a pre-defined time. Furthermore, for bacterial cellulose, if necessary, it is possible to genetically modify the bacteria in order to produce cellulose with specific properties for different uses (Mondal, 2017).

With regard to the type of BC cultivation, polymer biosynthesis can occur in statis or agitated conditions, and this reflects strictly on the physical and mechanical aspect, as well as on the morphological structure of the product obtained. The most common method is static culture, whereby at the airliquid interface, a uniform gelatinous film is formed with a configuration similar to the surface of the reactor in which it is produced (Gao *et al.*, 2020a). However, in this type of culture, microbiological growth is slower due to the difficulty in transferring oxygen and transporting nutrients to the interior of the bacterial cell. Therefore, the production of bacterial cellulose requires a longer incubation time (Cazón and Vázquez, 2021).

In order to overcome the problems with the cultivation time, some authors suggest the use of agitation to improve the oxygenation of the medium. In this culture mode, the bacterium grows faster and the cellulose is produced in suspension, pellets, or irregular spheres format (Blanco *et al.*, 2018).

However, although the fermentation process under agitation has some advantages and higher productivity, being more attractive at an industrial level due to the time and reaction space required, the membranes formed have a lower degree of polymerization, and consequently, a lower crystallinity index, modulus of elasticity, and mechanical strength.

Thus, the deficiency in such properties can be considered an inconvenience, depending on the application for which it is intended, and for this reason, the pulp produced under this cultivation may show a lower cost/benefit compared to the static ones (Wang *et al.*, 2019). Furthermore, these authors report that in agitated culture, another important aspect that should be highlighted concerns the formation of mutant cells originating from genetic instability that reduce the production.

Consequently, considering these factors, together with the ease of obtainment and biodegradability, the production of this type of cellulose came to be considered a promising and sustainable alternative to vegetable cellulose. Given its physicochemical characteristics, CB started to attract the attention of researchers who aim at the synthesis of biocomposites for use in the fields of medicine, cosmetics production, food packaging, and materials engineering (Ding *et al.*, 2021).

Among some of the characteristics of CB, its high mechanical and thermal stability, tensile strength, large surface area, and water absorption capacity, porosity, and high degree of purity can be highlighted (Gao *et al.*, 2019). About this, Inoue *et al.* (2020) emphasize that purity is one of the main characteristics of bacterial cellulose when compared to other sources of cellulose. This is because, unlike what occurs in the plant cell wall, the CB D-glucopyranose units are not associated with other molecules, such as hemicelluloses, lignin, and pectin.

Therefore, the polysaccharide produced by bacteria does not require chemical treatments for its purification. This occurs because these can cause contamination and damage its structure, generating an additional cost to the process and lead to the formation of harmful organochlorine compounds to the environment. Thus, different culture media have been used in the production of BC.

11.3.1 MEANS OF PRODUCTION

Wang *et al.* (2019) state that the yield and properties of the biopolymer are influenced by several factors such as the type of bacterial strain to be used, the composition of the culture medium, and the conditions applied during the fermentation process, which involve variables such as pH, temperature, time, and culture speed (static or agitated), cell concentration, and volume of the growing medium. The fermentation media commonly used in the production of BC have a defined or complex composition. Complex media usually come from natural substrates, whereas those with a defined chemical composition can be obtained from inorganic salts, for example (Duarte *et al.*, 2019).

For the synthesis of BC, microorganisms are inoculated in culture media that, in general, contain sources of carbon, nitrogen, and phosphorus, in addition to trace elements necessary for microbial growth. Among the most used culture media, there is the Hestrin-Schram (HS) medium, which contains glucose, yeast extract, and peptone as the main source of nitrogen, in addition to citric acid and sodium phosphate (Singhsa *et al.*, 2018).

However, alternatives have been researched to replace this synthetic medium, since it is composed of nutrients of high economic value, making the production process expensive (Revin *et al.*, 2018). Amorim *et al.* (2020) revealed that about 30% of the total cost of the process is associated with the culture medium. Thus, for cell suspension and production of BC, some authors have evaluated the use of culture media composed only of low-cost salts.

In this context, it was possible to observe that there was an increase in the application of residues and byproducts, agribusiness, and food as a nutrient source for BC production (Pacheco *et al.*, 2017). Thus, some scientific studies used residues such as orange peel (Tsouko *et al.*, 2020), cashew tree (Pacheco *et al.*, 2017), tobacco residue extracts (Ye *et al.*, 2019), and juice cashew nuts with soy molasses (Souza *et al.*, 2019), aiming to develop less costly alternative means of cultivation.

In the study developed by Souza *et al.* (2019), for example, the production of bacterial cellulose in alternative cultures was evaluated, using cashew juice with yeast extract (CYE), soy molasses (SJ), and a mixture of cashew juice with soy molasses (CSM) as the source of nutrients. Therefore, the yield of BC produced from the aforementioned culture media were compared with the yield of BC produced in Hestring (HS) medium. Thus, when evaluating the results of the yields obtained, 4.54 2.23, 4.50, and 4.04 g/L^{-1}, respectively, for the CYE, SJ, CSM, and HS media, it was noted that the alternative culture media, based on cashew, showed good yields in BC production, including better than the traditional Hestring medium.

Given the exposure of various means used in the production of bacterial cellulose, the possibility arises of applying the material in different areas and ways.

11.4 APPLICATION

Due to the aforementioned properties and advantages, CB is considered a material with great potential for application and added value. This material is widely used in areas such as food, biomedicine, pharmaceuticals, cosmetics, and bioengineering (Fernandes *et al.*, 2020).

11.4.1 APPLICATION OF BACTERIAL CELLULOSE IN THE FOOD INDUSTRY

Non-biodegradable plastic food packing is responsible for daily generating a large amount of waste that is harmful to the environment, if disposed of incorrectly. For this reason, BC is already being studied as an alternative material to be used in food packaging. Its use is considered thanks to characteristics such as high purity, biodegradability, and production from a biological source (Bandyopadhyay *et al.*, 2019).

In addition, some substances with antimicrobial and antioxidant characteristics can be added to the CB structure for the production of active packaging. As an example, it can be mentioned that the packages produced with bacterial cellulose incorporated with chitosan and polypyrrole (Gao *et al.*, 2020b), nisina (Gedarawatte *et al.*, 2021), and lauric acid (Zahan *et al.*, 2020).

Almeida *et al.* (2013) used blends of CB and potato starch to produce polymeric films and improve the water retention/absorption-desorption capacity, in order to be applied in food packages that need coverage and moisture maintenance. The authors obtained denser films with greater solubility. Rovera *et al.* (2020) used CB nano-crystals obtained by enzymatic hydrolysis to produce a coating with a high oxygen barrier to generate food packaging films.

11.4.2 APPLICATION OF BACTERIAL CELLULOSE IN THE MEDICAL FIELD AND COSMETICS INDUSTRY

The high biocompatibility of bacterial cellulose associated with its non-toxic, anti-allergic characteristics, and its good physical and mechanical parameters, allow use of BC in different biomedical applications, development of pharmaceutical products (Inoue *et al.*, 2020), and use in the cosmetics industry (Bianchet *et al.*, 2020). Since 1980, bacterial cellulose has been investigated and used in would healing, including tissue regeneration and the treatment of severe burns. It is used as a second skin or in the form of dressing and stands out for being able to guarantee an effective physical barrier against infection agents. At the same time, it promotes the maintenance of a humid environment that favors gas exchange and even allows the transfer of medica-tion to the injured tissue (Betlej *et al.*, 2021; Bianchet *et al.*, 2020).

Different authors also describe the potential of bacterial cellulose in drug con-trolled release systems, in which there is great interest in the complete investigation of the interaction between the drug used and the BC that acts as a vehicle (Inoue *et al.*, 2020; Adepu and Khandelwal, 2020; Solomevich *et al.*, 2020). In a study on the transdermal delivery on an anti-inflammatory, Silva *et al.* (2014) was able to obtain high rates of incorporation of sodium diclofenac into bacterial cellulose membranes with diffusion processes and release profiles similar to those found in commercial adhesives. In the work by Ullah *et al.* (2019), the loading and release efficiency of model drugs was evaluated, as the material produced led to high release rates and good drug activity.

In the cosmetics industry, CB is commonly used as an ingredient in skin and hair care products, and as a vehicle for the release of active compounds in facial masks. In this context, Amorim *et al.* (2019) evaluated the production of bacterial cellulose to be applied as facial masks, using tropical fruit residues as an alternative source of nutrients and as an agent capable of preventing skin free radicals. In addition, there are also stud-ies demonstrating its good application as a stabilizer in water–oil emulsions, which is one of the important resources for the cosmetics industry (Bianchet *et al.*, 2020).

11.4.3 APPLICATION OF BACTERIAL CELLULOSE TO PROTECT THE ENVIRONMENT FOR WASTEWATER TREATMENT

Problems related to environmental pollution of water, soil, and air are very present in many countries around the world. There are several substances that can contribute to this scenario, causing countless health problems in humans and animals, in addition to causing disturbances in the ecosystem and climate change. In this scenario, bacterial cellulose, which is a biodegradable and biocompatible polymer, emerges as an alternative in applications for solving environmental problems, such as, for example, in wastewater treatment technologies (Betlej *et al.*, 2021).

One of the attempts to use bacterial cellulose in wastewater treatment was the object of a study in the work developed recently by Silva *et al.* (2021). The researchers evaluated the development of a CB membrane filter for the purification of water contaminated with oily substances. In the experiments, in addition to achieving efficient separation, the CB membrane proved to be highly resistant and durable, enabling its use as a possible industrial biotechnological filter. Another interesting work was presented by Wang *et al.* (2021), who also reported the use of BC in the separation of water-oil systems. In this work, a CB composite was prepared with super-hydrophobic properties, from the addition of kraft pulp (rich in hemicellulose, with low lignin content and high density) bleached vegetable cellulose and copper hydroxide nanoparticles ($Cu[OH]_2$). The resulting composite exhibited excellent water-oil separation efficiency, in addition to good recyclability.

In addition to these applications, CB can be used as a polymer matrix in the formation of nanocomposites, incorporated with catalysts as reinforcement materials, to increase the efficiency of removal and/or degradation of contaminants present in industrial effluents.

11.4.3.1 Structural Modification for Nanocomposite Formation

As already mentioned, the CB allows modifications in its structure, aiming at the improvement of its characteristics, which can be carried out by the in situ and ex situ methods, represented schematically in Figure 11.2.

As shown in Figure 11.2A, the in situ method consists of modifying the structure as the BC is being built. That is, additives are added to the culture medium and as a result, there is the formation of a composite (whose polymeric matrix is bacterial cellulose) with altered morphology and physical-chemical properties (Amorim *et al.*, 2020).

In the ex situ method, unlike in the in situ method, CB modification occurs after its formation. To this end, the different types of additives are impregnated into the bacterial cellulose structure, as shown in Figure 11.2B. This can be done chemically (copolymerization or crosslinking) and physically (absorption). In the latter case, the chemical structure of cellulose is practically unchanged, as there is only an interaction between the polymer matrix and the additive used (Stumpf *et al.*, 2016).

In view of this and considering factors mentioned previously, CB seems to be an interesting alternative for the immobilization of metallic catalysts, which can occur through in situ and ex situ methods (Stumpf *et al.*, 2016). On the other hand, the addition of modifying agents can cause the adjustment of interactions within the polysaccharide chains, which also allows changing the degree of crystallinity and porosity of the BC (Sun *et al.*, 2019). By adding new characteristics and properties to CB forming nanocomposites, these new materials can then be used as catalysts for wastewater treatment processes from different industrial branches.

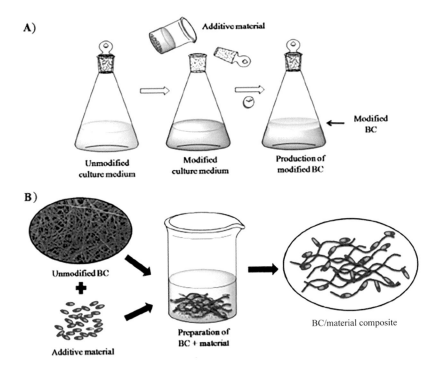

FIGURE 11.2 Schematic representation of the modification methods: A) in situ; B) ex situ of the CB (adapted from Stumpf *et al.* [2016]).

11.4.3.2 Application of Bacterial Cellulose-Based Nanocomposites for Wastewater Treatment

In addition to the aforementioned applications, CB can be used as a polymer matrix in the formation of nanocomposites, reinforced with substances capable of degrading and/or removing persistent pollutants present in industrial effluents. Textiles, food, pharmaceuticals, petrochemicals, paints, and mining can be mentioned as examples of industries which are responsible for generating large amounts of effluents containing different types of contaminants (Jain and Gogate, 2018; Chen *et al.*, 2019; Cristóvão *et al.*, 2019; Bhattacharjee *et al.*, 2020).

Among the pollutants that deserve attention are synthetic dyes, drugs, pesticides, polycyclic aromatic hydrocarbons, and heavy metals, which are difficult to remove and/or degrade, and therefore persistent in the environment. These pollutants are potentially toxic to the aquatic environment and to humans, as they are carcinogenic, mutagenic and, in the case of heavy metals, can also be bioaccumulative (Suganya *et al.*, 2020; Aruna *et al.*, 2021).

Therefore, the use of nanocomposite materials – in photocatalytic and/or adsorptive processes, for example – emerges as an interesting alternative, as it avoids the leaching of catalysts and the possible contamination of the aquatic environment by these substances. In addition, the use of nanocomposites as catalysts and adsorbents facilitates their removal from the medium and their subsequent reuse (Oyewo *et al.*, 2020).

Núñez *et al.* (2020) studied the development of a new composite biomaterial, by the in situ method, using CB nanofibers as polymeric matrix and hydroxyapatite nanocrystals, obtained from mollusks shells, as reinforcement material. This biomaterial aimed to remove the heavy metal Pb (II) from an aqueous solution, with an adsorption capacity of 192.3 mg/g^{-1}. This result was possible due to the structure of hydroxyapatite being a bioceramic material with a high content of calcium ions, which – due to its atomic radius similar to that of Pb (II) – is able to increase the availability of active sites through the exchange of cations.

Li *et al.* (2019), for example, developed a new multifunctional composite of CB, chitosan, and zeolitic-67 imidazolate, a type of organic metal structure, capable of adsorbing contaminants such as heavy metals, Cu (II), Cr (VI), and the wastewater red dye X-3B. The material presented adsorptive capacities (q), after 24 hours of testing, equal to 200.6 and 152.1 mg.g^{-1} for Cu (II) and Cr (VI), respectively, and a removal efficiency of almost 100% of the studied dye. In addition to this work, several studies are being carried out with CB-based nanocomposites, reinforced with different types of materials, aiming at their application in adsorptive processes, as shown in Table 11.1.

Another CB-based nanocomposite material, reinforced with polydopamine (PDA) and titanium dioxide (TiO$_2$), was developed by the ex situ method by Yang *et al.* (2020). This study aimed to evaluate the efficiency of the CB/PDA/TiO$_2$ nanocomposite for the adsorption and degradation of organic dyes such as methyl orange (MO), methylene blue (MB), and rhodamine B (RhB). It was verified that the addition of PDA increased the adsorptive capacity (q) of the CB/PDA/TiO$_2$ nanocomposite in relation to the CB/TiO$_2$ nanocomposite, obtaining values of q = 11.8, 13.4, and 28.9 mg.g^{-1} for LM, AM, and RhB, respectively. As for the heterogeneous photocatalysis, using UV radiation, a degradation efficiency above 95% was obtained for all studied dyes, in a maximum time of 60 min.

Li *et al.* (2017) developed a CB-based nanocomposite material, combining the biocatalytic properties of the laccase enzyme with the photocatalytic properties of TiO$_2$ for the degradation of the reactive X-3B textile dye. Given the best operating conditions (pH = 5 and temperature = 40°C), the authors evaluated the dye degradation

TABLE 11.1

Application of CB-Based Nanocomposites, with Different Types of Reinforcement Materials and Their Respective Contaminants

Reinforcement Material	Contaminant	Removal Efficiency	References
2-acetyl-1-pyrroline	Cr (VI)	83.06%	Muhamad *et al.* (2020)
Chitosan	Cu (II)	55.00%	Urbina *et al.* (2018)
Polyaniline	Cr (VI)	85.00%	Hosseini and Mousavi (2021)

when using the developed material via a photocatalytic process with UV-C radiation and obtained an efficiency of 95% after two hours of experimentation.

Wahid *et al.* (2019) used zinc oxide (ZnO) to produce nanocomposite materials based on bacterial cellulose (CB/ZnO), by the ex situ method, with photocatalytic and antibacterial activities. The authors verified that the CB/ZnO nanocomposites present a degradation efficiency of approximately 91% of the methyl orange dye, after two hours of experimentation and under UV irradiation. In addition, the material produced also demonstrated antibacterial activities for gram-negative and gram-positive bacteria.

In this context, it was observed that BC has favorable properties for the formation of nanocomposites, reinforced with catalysts capable of assisting in the process of removal/degradation of contaminants from industrial effluents. These materials have the advantages of being environmentally correct, avoiding costs with the post-use catalyst removal processes, and the high efficiency of pollutant removal, as shown in the aforementioned works.

11.5   CONCLUSION

Given the material presented in this chapter, it was possible to observe that bacterial cellulose (BC) is a material that has been widely explored in different areas of application. This is mainly due to the fact that BC has good compatibility with several substances, being a biodegradable polymer produced from biological sources. Therefore, the application of CB as a polymer matrix for the formation of nanocomposites, with catalytic and adsorptive properties, has increasingly attracted the attention of researchers in the environmental area. As stated in the aforementioned works, these nanocomposites have high efficiency of degradation and/or removal of contaminants present in industrial effluents, proving to be a material with great potential for this application. In addition, it is noteworthy that these nanocomposites facilitate the recovery of catalytic substances in wastewater decontamination processes, reducing the operational costs for removal of catalysts.

REFERENCES

Adepu, S, Khandelwal, M. (2020) Ex-situ modification of bacterial cellulose for immediate and sustained drug release with insights into release mechanism. *Carbohydrate Polymers*, v.249, 116816. DOI: 10.1016/j.carbpol.2020.116816.

Albuquerque, R. M. B, Meira, H. M, Silva, I. D. L, Silva, C. J. G, Almeida, F. C. G, Amorim, J. D. P, Vinhas, G. M, Costa, A. F. S, Sarubbo, L. A. Production of a bacterial cellulose/poly(3-hydroxybutyrate) blend activated with clove essential oil for food packaging. *Polymers & polymer composites*, v. 1, p.259–270, 2020. DOI: 10.1177/0967391120912098

Almeida, D. M, Woiciechowski, A. L, Wosiacki, G, Prestes, R. A, Pinheiro, L. A. Propriedades físicas, químicas e de barreira em filme formados por blenda de celulose bacteriana e fécula de batata. *Polímeros*, v.23, n 4, p.538–546, 2013. DOI: 10.4322/polimeros.2013.038

Amorim, J. D. P, Costa, A. F. S, Galdino, C. J. S, Santos, E. M. S, Sarubbo, I. A. Bacterial cellulose production using fruit residues as substract to industrial application. *Chemical Engineering Transactions*, v.74, p.1165–1170, 2019. DOI: 10.3303/CET1974195.

Amorim, J. D. P, Souza, K. C, Duarte, C. R, Duarte, I. S, Ribeiro, F. A. S, Silva, G. S, Farias, P. M. A, Stingl, A, Costa, A. F. S, Vinhas, G. M, Sarubbo, L. A. Plant and bacterial nanocellulose: production, properties and applications in medicine, food, cosmetics, electronics and engineering. A review. *Environmental Chemistry Letters*, v. 18, p. 851–869, 2020. DOI:10.1007/s10311-020-00989-9.

Aruna, Bagotia, N, Sharma, A. K, Kumar, S. A review on modified sugarcane bagasse biosorbent for removal of dyes. *Chemosphere*, v.268, 129309, 2021. DOI: 10.1016/j.chemosphere.2020.129309.

Azeredo, H, Barud, H, Farinas, C. S, Vasconcellos, V. M, Claro, A. M. Bacterial cellulose as a raw material for food and food packaging applications. *Frontiers in Sustainable Food Systems*, v. 3, p. 1–14, 2019. DOI:10.3389/fsufs.2019.00007.

Bandyopadhyay, S, Saha, N, Brodnjak, U. V, Sáha, P. Bacterial cellulose and guar gum based modified PVP-CMC hydrogel films: Characterized for packaging fresh berries. *Food Packaging and Shelf Life*, v.22, 100402, 2019. DOI: 10.1016/j.fpsl.2019.100402.

Betlej, I, Zakaria, S, Krajewski, K. J, Boruszewski, P. Bacterial Cellulose – Properties and Its Potential Application. *Sains Malaysiana*, v. 50, n. 2, p. 493–505, 2021. DOI: 10.17576/jsm-2021–5002–20.

Bhattacharjee, C, Dutta, S, Saxena, V. K. A review on biosorptive removal of dyes and heavy metals from wastewater using watermelon rind as biosorbent. *Environmental Advances*, v.2, 100007, 2020. DOI: 10.1016/j.envadv.2020.100007.

Bianchet, R. T, Cubas, A. L. V, Machado, M. M, Moecke, E. H. S. Applicability of Bacterial Cellulose in Cosmetics – Bibliometric Review. *Biotechnology Reports*, v. 27, p. 1–25, 2020. DOI: 10.1016/j.btre.2020.e00502.

Blanco, A, Monte, M.C, Campano, C. Balea, A. Merayo, N. Negro, C. Nanocellulose for industrial use: Cellulose nanofibers (CNF), cellulose nanocrystals (CNC), and bacterial cellulose (BC) *Handbook of Nanomaterials for Industrial Applications*, Elsevier: Newark, NJ, USA, p. 74–126, 2018. DOI: 10.1016/B978-0-12-813351-4.00005–5.

Brown, A. J. An acetic ferment which forms cellulose. *Journal of the Chemical Society*, v.49, p.432–439, 1886. DOI: 10.1039/CT8864900432.

Bundjaja, V. Santoso, S. P. Angkawijaya, A. E. Yuliana, M., Soetaredjo, F. E., Ismadji, S., Ayucitra, A., Gunarto, C., Ju, Y., Ho, M. H. Fabrication of cellulose carbamate hydrogel-dressing with rarasaponin surfactant for enhancing adsorption of silver nanoparticles and antibacterial activity. *Materials Science and Engineering: C*, v. 118, p. 111542, 2021. DOI: 10.1016/j.msec.2020.111542.

Callister, WD (2008) *Ciências e engenharia de materiais: uma introdução*. LTC, 7th ed., 705. ISBN:978-85-216-1595-8.

Cazón, P., Vázquez, M. Improving bacterial cellulose films by ex-situ and in-situ modifications: a review. *Food Hydrocolloids*, v. 113, p. 106514, 2021. DOI: 10.1016/j.foodhyd.2020.106514.

Chen, X., Zhou, Q., Liu, F., Peng, Q., Teng, P. Removal of nine pesticide residues from water and soil by biosorption coupled with degradation on biosorbent immobilized laccase. *Chemosphere*, v.233, p.49–56, 2019. DOI: 10.1016/j.chemosphere.2019.05.144.

Cristóvão, M. B., Torrejais, J., Janssens, R., Luis, P., Der Bruggen, B. V., Dubey, K. K., Mandal, M. K., Bronze, M. R., Crespo, J. G., Pereira, V. J. Treatment of anticancer drugs in hospital and wastewater effluents using nanofiltration. *Separation and Purification Technology*, v.224, p.273–280, 2019. DOI: 10.1016/j.seppur.2019.05.016.

Ding, R., Hu, S., Xu, M., Hu, Q., Jiang, S., Xu, K., Tremblay, P., Zhang, T. The facile and controllable synthesis of a bacterial cellulose/polyhydroxybutyrate composite by co-culturing Gluconacetobacter xylinus and Ralstonia eutropha. *Carbohydrate Polymers*, v. 252, p. 117137, 2021. DOI: 10.1016/j.carbpol.2020.117137.

Duarte ÉB, Andrade FK, Lima HLS, Nascimento ES, Carneiro MJM, Borges MF, Luz EPCG, Chagas BS, Rosa MF (2019) *Celulose bacteriana propriedades, meios fermentativos e aplicações*. Embrapa Agroindústria Tropical, 1st ed., 35. ISSN:2179-8184.

Fernandes, I. A. A., Pedro, A. C., Ribeiro, V. R., Bortolini, D. G., Ozaki, M. S. C., Maciel, G. M., Haminiuk, C. W. I. Bacterial cellulose: From production optimization to new applications. *International Journal of Biological Macromolecules*, v.164, p.2598–2611, 2020. DOI: 10.1016/j.ijbiomac.2020.07.255.

Fonseca, D. F., Vilela, C., Pinto, R. J., Bastos, V., Oliveira, H., Catarino, J., Faísca, P., Rosado, C., Silvestre, A. J. D., Freire, C. S. Bacterial nanocellulose-hyaluronic acid microneedle patches for skin applications: In vitro and in vivo evaluation. *Materials Science and Engineering: C*, v. 118, 111350, 2021. DOI:10.1016/j.msec.2020.111350.

Fontana, J. D., Koop, H. S., Tiboni, M., Grzybowski, A., Pereira, A., Kruger, C. D., Silva, M. G. R., Wielewski, L. P. New Insights on Bacterial Cellulose. *Food Biosynthesis*, p. 213–249, 2017. DOI: 10.1016/B978-0-12-811372-1.00007–5.

Gallegos, A. M. A., Carrera, S. H., Parra, R., Keshavarz, T., Iqbal, H. M. N. Bacterial Cellulose: A Sustainable Source to Develop Value-Added Products – A Review. *BioResources*, v. 11, n. 2, p. 5641–5655, 2016. DOI: 10.15376/biores.11.2.Gallegos.

Gao, H., Sun, Q., Han, Z., Li, J., Liao, B., Hu, L., Huang, J., Zou, C., Jia, C., Huang, J., Chang, Z., Jiang, D., Jin, M. Comparison of bacterial nanocellulose produced by different strains under static and agitated culture conditions. *Carbohydrate Polymers*, v.227, 115323, 2020 (a) DOI: 10.1016/j.carbpol.2019.115323.

Gao, M., Li, J., Bao, Z., Hu, M., Nian, R., Feng, D., An, D., Li, X., Xian, M., Zhang, H. A natural in situ fabrication method of functional bacterial cellulose using a microorganism. *Nature Communications*, v. 10, n. 437, 2019. DOI: 10.1038/s41467-018-07879-3.

Gao, Q., Lei, M., Zhou, K., Liu, X., Wang, S., Li, H. Preparation of a microfibrillated cellulose/chitosan/polypyrrole film for Active Food Packaging. *Progress in Organic Coatings*, v.149, 105907, 2020 (b) DOI: 10.1016/j.porgcoat.2020.105907.

Garba, Z. N., Lawan, I., Zhou, W., Zhang, M., Wang, L., Yuan, Z. Microcrystalline cellulose (MCC) based materials as emerging adsorbents for the removal of dyes and heavy metals-A review. *Science of the Total Environment*, v.717, 135070, 2019. DOI: 10.1016/j. scitotenv.2019.135070 .

Gedarawatte, S. T., Ravensdale, J. T., Al-Salami, H., Dykes, G. A., Coorey, R. Antimicrobial efficacy of nisin-loaded bacterial cellulose nanocrystals against selected meat spoilage lactic acid bacteria. *Carbohydrate Polymers*, v. 251, 117096, 2021. DOI: 10.1016/j. carbpol.2020.117096.

Gromovykh, T. I., Sadykova, V. S., Lutcenko, S. V., Dmitrenok, A. S., Feldman, N. B., Danilchuk, T. N., Kashirin, V. V. Bacterial Cellulose Synthesized by *Gluconacetobacter hansenii* for Medical Applications. *Applied biochemistry and Microbiol*, v. 53, n. 1, p. 60–67, 2017. DOI: 10.1134/s0003683817010094.

Hosseini, H., Mousavi, S. M. Bacterial cellulose/polyaniline nanocomposite aerogels as novel bioadsorbents for removal of hexavalent chromium: Experimental and simulation study. *Journal of Cleaner Production*, v.278, 123817, 2021. DOI: 10.1016/j. jclepro.2020.123817.

Inoue, B. S., Streit, S., Schneider, A. L. S., Meier, M. M. Bioactive bacterial cellulose membrane with prolonged release of chlorhexidine for dental medical application. *International Journal of Biological Macromolecules*, v. 148, p. 1098–1108, 2020. DOI: 10.1016/j.ijbiomac.2020.01.036.

Jain, S. N., Gogate, P. R. Efficient removal of Acid Green 25 dye from wastewater using activated Prunus Dulcis as biosorbent: Batch and column studies. *Journal of Environmental Management*, v. 210, p.226–238, 2018. DOI: 10.1016/j.jenvman.2018.01.008.

Klinthoopthamrong, N., Chaikiawkeaw, D., Phoolcharoen, W., Rattanapisit, K., Kaewpungsup, P., Pavasant, P., Hoven, V. P. Bacterial cellulose membrane conjugated with plant-derived osteopontin: Preparation and its potential for bone tissue regeneration. *International Journal of Biological Macromolecules*, v. 149, p. 51–59, 2020. DOI: 10.1016/j.ijbiomac.2020.01.158.

Li, D., Tian, X., Wang, Z., Guan, Z., Li, X., Qiao, H., Ke, H., Luo, L., Wei, Q. Multifunctional adsorbent based on metal-organic framework modified bacterial cellulose/chitosan composite aerogel for high efficient removal of heavy metal ion and organic pollutant. *Chemical Engineering Journal*, v.383, 123127, 2019. DOI: 10.1016/j.cej.2019.123127.

Li, G., Nandgaonkar, A. G.; Wang, Q., Zhang, J., Krause, W. E., Wei, Q., Lucia, L. A. Laccase-immobilized bacterial cellulose/TiO₂ functionalized composite membranes: Evaluation for photo- and bio-catalytic dye degradation. *Journal of Membrane Science*, v.525, p.89–98, 2017. DOI: 1016/j.memsci.2016.10.033.

Liu, H., Hu, Y., Zhu, Y., Wu, X., Zhou, X., Pan, H., Chen, S., Tian, P. A simultaneous grafting/vinyl polymerization process generates a polycationic surface for enhanced antibacterial activity of bacterial cellulose. *International Journal of Biological Macromolecules*, v.143, p. 224–234, 2019. DOI: 10.1016/j.ijbiomac.2019.12.052.

Meneguin, A. B., Barud, H. S., Sábio, R. M., Sousa, P. Z., Manieri, K. F., Freitas, L. A. P., Pacheco, G., Alonso, J. D., Chorilli, M. Spray-dried bacterial cellulose nanofibers: A new generation of pharmaceutical excipient intended for intestinal drug delivery. *Carbohydrate Polymers*, v. 249, 116838, 2020. DOI: 10.1016/j.carbpol.2020.116838.

Mondal, S. Preparation, properties and applications of nanocellulosic materials. *Carbohydrate Polymers*, v. 163, p. 301–316.2017. DOI: 10.1016/j.carbpol.2016.12.050.

Muhamad, I. I., Muhamad, S. N. H., Salehudin, M. H., Zahan, K. A., Tong, W. Y., Pa'e, N. Effect of pandan extract concentration to chromium (IV) removal using bacterial cellulose-pandan composites prepared by in-situ modification technique. *Materials Today: Proceedings*, v.31, p.89–95, 2020. DOI: 10.1016/j.matpr.2020.01.204.

Núñez, D., Cáceres, R., Ide, W., Varaprasad, K., Oyarzún, P. An ecofriendly nanocomposite of bacterial cellulose and hydroxyapatite efficiently removes lead from water. *International Journal of Biological Macromolecules*, v.165B, p.2711–2720, 2020. DOI: 10.1016/j.ijbiomac.2020.10.055.

Oyewo, O. A., Elemike, E. E., Onwudiwe, D. C., Onyango, M. S. Metal oxide-cellulose nanocomposites for the removal of toxic metals and dyes from wastewater. *International Journal of Biological Macromolecules*, v.164, p.2477–2496, 2020. DOI: 10.1016/j.ijbiomac.2020.08.074.

Pacheco, G., Nogueira, C. R., Meneguina, A. B., Trovatti, E., Silva, M. C. C., Machado, R. T. A., Ribeiro, S. J. L., Filho, E. C. S., Barud, H. S. Development and characterization of bacterial cellulose produced by cashew tree residues as alternative carbon source. *Industrial Crops & Products*, v.107, p.13–19, 2017. DOI: 10.1016/j.indcrop.2017.05.026.

Parte, F. G. B., Santoso, S. P., Chou, C. C., Verma, V., Wang, H. T., Ismadji, S., Cheng, K. C. Current progress on the production, modification, and applications of bacterial cellulose. *Critical Reviews in Biotechnology*, v. 40, n. 3, p. 397–414, 2020. DOI: 10.1080/07388551.2020.1713721.

Revin, V., Liyaskin, E., Nazarkina, M., Bogatyreva, A., Shchankin, M. Cost-effective production of bacterial cellulose using acidic food industry by-products. *Brazilian Journal of Microbiol*, v. 49, p. 151–159, 2018. DOI: 10.1016/j.bjm.2017.12.012.

Ross, P., Mayer, R., Benziman, M. Biossíntese de celulose e função em bactérias. *Microbiol and Molecular Biology Reviews*, v. 55, n. 1, p. 35–58, 1991. DOI: 10.1128/mr.55.1.35–58.1991

Rovera, C., Fiori, F., Trabattoni, S., Romano, D., Farris, S. Enzymatic Hydrolysis of Bacterial Cellulose for the Production of Nanocrystals for the Food Packaging Industry. *Nanomaterials*, v. 10, n. 4, p. 735, 2020. DOI: 10.3390/nano10040735.

Silva, C.J.G., Medeiros, A. D. M., Amorim, J. D. P., Nascimento, H. A., Converti, A., Costa, A.F. S., Sarubbo, L. A. Bacterial cellulose biotextiles for the future of sustainable fashion: a review. *Environmental Chemistry Letters*, 2021. DOI: 10.1007/s10311–021–01214-x

Silva, N. H. C. S., Rodrigues, A. F., Almeida, I. F., Costa, P., Rosado, C., Neto, C. P., Silvestre, A. J. D., Freire, C. S. R. Bacterial cellulose membranes as transdermal delivery systems for diclofenac: In vitro dissolution and permeation studies. *Carbohydrate Polymers*, v. 106, p. 264–269, 2014. DOI: 10.1016/j.carbpol.2014.02.014.

Singhsa, P., Narain, R., Manuspiya, H. Physical structure variations of bacterial cellulose produced by different Komagataeibacter xylinus strains and carbon sources in static and agitated conditions. *Cellulose*, v. 25, p. 1571–1581, 2018. DOI: 10.1007/s10570-018-1699-1.

Solomevich, S. O., Dmitruk, E. I., Bychkovsky, P. M., Nebytov, A. E., Yurkshtovich, T. L., Golub, N. V. Fabrication of oxidized bacterial cellulose by nitrogen dioxide in chloroform/cyclohexane as a highly loaded drug carrier for sustained release of cisplatin. *Carbohydrate Polymers*, v.248, 116745, 2020. DOI: 10.1016/j.carbpol.2020.116745.

Souza, E. F., Furtado, M. R., Carvalho, C.W.P., Freitas-Silva, O., Gottshalk, L.M.F. Production and characterization of Gluconacetobacter xylinus bacterial cellulose using cashew apple juice and soybean molasses. *International Journal of Biological Macromolecules*, v.146, p.285–289, 2019. DOI: 10.1016/j.ijbiomac.2019.12.180.

Stumpf, T. R., Yang, X., Zhang, J., Cao, X. In situ and ex situ modifications of bacterial cellulose for applications in tissue engineering. *Materials Science & Engineering C*, v.82, p. 372–383, 2016. DOI: 10.1016/j.msec.2016.11.121.

Suganya, E., Saranya, N., Sivaprakasam, S., Varghese, L. A., Narayanasamy, S. Experimentation on raw and phosphoric acid activated Eucalyptuscamadulensis seeds as novel biosorbents for hexavalent chromium removal from simulated and electroplating effluents. *Environmental Technology & Innovation*, v.19, 100977, 2020. DOI: 10.1016/j.eti.2020.100977.

Sun, B., Zhang, L., Wei, F., Al-Ammari, A., Xu, X., Li, W., Chen, C., Lin, J., Zhang, H., Sun, D. In situ structural modification of bacterial cellulose by sodium fluoride. *Carbohydrate Polymers*, v.23, 115765, 2019. DOI: 10.1016/j.carbpol.2019.115765.

Tsouko, E., Maina, S., Ladakis, D., Kookos, I. K., Koutinas, A. Integrated biorefinery development for the extraction of value-added components and bacterial cellulose production from orange peel waste streams. *Renewable Energy*, v.160, p.944–954, 2020. DOI: 10.1016/j.renene.2020.05.108.

Ullah, H., Badshah, M., Correia, A., Wahid, F., Santos, H. A., Khan, T. Functionalized Bacterial Cellulose Microparticles for Drug Delivery in Biomedical Applications. *Current Pharmaceutical Design*, v. 25, p. 3692–3701, 2019. DOI: 10.2174/138161282 5666191011103851

Urbina, L., Guaresti, O., Requies, J., Gabilondo, N., Eceiza, A., Corcuera, M. A., Retegi, A. Design of reusable novel membranes based on bacterial cellulose and chitosan for the filtration of copper in wastewaters. *Carbohydrate Polymers*, v.193, p.362–372, 2018. DOI: 10.1016/j.carbpol.2018.04.007.

Wahid, F., Duan, Y., Hu, X., Chu, L., Jia, S., Cui, J., Zhong, C. A facile construction of bacterial cellulose/ZnO nanocomposite films and their photocatalytic and antibacterial properties. *International Journal of Biological Macromolecules*, v. 132, p. 692–700, 2019. DOI: 10.1016/j.ijbiomac.2019.03.240.

Wang, F., Zhao, X., Wahid, F., Zhao, X., Qin, X., Bai, H., Xie, Y., Zhong, C. Sustainable, Superhydrophobic Membranes Based on Bacterial Cellulose for GravityDriven Oil/Water Separation. *Carbohydrate Polymers*, v. 253, p. 117–220, 2021. DOI:10.1016/j.carbpol.2020.117220.

Wang, J., Tavakoli, J., Tang, Y. Bacterial cellulose production, properties and applications with different culture methods – A review. *Carbohydrate Polymers*, v. 219, p. 63–76, 2019. DOI: 10.1016/j.carbpol.2019.05.008.

Xie, Y.-Y., Hu, X.-H., Zhang, Y.-W., Wahid, F., Chu, L.-Q., Jia, S.-R., Zhong, C. Development and antibacterial activities of bacterial cellulose/graphene oxide-CuO nanocomposite films. *Carbohydrate Polymers*, v. 229, 115456, 2020. DOI: 10.1016/j.carbpol.2019.115456.

Yang, H. J., Lee, T., Kim, J. R., Choi, Y., Park, C. Improved production of bacterial cellulose from waste glycerol through investigation of inhibitory effects of crude glycerol-derived compounds by *Gluconacetobacter xylinus*. *Journal of Industrial and Engineering Chemistry*, v. 75, p.158–163, 2019. DOI: 10.1016/j.jiec.2019.03.017.

Yang, L.; Chen, C., Hu, Y., Wei, F., Cui, J., Zhao, Y., Xu, X., Chen, X., Sun, D. Three-dimensional bacterial cellulose/polydopamine/TiO$_2$ nanocomposite membrane with enhanced adsorption and photocatalytic degradation for dyes under ultraviolet-visible irradiation. *Journal of Colloid and Interface Science*, v.562, p.21–28, 2020. DOI: 10.1016/j.jcis.2019.12.013.

Ye, J., Zheng, S., Zhang, Z., Yang, F., Ma, K., Feng, Y., Zheng, J., Mao, D., Yang, X. Bacterial cellulose production by Acetobacter xylinum ATCC 23767 using tobacco waste extract as culture médium. *Bioresource Technology,* v.274, p.518–524, 2019. DOI: 10.1016/j.biortech.2018.12.028.

Yue, L., Xie, Y., Zheng, Y., He, W., Guo, S., Sun, Y., Zhang, T., Liu, S. Sulfonated bacterial cellulose/polyaniline composite membrane for use as gel polymer electro-lyte. *Composites Science and Technology*, v. 145, p. 122–131, 2017. DOI: 10.1016/j.compscitech.2017.04.002.

Zahan, K. A., Azizul, N. M., Mustapha, M.; Tong, W. Y., Abdul Rahman, M. S., Sahuri, I. S. Application of bacterial cellulose film as a biodegradable and antimicrobial packaging material. *Materials Today: Proceedings*, v.31, p.83–88, 2020. DOI: 10.1016/j.matpr.2020.01.201.

12 Application of Marine Biofilms

An Emerging Thought to Explore

Md. Foysul Hossain, Jakir Hossain
and Roksana Jahan

CONTENTS

DOI: 10.1201/9781003184942-14

12.1   INTRODUCTION

Marine environments are the most diversified and dynamic habitats for the survival of marine biota. Marine microorganisms, the microbial warriors, establish their survival in marine life through forming communities or colonization commonly termed as biofilms. Microorganisms such as bacteria, diatoms, algae, and protozoa frequently form biofilms, which are high density group cells contained in an extra-cellular matrix and possessing microcolony structures or other multicellular arrange-ments, an adaptive way to meet the environmental challenges (Egan *et al.*, 2008). Biofilm formation is a protective skill for colony-forming microorganisms, and also it has ecological and biogeochemical aspects in the changing marine environment. Though having some negative effects, marine biofilm is very promising for good pur-poses in some sectors like marine aquaculture, wastewater treatment, bioremediation of pollutants, antibiofouling, biotechnology, medicine, ecotoxicology, and so on – but some problems are still to be resolved and new sectors, along with problems, are aris-ing day by day. Conventional microbes or biofilms-based technologies have short-comings to address those challenges. To settle the existing problems and address the upcoming issues, it is crucial to find out those types of microbes containing biofilm that have diverse characteristics and can prevail in unfavorable conditions. As is well known, due to physical and chemical differences, such as temperature – with a range that goes from the extreme low values of Antarctic waters to the 350°C of hydrother-mal vents – and pressure (1–1,000 atm), the biofilms found in marine environments have intriguing properties to adapt and flourish in these adverse conditions (Jha and Zi-Rong, 2004). Diverse marine biofilms form protective microenvironments in these dynamic and incessantly changing environments and support a variety of microbial processes (Decho, 2000). For the aforementioned properties, it is safe to say that application of marine biofilm could be the best solution to tackle the chal-lenges. It has been evidently realized that marine biofilms have the upper hand over biofilms in particular sectors due to their diversity and ability to thrive in the most adverse conditions on the planet.

   As of now, marine biofilm has been applied in distinctive sectors, some places in full scale and limited scale in other places, but is yet to be materialized in full-blown scale. Already, several experiments have been conducted and many more are being carried out to explore the potential of marine biofilm by applying it in different arenas. Marine biofilms are being widely used in marine aquaculture for feedstock produc-tion, wastewater treatment and disease control (Thompson *et al.*, 2002; Rusten *et al.*, 2006; Azad *et al.*, 2000). The outcomes of applying marine biofilms in these sectors are highly impressive. Moreover, biofilm is being considered prominently for pro-ducing antifouling metabolites to control marine biofouling on natural and human-made marine structures like nets, ship hulls and ballast tanks, etc. (Qian *et al.*, 2009;

Wang *et al.*, 2017b; Cho and Kim, 2012). Besides, biodegradation and bioremediation of persistent organic pollutants, wastewater treatment including anammox, removing petroleum hydrocarbons, ameliorating ecosystems by protecting bacteria from harsh environmental conditions, and removing pollutants from open waters can be done through application of marine biofilm (Abdel-Raouf *et al.*, 2012; Tyagi *et al.*, 2011; de Carvalho and Caramujo, 2012; Zhu *et al.*, 2011). So, marine biofilm has a great potential to flourish in these sectors. The main focus of this chapter, therefore, is to explore the immense prospects and potentiality of the application of marine biofilms in marine aquaculture, marine antifouling, environmental bioremediation, wastewater treatment and environmental restoration for addressing several issues.

12.2 APPLICATION OF MARINE BIOFILMS IN AQUACULTURE

12.2.1 MARINE BIOFILMS AS FOOD SOURCES

Although there are many successful applications of biofilm in freshwater aquaculture, biofilm acts as major food source in marine aquaculture, as well (Azim and Wahab, 2005; Asaduzzaman *et al.*, 2008). Microorganisms in the biofilm consumed by marine organisms provide essential nutrients (i.e. sterols, essential fatty acids, amino acids and vitamins) that make them an important complementary food source (Thompson *et al.*, 2002). For instance, microalgae (*Thalassiosira*, *Chaetoceros*, *Navicula*, etc.) and bacteria – major components of biofilm – are familiar for their nutritional benefits (i.e. vitamins, carotenoids, etc.) and widely used as dietary catalysts to enhance shrimp growth, including stimulating immune systems, increasing stress tolerance and enhancing embryonic development, growth and gonadal maturation (Linan-Cabello *et al.*, 2002; Ju *et al.*, 2009). Beside this, epiphytes also enhanced shrimp growth by supplementing carbon and nitrogen in their diets (Burford *et al.*, 2004).

The three most abundant bacterial phyla in the biofilm are Proteobacteria, Bacteroidetes and Planctomycetes, associated with nine orders: Rhizobiales, Clostridiales, Cytophagales, Actinomycetales, Vibrionales, Flavobacterales, Planctomycetales, Chlamydiales and Rhodobacterales (Ortiz-Estrada *et al.*, 2019; Lee *et al.*, 2008; Gao *et al.*, 2016). Marine microalgae (*Amphora*, *Campylopyxis*, *Navicula*, *Synedra*, *Hantschia* and *Cylindrotheca*) were also important component of biofilm including filamentous cyanobacteria (*Oscillatoria* and *Spirulina*), bacteria, flagellates and ciliates in intensive marine aquaculture systems (i.e. *Farfantepenaeus paulensis*) (Thompson *et al.*, 2002). Moreover, centric diatom and nematodes could be important sources of lipids and protein in the biofilm (Abreu *et al.*, 2007). The highest growth (1.03–1.72 g/week^{-1}) of juvenile shrimp (*F. paulensis)* could be achieved where nematodes were the major components in marine biofilms (Pissetti, 2004).

12.2.2 MARINE BIOFILMS FACILITATE LARVAL SETTLEMENT

Biofilms play an important role on the larval settlement processes of marine invertebrate taxa such as polychaetes, mollusks, sponges, annelids, echinoderms, phoronids, corals, etc. (Woollacott and Hadfield, 1996; Unabia and Hadfield, 1999).

Bacteria and algae in biofilms have been shown to be responsible for the induction of larval settlement and metamorphosis in some species (Kirchman and Mitchell, 1983; Unabia and Hadfield, 1999). For instance, coralline red algae (i.e. Corallinales, Rhodophyta) can enhance larval settlement of some species of abalone (i.e. *Haliotis rufescens, H. iris, H. virginea* and *H. rubra*) (Daume *et al.*, 1997, 1999); bacteria diatoms and mucus, living on those algae, could enhance settlement. Biofilms also induced larval settlement of the pearl oyster *Pinctada maxima* and *Crassostrea virginica*, although it is opaque whether inductive cues initiated from diatoms or bacteria, or both (Weiner, 1993; Zhao *et al.*, 2003). Note that, oyster larvae can respond similarly to waterborne substances released by either adult oysters or bacterial biofilms (Weiner, 1993).

The composition of biofilms varies in accordance with several environmental factors shown in Figure 12.1 like temperature, salinity, nutrients, season, tidal level, biofilm age, types of substratum, etc. (Lee *et al.*, 2014; Wang *et al.*, 2012; Chung *et al.*, 2010). For instance, biofilm age and type of substratum alter microbial community composition and chemical profiles of the biofilms (Chung *et al.*, 2010). Older biofilms induced comparatively higher levels of settlement and metamorphosis of pediveliger larvae of *Mytilus coruscus* because biofilm age was positively correlated with bacterial and diatom densities that had a vital role in larval settlement (Wang *et al.*, 2012). Note that some marine matured biofilms can help to form a pattern of complex molecules and morphogenic signaling compounds that might have a greater role on the larval settlement in marine invertebrates (Hadfield, 2011). Elevated seawater temperature associated with biofilm age had remarkable positive influence on biofilm community composition that ultimately contributed to settlement for the coral reef sponge larvae *Rhopaloeides odorabile* (Whalan and Webster, 2014). Sea urchin (*Heliocidaris erythrogramma* and *H. purpurascens*) larval settlement depended on epiphytic bacterial community composition on the benthic coralline algae (Nielsen *et al.*, 2015). Settlement of the barnacle (*Balanus amphitrite*) preferred on the intertidal rather than subtidal biofilm that were correlated with the composition of biofilm communities such as Proteobacteria and Cyanobacteria (Lee *et al.*, 2014). In addition, larval settlement influenced by the presence of pollutants (i.e. heavy metals) on the biofilm that had remarkable influence on microbial community composition (Figure 12.1). For instance, copper could stress on larval settlement of the polychaete (*Hydroides elegans*) on biofilms because it could suppress on biofilms development processes. Yong biofilms were comparatively more vulnerable to pollution because pollutants could change in bacterial community composition, survival and the chemical profiles of biofilms (Bao *et al.*, 2010).

12.2.3 AQUACULTURE WASTEWATER TREATMENT BY MARINE BIOFILM

Biofilm processes are widely used in aquaculture wastewater treatment (Li *et al.*, 2017; Zhang *et al.*, 2020; Natrah *et al.*, 2014). There are many successful applications shown in marine aquaculture industries through different processes such as microalgal biofilms, moving-bed biofilm reactors systems (MBBRs), fixed bed biofilm systems, trickling biofilters, etc.

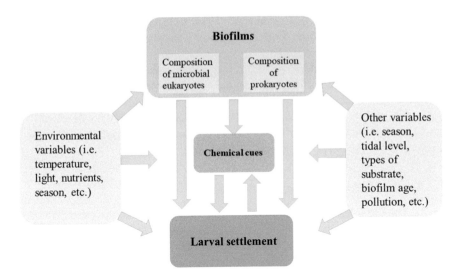

FIGURE 12.1 Interactions among biofilms, their chemical compounds and larvae of marine organisms at various environmental conditions, and other variables.

Algal-bacterial biofilms have an excellent capacity to reduce the level of nitrogen and phosphate from aquaculture wastewater (Liu *et al.*, 2019; Han *et al.*, 2020). Microalgae can transform nutrients, CO_2 and other substances into proteins, carbohydrates, lipids and other organic components by photosynthesis (Ruiz *et al.*, 2016). Algal-bacterial biofilms can remove nitrogen and phosphorus from aquaculture wastewater was up to 89% and 99.8%, respectively (Han *et al.*, 2020). Note that microalgae can remove nutrients through absorption and some chemical processes such as ammonia stripping and chemical precipitation of phosphates (Christenson and Sims, 2011). Microalgal biofilms have also great capacity to remove antibiotic substances from aquaculture wastewater through several processes such as abiotic processes including microalgae adsorption, bioaccumulation and biodegradation (Xiong *et al.*, 2018). For instance, cephalosporin antibiotics were mainly removed by microalgae adsorption, hydrolysis and photolysis reactions (Gao *et al.*, 2016).

Moving-bed biofilm reactors systems (MBBRs) remove solid waste and particulate matter, including organic and inorganic substances, from marine aquaculture wastewater through nitrification and denitrification (Rusten *et al.*, 2006). Nitrification rates – an aerobic process of converting ammonia to nitrate by two major groups of bacteria: (1) converting ammonia to nitrite (AOB); and (2) converting nitrite to nitrate (NOB) – are influenced by organic load, dissolved oxygen, total ammonium nitrogen (TAN), temperature, pH and alkalinity, and history of the biofilm (Rusten *et al.*, 2006). For instance, the removal rate of total ammonium nitrogen was 1 g NH_4^--N/ (m^2 d) when dissolved oxygen and organic load was 5 mg/L and 1 g BOD^5/m^2 biofilm surface area/d, respectively (Rusten *et al.*, 2006). Denitrification has been implemented by using both internal and external carbon sources (i.e. methanol, ethanol, monopropylene glycol MPG) (Rusten *et al.*, 1994, 1995). Furthermore, MBBRs also used for removal of antibiotics from aquaculture wastes such as norfloxacin, ciprofloxacin, enrofloxacin (Li *et al.*, 2017).

Trickling filters are widely applied in marine recirculatory aquaculture systems (RASs). Like MBBRs, these systems can also remove total ammonia nitrogen, biological and chemical oxygen demand, total suspended solids and carbon dioxide from aquaculture effluents through nitrification and denitrification. They are used as the major biofilter system or combined with granular type of biological filters to increase their denitrification efficiency (Timmons *et al.*, 2006).

12.2.4 VACCINATION IN MARINE AQUACULTURE

In 1942, Duff first developed the oral/mucosal vaccine for fish against marine bacteria, *Aeromonas salmonicida* (Duff, 1942). This result revealed that fish mortality was remarkably reduced after prolonged feeding (64–70 consecutive days) with feed containing *Aeromonas salmonicida*, and the protection appeared to correlate with antibody production. The use of bacterial biofilms as vaccine candidates for aquaculture species dates back to the late 1990s. To develop biofilms, microorganisms are cultured on particular substrates, on which they aggregate. The mature biofilms were harvested and their pathogenicity inactivated and their antigenicity maintained and used as oral vaccine candidates by mixing in the feed for fish in different studies (Vinay *et al.*, 2016). A schematic representation of application of marine bacterial biofilm as an oral vaccination shown in Figure 12.2.

There are some examples of oral vaccination in case of freshwater fishes. For instance, chitin-based, heat-inactivated *Aeromonas hydrophila* biofilm delivered to

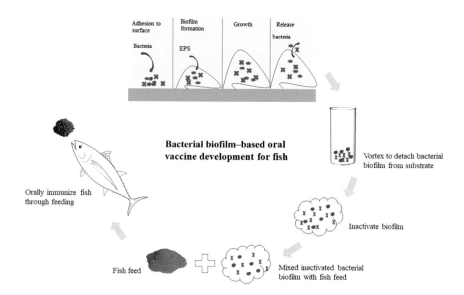

FIGURE 12.2 Application of biofilms as oral vaccine candidates for aquaculture species.

catla, rohu, common carp and walking catfish (*Clarias batrachus*) via feed provided significant protection that generates higher antibody responses than normally cultured bacterial cells (Azad *et al.*, 1999, 2000; Nayak *et al.*, 2004). For marine fishes, biofilms (BF) of *Vibrio anguillarum* were used for oral vaccination in Asian seabass, *Lates calcarifer* (Ram *et al.*, 2019). Here, BF at 10^{10} cfu g^{-1} fish d^{-1} was proper dose for vaccination where higher relative percentage survival (RPS) was observed in vaccinated fish (85.4%) compared to that with free cells (27.0%). Vaccinated sea bass showed higher protection to the homologous injection challenge (Ram *et al.*, 2019).

Initially, the major constraint of application of oral vaccination on fish was the breaking down of antigenic epitopes in the stomach and foregut of the fish before the vaccine reaches immune-responsive sites in the hindgut (Rombout *et al.*, 1985). Strategies to protect antigen breakdown are the investigation of uptake of antigens, immune responses and efficiency of vaccine after oral or anal administration (Stroband and Kroon, 1981; Rombout *et al.*, 1986), as well as the use of encapsulated antigen microspheres were needed (Dalmo *et al.*, 1995). For instance, compared with oral and batch vaccination, intubation showed better protection for marine fish (i.e. salmonids) against *Vibrio anguillarum* and *Yersinia ruckeri* (Johnson and Amend, 1983) and for carp against *V. anguillarum* (Rombout *et al.*, 1986). Furthermore, gastrointestinal stability of the biofilm-based oral vaccines was recently developed by employing monoclonal antibody–based immunofluorescence and immunoperoxidase assays (Azad *et al.*, 2000). The glycocalyx of the biofilm vaccine is believed to protect the antigen against destruction in the gut, thus facilitating its transport in intact condition to the immune-responsive areas (Azad *et al.*, 1999, 2000).

Microbes of marine biofilms may suppress microbial infection and could be useful in preventing diseases. For instance, the bacterium *Alteromonas sp.* in biofilm often colonized on the surface of estuarine shrimp embryos (*Palaemon macrodactylus*) that were resistant to fungal infection (*Lagenidium callinectes*) (Gil-Turnes and Fenical, 1992). When the embryos were exposed to the pathogenic fungus, bacteria-free embryos immediately expire, whereas similar embryos coated with *Alteromonas sp.* were comparatively well. The underlying mechanism was that bacteria protect embryos from fungal infection by producing and liberating the antifungal metabolite 2,3-indolinedione (Gil-Turnes and Fenical, 1992). Similarly, a gram-negative bacterium living on the surface of the lobster embryos *Homarus americanus* suppressed the growth of the pathogenic fungi *Lagenidium callinectes* in vitro by producing the antifungal compound 4-hydroxyphenethyl alcohol (tyrosol) (Gil-Turnes and Fenical, 1992). Moreover, marine bacteria *Alteromonas haloplanktis* also control the mortalities in larval culture of scallops *Agropecten purpuratus* by producing the active inhibitory compounds that suppressed the growth of pathogenic strains such as *Vibrio alginolyticus* and *V. anguillarum* (Riquelme *et al.*, 1996).

12.3 MARINE ANTIFOULING: AN ENVIRONMENTALLY FRIENDLY SOLUTION TO PROBLEMATIC BIOFOULING

Marine biofouling, a biological phenomenon, is the undesirable growth and accumulation of microorganisms, plants, algae or animals on submersed and immersed human-made objects such as ship hulls and pier pylons, resulting biodeterioration and increased roughness and drag on ships, which leads to economic loss, environmental degression or safety-related adversities (Callow and Callow, 2011; Schultz *et al.*, 2011; Dang and Lovell, 2016). Biofouling organisms, also known as aquatic invasive ones, can destructively hamper shipping operations, and so are presently considered as a threat to the marine environment. Thus, biofouling has long been seen as one of the major problems in marine shipping and naval activities. The key to controlling biological accumulation is to block the biological attachment from the source. However, there is a solution to this problem called antifouling. Antifouling is the prevention or inhibition of the growth or accumulation of barnacles and other marine organisms on ships' hulls or any other human-made surfaces with particular designed substances like special paints that block the organisms from piling up in the hull.

Stopping or reducing biofouling on submerged and immersed surfaces and ship hulls requires a coating system that will make a clean and smooth hull; is resistant to slimes, roughness and fouling; and has an insignificant destructive effect on the marine and coastal environments. There are some techniques used to control these, including copper, which was observed to have a shortly effective period and also be toxic to various marine organisms. With much attentive work and little progress in this field, there is still hardly any viable non-toxic antifouling agent (AF) which can satisfy these constraints.

12.3.1 ANTIFOULING STRATEGIES

An outstanding approach being considered to remedy to biofouling is with the use of marine biotechnologies in marine antifouling coatings. Some natural products having antifouling activities have been extracted from marine macroorganisms and microorganisms like bacteria, fungi, plants, seaweed, seagrasses, sponges and soft corals etc. Antifouling (AF) coatings, special types of marine paint, contain antifouling agents are used for coating the ship's hull to control marine biofouling. Modern AF coatings are of two types: biocidal and non-biocidal. There are some traditional antifouling strategies that include the use of biocidal tin oxide marine paints specially on the hulls of vessels – and these paints effectively block biofouling for a couple of years, but contaminate the surrounding waters with tributyltin (TBT), which is highly toxic to many marine and coastal living organisms. After the international banning of TBT in antifouling painting materials, copper has been focused on as the major AF biocide that needs additional 'booster biocides' to cover copper-tolerant marine species (Dafforn *et al.*, 2011). Control depletion polymers, self-polishing copolymers and hybrid or self-polishing AF coatings are currently three common biocidal antifouling approaches. On the other hand, fouling release coatings and hard inert coatings are the non-biocidal ones. These are considered non-toxic because they contain no biocides to control fouling, and their development includes very complex and sophisticated processes.

Mechanisms or mode of actions followed by certain types of marine organisms against fouling are of four categories: physical, chemical, mechanical and behavioral. Among these, physical, chemical and mechanical have gained much interest in marine biotechnology and have been the key topics of research on marine antifouling agents and micro-texturing of surfaces. After the international banning of TBT in antifouling painting materials, more concern is being given to establish more efficient and environmentally friendly replacements. Though a clear explanation of the biofouling process is not fully done yet, marine biotechnology has found some fruitful outputs for developing antifouling coatings. Some promising important sources and antifouling compounds have been discovered, though a budget-friendly solution has yet to be found.

12.3.2 ANTIBIOFILM METABOLITES FROM MARINE ORGANISMS

Many marine biota like microorganisms, macroorganisms, invertebrates, sponges and seaweeds have been explored to find out some advanced bioactive compounds having antibiofilm potentiality; among these organisms, marine microorganisms are getting much concern for having some effective metabolites, along with antibiofilm properties. Some of the marine organisms and their derived antibiofilm compounds and the mechanisms of biofilm inhibition are listed in Table 12.1. Interestingly, the microbial metabolic compounds of these species are observed to block the biochemical compounds responsible for biofilm development, and they also are proved to prevent the attachment, biofilm formation and quorum sensing activities among the biofilm-capable bacteria community (Adnan et al., 2018). Marine macroorganisms and microorganisms – for example, red seaweed, brown seaweed, microalgae, macroalgae, cyanobacteria, crab shells, sponges, soft corals, fungi, bacteria, etc. – have been found to produce some important metabolites like sesquiterpenes, fucoidan and other polysaccharides, chitosan, ianthellin, brominated tyrosine, lectins, surfactants, enzymes, culture filtrates, etc., that can act against biofilm formation of certain bacteria communities. The antibiofilm activities of these marine organisms and metabolites include antimicrobial activities, destabilization of plasma membranes, inhibiting primary attachment, regulating virulence indicators, inhibiting swarming motilities, quorum sensing inhibition, interruption of cell surface hydrophobicity, inhibiting resistant characters, degradation of cell walls, etc.

12.3.3 FOULING ORGANISMS AND ANTIFOULING COMPOUNDS

There are more than 4,000 diversified biofouling organisms, including microorganisms, macroorganisms, algae, plants, animals, etc., present in the marine and coastal environments (Shao et al., 2011a). Most of them are found to live in nutrient-rich shallow-water marine regions like coasts and harbors (Dang and Lovell, 2016; Wang et al., 2017a). Categorically, the microfoulers or micro biofilm organisms involve diversified marine bacteria, algae and protozoa, and macrofouler organisms are marine barnacles, bryozoans, tubeworms, mussels, polychaete worms and seaweed, which can cause severe biofouling (Shao et al., 2011b; Yebra et al., 2004).

Several antifouling compounds are used to control biofouling internationally. Among these, metal-based ones like tributyltin and cuprous oxide and synthetic organic biocide ones like igarol and diuron are prominently known as antifouling compounds to reduce biofouling in marine environments (Yebra et al., 2004), but many of them cause high toxicity on non-target organisms, and thus show adverse

TABLE 12.1
Marine Organisms and Their Extracted Antibiofilm Metabolites/Compounds

Marine Organisms	Antibiofilm Compounds	Targeting Biofilm Organisms	Possible Mechanism/ Implication	References
Red seaweed, *Laurencia dendroidea*	Elatol, sesquiterpene	*Leishmania amazonensis*	Disrupts plasma membrane	Santos *et al.*, 2018
Seaweed, *Fucus vesiculosus*	Fucoidan	*Streptococcus mutans* and *S. sobrinus*	Antimicrobial activities	Jun *et al.*, 2018
Marine brown alga, *Halidrys siliquosa*	Partly fractionated methanolic extract	*Staphylococcus aureus* (MRSA)	Antimicrobial activities	Busetti *et al.*, 2015
Cyanobacteria, *Westiellopsis prolifica*	Acetone extract	*Bacillus subtilis*, *Shigella* sp. and *Proteus* sp.	Antimicrobial activities	Al-Tmimi *et al.*, 2018
Crab shells (biowastes), *Portunus sanguinolentus*	Chitosan	*Staphylococcus aureus*	Lowers the staphyloxanthin pigmenting compound	Rubini *et al.*, 2018
Soft coral, *Eunicea* sp.	Batyl alcohol (1) and fuscoside E peracetate (6)	*Pseudomonas aeruginosa* ATCC 27853 and *Staphylococcus aureus* ATCC 25923	Inhibits specific biofilm with low antimicrobial effects	Díaz *et al.*, 2015
Marine sponge, *Aplysina fulva*	Mucin-binding lectin	*Staphylococcus aureus, S. epidermidis* and *Escherichia coli*	Decreases the biomass of biofilm	Carneiro *et al.*, 2019
Actinomycetes Nocardiopsis sp.	Pyrrolo[1,2-a] pyrazine-1,4-dione, hexahydro-3-(2-methylpropyl)	*Proteus mirabilis* and *E. coli.*	Degrades cell wall in treated cells of the bacteria	Rajivgandhi *et al.*, 2018
Deep-sea bacterium, *Bacillus cereus*	Amylase enzyme	*Pseudomonas aeruginosa* and *Staphylococcus aureus*	Blocks the complete biofilm formation	Vaikundamoorthy *et al.*, 2018
Marine-derived fungus of *Emericella variicolor*	Sesterterpenes	*Mycobacterium smegmatis*	Antimicrobial activity	Arai *et al.*, 2013

effects on marine communities and the environment, thereby getting banned and restricted in developing antifouling substances (Yebra *et al.*, 2004; Guerin *et al.*, 2007; Wang *et al.*, 2016). Therefore, there is growing motivation to find effective and environment-friendly antifouling agents to control marine biofouling.

12.3.4 Antifouling Natural Products from Marine Microorganisms

The metabolites extracted from various marine microorganisms – namely, bacteria and fungi, having antifouling activities – are being revealed as active inhibitors of growing and colonizing biofouling organisms with lower and non-toxic property (Qian *et al.*, 2009). Polyketides, lactones, nucleosides, peptides, phenyl-ethers, fatty acids, steroids, benzenoids, alkaloids, terpene, etc., are some natural products produced from marine microorganisms which act as antifouling agents (Dang and Lovell, 2016; Wang *et al.*, 2017a, 2017b). Importantly, some strains of *Bacillus* and *Streptomyces* are good sources of antifouling metabolites, and also, some symbiotic bacteria – along with different marine communities – can potentially produce strong compounds against biofouling (Table 12.2). For instance, two important fatty acids (FA) from marine *Shewanella oneidensis* were found to inhibit the germination of green alga (*Ulva pertusa*) spores and a prominent non-toxic antifouling metabolite called diterpene that was extracted from a marine bacteria *Streptomyces cinnabarinus* showed strong antifouling activities over the diatom *Navicula annexa* and the marine macroalga *U. pertusa* (Bhattarai *et al.*, 2007; Cho and Kim, 2012).

Moreover, there are some groups of marine fungi that produce various chemical antifouling metabolites. A few of them are listed in Table 12.2. A notable fungi, *Aspergillus* sp., was found to produce substantial amount of antifouling metabolites (Table 12.2). Bisabolane-type sesquiterpenoids (Li *et al.*, 2012) and aspergilone A (Shao *et al.*, 2011a, 2011b) found in *Aspergillus* sp., eurotiumides A–D extracted from *Eurotium* sp. (Chen *et al.*, 2014) and benzenoids in *Ampelomyces* sp. (Kwong *et al.*, 2006) are some proven antifouling compounds that expressed antifouling activities over the larval attachment of *Balanus amphitrite*, even at the lower concentration. Another study showed pestalachlorides found in marine *Pestalotiopsis* sp. displayed higher antifouling potentiality over the larval attachment of *B. amphitrite* with zero toxicity level (Xing *et al.*, 2016).

12.3.5 Antifouling Natural Products from Marine Macroorganisms

Some marine invertebrates – such as sponges, gorgonians and soft corals – are well-known macroorganisms for extracting antifouling compounds. A study conducted by Qi and Ma (2017) reported around 198 antifouling compounds found in diversified marine invertebrates: those are diterpenoids, sesquiterpenoids, prostanoids, alkaloids and steroids. Cembranoid epimer, a prominent antifouling compound extracted from the Caribbean gorgonian *Pseudoplexaura flagellosa*, is observed to block the formation and maturation of biofilm of some bacteria like *Pseudomonas aeruginosa*, *Vibrio harveyi* and *Staphylococcus aureus*, with no interruption of their growth (Tello *et al.*, 2011).

In addition, some marine macroalgae – especially green, brown and red algae – have shown their potential of having antifouling properties (Dahms and Dobretsov, 2017). Organic compounds and metabolites extracted from *Ulva reticulata* were found to be capable of sensible antifouling activities and antibiofilm potential (Prabhakaran *et al.*, 2012). Moreover, some non-polar compounds of invasive *Sargassum* sp. (Schwartz *et al.*, 2017) and extracted metabolites (ethanol and dichloromethane) from *Sargassum muticum* (Silkina *et al.*, 2012) restricted the growth of diatoms and their attachment.

TABLE 12.2

List of a Few Marine Bacterial and Fungal Strains and Their Derived Antifouling Compounds

Bacterial/Fungal Strain	Source	Antifouling Compound	Target Fouling Organism	References
Bacteria, *Streptomyces violaceoruber*	Seaweed	Butenolides	*Ulva pertusa*	Hong and Cho, 2013
Bacteria, *Bacillus cereus*	Sponge, *Sigmadocia* sp.	Culture extracts	High inhibiting activity of microalgal settlement	Satheesh *et al.*, 2012
Bacteria, *Pseudoalteromonas haloplanktis*	Ascidian	Culture extracts	Spore germination of *Ulva pertusa*	Ma *et al.*, 2010
Bacteria, *Leucothrix mucor*	Red algae	Steroids	Larval settlement of diatoms and B. amphitrite	Cho and Kim, 2012
Bacteria, *Streptomyces* sp.	Deep-sea sediment	12-Methyltetradecanoic Acid	*Hydroides elegans* larvae	Xu *et al.*, 2009
Marine cyanobacterium, *Kyrtuthrix maculans*	Exposed to sheltered rocky shores	Maculalactone A	Naupliar larvae of the barnacles *B. amphitrite*, *Tetraclita japonica*, and *Ibla cumingii*	Brown *et al.*, 2004
Fungi, *Sarcophyton* sp.	Soft coral	Amibromdole	Larvae of *B. amphitrite*	Xing *et al.*, 2016
Fungi, *Scopulariopsis* sp.	Gorgonian coral	Alkaloids	Larval settlement of barnacle *B. amphitrite*	Shao *et al.*, 2015
Fungi, *Cochliobolus lunatus*	Sea anemone	Cochliomycins D–F	Larval settlement of *B. amphitrite*	Liu *et al.*, 2019
Fungi, *Aspergillus elegans*	Soft coral, *Sarcophyton* sp.	Phenylalanine derivatives and cytochalasins	Larvae attachment of *B. amphitrite*	Zheng *et al.*, 2013
Fungi, *Aspergillus* sp.	Gorgonian coral	Polyketides	Antifouling activity	Bao *et al.*, 2017

12.4 APPLICATION OF MARINE BIOFILM FOR BIOREMEDIATION OF PERSISTENT ORGANIC POLLUTANTS (POPS)

Persistent organic pollutants (POPs) basically indicates a group of compounds that can be present in the environment for a long period of time and cannot be biode-graded naturally due to their nature. These include polychlorinated biphenyls (PCBs), dioxins, chlorinated ethenes, polycyclic aromatic hydrocarbons (PAHs) and brominated flame retardants (PBDEs) that have been contaminating bodies of water, sediments and terrestrial soil all over the world for decades. As they have the ability to remain in the environment for unimaginably longer time periods and ultimately impair human health by entering into the food chain, the removal of these pollutants from the environments is of utmost importance (Kumar *et al.*, 2014).

Various approaches have been practiced to spur biodegradation of POPs, but most of these methods have been proven ineffective and costly. Some methods like dredging and capping are not only expensive, but can pose other threats such as bioaccumulation of contaminated particles and hazardous particles remaining to some extent (Martins *et al.*, 2012). Our traditional waste removal systems have several drawbacks to removing POPs from the environment. To addressing these issues, a solution would be to introduce microorganisms on carbon surfaces and expose these biofilm communities regardless of the polluted sites. Therefore, scientists all across the world are trying to establish a sustainable and effective approach to wrestle with this long-standing problem of POPs. The cutting-edge microbes-based bioremediation of persistent pollutants, popularly known as biofilm-based solutions, are becoming popular. Although most of the experiments for removing pollutants have been conducted in laboratory conditions, for several, prospects and potential biofilm-based approaches are becoming more common. The majority of those, especially using marine biofilms, appear promising. Therefore, it is not unsurprising that studies and experiments regarding the mechanisms of forming biofilms by microbes are being extended and have gained attention.

12.4.1 POLYCHLORINATED BIPHENYLS (PCBS)

Polychlorinated biphenyls (PCBs) are one of the most prevalent pollutants found in marine – even in polar – regions. Quite a number of bacteria available in marine regions have the capacity to degrade PCBs to innocuous materials. and these microbial communities are well known as plausible factors for bioremediation of PCBs. The most prominent marine bacteria having degradation capability of PCBs, are *Psychrobacter* sp., *Pseudoalteromonas* sp. and *Arthrobacter* sp. Those species are isolated from the Ross Sea, Antarctica (Michaud *et al.*, 2007). Apart from this, biofilm-forming marine bacteria having promising ability to biodegrade PCBs found in cryoconites are *Pseudomonas*, *Shigella*, *Subtercola*, *Chitinophaga* and *Janthinobacterium* species (Weiland-Bräuer *et al.*, 2017).

12.4.2 DIOXINS

Biodegradation of dioxins has been observed in the same way as that of PCBs. Basically, there are three types of dioxins that contaminate environment to a great extent: chlorinated biphenyl, dibenzofuran and dibenzo-p-dioxin. *Alteromonas macleodii*, *Neptunomonas naphthovorans* and *Cycloclasticus pugetii* are marine bacteria which can biodegrade those pollutants to clean the environment. However, their modes of action are somewhat different and pollutant specific. Some species like *Alteromonas macleodii* partially degrade pollutants, whereas the latter two degrade completely through oxygenation processes and it has been observed that the presence of more than one bacterial species in marine biofilm increased the degradation efficiency for dioxins (Yoshida *et al.*, 2009).

12.4.3 CHLORINATED ETHENES

In marine environment, the two most frequently occurring chlorinated ethenes are tetrachloroethene (PCE) and trichloroethene (TCE), which originate naturally from marine algae and pollute marine environments alongside others which are basically used in terrestrial environments, but their ultimate fates are in marine environments (Abrahamsson et al., 1995). Vinyl chloride (VC) and cis-1,2-dichloroethene (cDCE) are two other notorious pollutants of the group of chloroethene found in many contaminated spots. Chlorinated ethenes are basically degraded by two distinguished pathways, aerobic co-metabolism and anaerobic oxidation. Most of the chlorinated ethenes are, however, degraded anaerobically rather than aerobically (Dolinová et al., 2017). There are several biofilm-forming bacteria found in marine environments which have the capability to degrade different types of chlorinated ethenes in efficient ways. Specific bacteria act on definite pollutants and biodegrade them via respective pathways (Table 12.3).

TABLE 12.3
Chlorinated Ethenes Degrading Marine Bacteria

Type of POPs	Pathways	Biodegraded Chloroethenes	Microbes	References
Chlorinated ethenes	Anaerobic	cDCE, VC	Clostridium sp.	Kim et al., 2006
		PCE, TCE,	Clostridium bifermentans	Chang et al., 2001
		PCE	Desulfomonile tiedjei	DeWeerd et al., 1990
		PCE, TCE	Desulfuromonas chloroethenica	Krumholz, 1997
		PCE, TCE	Enterobacter agglomerans	Sharma and McCarty, 1996
		PCE	Methanosarcina sp.	Fathepure and Boyd, 1988
		PCE	Shewanella sediminis	Lohner and Spormann, 2013
	Aerobic	VC	Pseudomonas putida strain	Danko et al., 2004
		TCE, cDCE,	Pseudomonas putida strain	Wackett and Gibson, 1988
		TCE, cDCE	Rhodococcus sp.	Suttinun et al., 2009
		TCE, cDCE	Rhodococcus erythropolis	Dabrock et al., 1992

Notes: tetrachloroethene (PCE); trichloroethene (TCE); vinyl chloride (VC); cis-1,2-dichloroethene (cDCE)

12.4.4 POLYCYCLIC AROMATIC HYDROCARBONS (PAHS)

PAHs are highly toxic and ubiquitous in marine environments. However, a matter for optimism is that marine bacteria itself belongs to the genera *Cycloclasticus*, which can degrade these pollutants. Taking *Cycloclasticus* sp. is an example of that genera isolated from the deep-sea sediment that can biodegrade PAHs such as naphthalene, phenanthrene, pyrene, etc. It has been already established as a prominent PAHs degrader by enzymatic assays (Wang *et al.*, 2018). Moreover, a group of marine biofilm–forming microorganisms has already been identified which can effectively degrade PAHs such as bacteria *Pseudomonas* spp., *Rhodococcus spp.*, *Mycobacterium* spp., *Sphingomonas* spp. (Walter *et al.*, 1991; Uyttebroek *et al.*, 2007), and cyanobacteria and algae (*Corynebacterium* spp.; *Cyanobacterium Oscillatoria, Agmenellum quadruplicatum, Chorella* sp., *Coccochloris* sp., *Oscillatoria* sp., *Porphyridium* sp.) (Narro *et al.*, 1992a, 1992b; Cerniglia *et al.*, 1985). In another experiment, utilizing moving-bed biofilm reactor showed that nitrates also reduced, together with PAH, in a reduced and anoxic environment that underscores the significance of co-metabolism in biofilm-based remediation (Lolas *et al.*, 2012).

12.4.5 BROMINATED FLAME RETARDANTS (PBDES)

Brominated flame retardants (e.g. PBDEs) can be degraded both aerobically and anaerobically. Anaerobic biodegradation can be carried out in an experiment which reveals that biofilm-forming bacteria can remove these types of pollutants. Anaerobic bacteria (*Sulfurospirillum, Sulfurimonas, Nautilia*) are generally used in this system isolated from sulphidic terrestrial and marine habitats (Campbell *et al.*, 2006). PBDEs were basically observed to degrade anaerobically in sediment (Tokarz *et al.*, 2008). Moreover, longer anoxic periods favored degradation of PBDEs (Bartrons *et al.*, 2011). Besides that, aerobic degradation is also possible by using bacteria such as *Pseudomonas citronellolis, Microbacterium oxydans* and *Delftia tsuruhatensis*, which are isolated from sediment of ground water, yet also found in marine and coastal regions. They can effectively degrade two types of brominated flame retardants (BFR): dibromoneopentyl glycol (DBNPG) and tribromoneopentyl alcohol (TBNPA), those which are responsible for polluting water (Balaban *et al.*, 2021). Generating the proper conditions for degradation of these organic pollutants could be challenging. To solve this problem, application of biofilm for creating supportive situations for debromination could be the most feasible option (Langford *et al.*, 2007).

12.4.6 ACTIVATED SLUDGE BIOFILMS

To enhance POP bioremediation by biofilm, one of the auspicious strategies is to harness the ability of activated sludge to co-metabolize. Kurane Ryuichiro (1997) studied the ability of activated sludge by inoculating bacteria available in marine environments like *Rhodococcus erythropolis* and *Pseudomonas* sp. to activate sludge for removing a group of PAHs and got an optimistic outcome. Besides, it has been observed in the case of a broad range of PAH substrates that mixed activated sludge biofilm performed better (Rodriguez and Bishop, 2008). On top of that, Falas *et al.*

(2013) demonstrated that a biofilm delivery system with activated sludge efficiently removed micropollutants such as diclofenac, which emphasized the effect of using the combined hybrid biofilm-activated sludge process rather than separate processes.

12.4.7 Marine Biofilm Barriers

Use of biofilm barriers – combinations of biobarriers with biofilm – is an approach that can be efficiently used to prevent spreading of toxic and long-lasting POPs and concomitantly ensuring better management for larger areas like open bodies of water. It was recorded that by using a mulch biofilm barrier, elimination of phenanthrene and pyrene is successfully possible; even up to 99% of removal was feasible within 150 days (Seo *et al.*, 2009). Experimentally, it has been proven that genus *Arthrobacter* spp. has the capability to tackle the contamination of PAH as well as trichloroethylene (TCE)-contaminated sites. One species of that genus associated with the marine sponge, *Arthrobacter ilicis*, can play a pivotal role keeping marine environments safe from those pollutants (Mohapatra and Bapuji, 1998). A study showed that by applying a combination of substrate, tetrachloroethylene (PCE) can be decayed by indigenous bacteria, ensuring constant electron donors (Kao *et al.*, 2001).

12.5 APPLICATION OF MARINE BIOFILM IN WASTEWATER TREATMENT

For treating wastewater, biofilm-based technology is widely practiced where suspended growth reactors are enriched with solid media to provide attachment surfaces for microorganisms. Pollutants removal from the water is faster when the biofilms concentration is higher. In this regard, marine algal biofilm can play a central role in removing different nutrients, carbonaceous materials and heavy metals accompanying trapped pathogens from the wastewater (Abdel-Raouf *et al.*, 2012). Furthermore, heavy metal chelation polysaccharides are produced from marine bacteria. *Enterobacter cloaceae*, a marine bacterium isolated from marine sediment having the ability to produce exopolysaccharides which can chelate heavy metals, was used in the experiments whereby effectively chelated chromium (Cr), cadmium (Cd), cobalt (Co), mercury (Hg) and copper (Cu) up to 70% (Iyer *et al.*, 2005). On top of that, other marine biofilms forming microorganisms like *Zoogloea* sp. and *chlorella* sp. can also remove pollutants from water, and have huge potentiality using in wastewater treatment at large scale. (Kaplan *et al.*, 1987; Matsuda *et al.*, 1992; Kong *et al.*, 1998).

12.5.1 Anammox Process in Marine Environments

A process of anaerobic ammonium oxidation commonly known as anammox is a watershed technological advancement that combines ammonia and nitrite directly into dinitrogen gas without passing through a two-stage process of aerobic nitrification and anaerobic denitrification (Shivaraman and Shivaraman, 2003). The discovery of this process has a great deal of significance for removing ammonium in wastewater and changes the course of wastewater treatment. The marine genera of anammox bacteria has a huge potential, as it is responsible for almost half of the total nitrogen yield in marine environments. In comparison with conventional methods, anammox removes the obligation of a carbon source for the process. In addition, it lowers the necessity of energy requirements for oxygenation and generates a negligible amount of surplus waste. It also emits lesser amount of CO_2 that can be a viable option to combat climate change (Bertino, 2011). The simplest representation of anammox reaction can be as $NH_4^- + NO_2^- = N_2 + 2H_2O$ (Figure 12.3), which is executed by special type of bacteria of the group planctomycete, popularly known as anammox bacteria (*Candidatus Brocadia anammoxidans*).

Annamox bacteria have shifted the approach of wastewater treatment to a large extent. It is considered the paradigm in wastewater treatment, particularly in marine contexts. This process is extensively used to tackle excess ammonia problems in water, particularly in intensive culture systems. This is the two-stage process whereby in the preliminary stage. a specific type of bacteria oxidizes ammonium and produces nitrite.

$$2NH_4^- + 3O_2 \rightarrow 2NO_2^- + 4H^+ + 2H_2O$$

The next stage is known as anammox process, where rest of the ammonium, along with nitrite, turns into dinitrogen. The bacteria in this system is popularly known as anammox bacteria.

$$NH_4^+ + NO_2^- \rightarrow N_2 + 2H_2O$$

Both parts of the process can occur in one-stage reactor, effectively eliminating ammonium from water with the help of the aforementioned type of bacteria. Application of the anammox process literally puts an end to conventional methods due to its various advantages in terms of cost, management and environmental

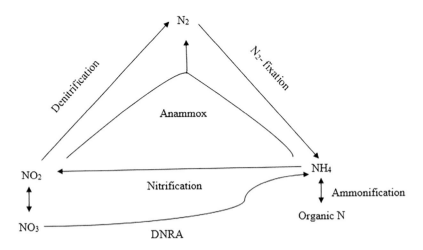

FIGURE 12.3   The biological nitrogen cycle. DNRA: Dissimilatory Nitrate Reduction to Ammonium.

matters (Daigger, 2014). Needless to say, annamox bacteria in biofilms are being used in wastewater treatment exclusively in different reactors like moving-bed reactors and sequencing batch reactors. Though anammox bacteria are extensively diverse, as of now, ten species of anammox have been identified and most of them enriched in the laboratory (Kartal *et al.*, 2013). All possess the same taxonomical stature *Candidatus*. Scientists have classified all ten species into five genera, where four genera – *Kuenenia* (*Kuenenia stuttgartiensis*), *Brocadia* (*B. anammoxidans, B. fulgida* and *B. sinica*), *Anammoxoglobus* (*A. propionicus)* and *Jettenia* (*J. asiatica)* – are found in treatment plant and fresh water, whereas the fifth genus *Scalindua* is predominantly marine (*S. brodae, S. sorokinii, S. wagneri* and *S. profunda*). In recent days, for the expansion of intensive marine RAS and treating of marine water for different purposes, a holistic quest of marine biofilms or biofilm-forming bacteria are being proceed. As marine anammox bacteria are diverse, so it presumably can be in treating of marine water. Already two experiments, one in an RAS and another one in membrane bioreactors to treat wastewater, have been carried out by using *Candidatus Scalindua* sp., a unique marine anammox bacteria. In both cases, more than 90% of nitrogen has been removed, making *Candidatus Scalindua* sp. a stable ammonium or nitrogen remover (Roques *et al.*, 2021; Ali *et al.*, 2020).

12.6   APPLICATION OF MARINE BIOFILMS IN REMOVING PETROLEUM HYDROCARBONS

Microorganisms – in other words, biofilms – have the capability to degrade petroleum hydrocarbons, which are responsible for contamination of aquatic sites, especially in marine (Atlas and Hazen, 2011; Tyagi *et al.*, 2011). In the recent past, marine microbes played an important role in attenuating the impact of two shocking accidents of the oil industry in the United States, the Exxon Valdez in 1989 and the BP Deepwater Horizon spill in 2010. It is well known, for valid reasons, that contaminated spots experienced lower amounts of essential nutrients like nitrates, phosphates and iron required to grow microbes. There are a number of marine biofilm–forming microbes shown in Table 12.4 which can degrade specific petroleum components like naphthalene, anthracene, phenanthrene, pyrene, and so on. Therefore, for removal of oil released by the Exxon Valdez oil spill, two fertilizers were used to boost up the growth of hydrocarbon-degrading microbes. In the preliminary stage, naturally oil-degrading bacteria increased around 30–40%; for instance, in the beginning the bacteria present was $1–5 \times 10^3$ cells/mL of seawater (1–10% of total heterotrophic population), while by late 1989, it was 1×10^5 cells/mL (40% of the total heterotrophic population). Around 49 tons of fertilizers were applied during the period of 1989–1991, and the cleanup was considered concluded by June 1992.

On the other hand, for eliminating the crude oil spilled by the accident in BP's Deepwater Horizon, at a depth of more than 5,000 m, one of the tactics that was used to wrestle with this situation was watershed at that moment (Atlas and Hazen, 2011). Dispersant was directly injected at the wellhead, this ultimately resulted in a 'cloud' of dispersed oil having 10^5 cells per mL. At the time when oil was being released, the half-life of alkanes ranged between 1.2 and 6.1 days in the 'cloud'. A marine

TABLE 12.4

Petroleum Hydrocarbons Degrading Marine Microorganisms and Their Metabolites

Phylum	Species	Metabolites	References
Bacteria	*Pseudomonas, Corynebacterium, Bacillus, Corynebacterium*	Aliphatic hydrocarbon, cyclohexane, alkyl benzene, and dicycloalkane	Yuan *et al.*, 2003
	Vibrio cyclotrophicus sp. nov.	Naphthalene, 1-methylnaphthalene, 2-methylnaphtha-lene, 2,6- dimethylnaphthalene, and phenanthrene	Harris *et al.*, 2002
	Aeram onas punctata TII	Naphthalene, anthracene, phenanthrene and pyrene	Page *et al.*, 2000; Bagi *et al.*, 2014
	Vibrio, Pseudoalteromonas, Marinomonas	Phenanthrene and Chrysene	Harayama *et al.*, 2004
	Cycloclasticus oligotrophus	n-Alkanes aromatic hydrocarbon, naphthalene, phenanthrene and anthracene	Dyksterhouse *et al.*, 1995; Wang *et al.*, 1996
	Alcanivorax sp.	n-Alkanes, branched alkanes and alkylbenzenes	Head *et al.*, 2006; Dutta and Harayama, 2001
	Oleiphilusand, Oleispira	Aliphatic hydrocarbons, alkanoles and alkanoates.	Golyshin *et al.*, 2002; Yakimov *et al.*, 2007
	Desulfococcus	Alkanes	Aeckersberg *et al.*, 1991
	Cycloclasticus	PAHs	Iwabuchi *et al.*, 2002; Kasai *et al.*, 2003
	Mycobacterium	Pyrene	Kanaly and Harayama, 2010; Gallego *et al.*, 2014
Fungus	*Aspergillius* sp., *Penicillium* sp.	Long carbon chain hydrocarbon component	Swannell *et al.*, 1999

bacterium, *Mycobacterium vanbaalenii*, is one of the most significant degraders of hydrocarbons. Considering crude oil components of this event, it was used as an experiment to prevent the dispersion of oil and clear the marine environment as early as possible. It was observed that shorter length alkenes were degraded faster, while complete degradation of phenanthrene and pyrene were noticed, along with partial degradation of fluoranthene (Kim *et al.*, 2015).

12.7   MARINE BIOFILMS ACTING AS A SAFEGUARD OF BENEFICIAL MICROBES IN ADVERSE CONDITIONS

Beneficial microbes (bacteria, fungi, algae) play pivotal roles in the wellbeing of an ecosystem. But in the undesired environmental conditions that are a very common phenomenon in the ever-changing dynamic marine ecosystem, they can be exposed to excess heat and ultraviolet (UV) radiation, causing transient desiccation. Alongside this, the engaging of competition for essential but limited nutrients cannot be denied. It is worth mentioning that the vivid instance of shielding rendered by biofilms ensures a higher survival percentage of microbes living near shore. Biofilms actually confer safety for beneficial microbes against hassles incurred by different harsh environmental conditions (Ortega-Morales *et al.*, 2010). Fluctuations of light intensity, temperature, salinity, radiation exposure, acidity, alkalinity, dissolved oxygen and other gases are frequent events in most aquatic ecosystems. In that situation, whereby cells experienced abrupt fluctuations of various environmental parameters, they utilize their inherent capability to sustain and preserve a sufficient fluidity of the cellular membrane. In doing so, cells generally change the profile of fatty acids associated with phospholipids (de Carvalho and Caramujo, 2012). Not only this, but also, some biofilm forming marine bacteria like *Shewanella colwelliana* and *Vibrio splendidus* produce polyunsaturated fatty acids (PUFA) (Freese *et al.*, 2009). In addition, for shielding themselves from harsh environmental conditions, organisms present in biofilms and mats produce secondary metabolites and carotenoid pigments, as well. These small secondary metabolites are popularly known as mycosporine-like amino acids (MAAs).

It is believed that the biofilms matrix, an integral part of biofilms, together with microbial mats, protect beneficial microorganisms from intense sunlight and harmful radiation (de Carvalho and Caramujo, 2017). It is realistic to assume that this process of safeguarding could have elevated the protection from the extreme variable parameters of the environment (Hall-Stoodley *et al.*, 2004). Another significant thing is that the bacteria which are found in extreme unfavorable conditions such as in polar or hydrothermal vent regions, and chemically disagreeable environments, engender specialized lipids by extremophiles. It also plays an crucial role to survive in those harsh conditions and renders protection to the other organisms (de Carvalho and Caramujo, 2012). Because of the presence of polyketide synthase, various psychrophilic or cold-loving bacteria can able to produce PUFA, which is crucial to thrive in freezing temperatures. On the other hand, thermophilic bacteria like *Thermotoga maritima* possess a glycerol ether lipid that enables them to flourish in hot environments (Metz *et al.*, 2001). It is assumed that marine biofilm is not only a noble way to survive microbes themselves; rather, it also offers protection for other microorganisms that play a greater role in an ecosystem.

12.8   POTENTIAL APPLICATION OF MARINE BIOFILM AND BIONIC PLANTS IN OPEN WATERS

There are a number of downsides of our commonly practiced way of treating open water to restore it. To overcome those drawbacks, a state-of-the-art technological way can be introduced. Bionic plants, which are basically 'artificial aquatic plants' or 'ecological fillers', emulate the form of actual aquatic plants, as well as their main functions in water purification. The most significant thing is that the bionic plants actually combine two technologies. One is the biofilm technology of water bodies and another is the filler technology for treating wastewater. The combined technology has already drawn the attention of scientific communities. It basically mimics a forest-like structure below water to provide a greater attachment area for microorganisms for growing and multiplying to form biofilms. By doing so, this technology ultimately assists microorganisms to degrade pollutants. As a result, water is purified, which is the most significant part of ecological restoration. Bionic plants combined with biofilm confers several advantages to restore ecosystems over conventional ways, such as huge specific surface area, season independency, withstanding water pollution and reusability (Zhu *et al.*, 2011). All of these supremacies can be a paradigm to restore open waters ecosystems like rivers, estuaries and coastal zones of marine environments.

Because of their larger surface areas, bionic plants fundamentally provide suitable habitats for organisms that determine the abundance and degradation of pollutants. However, attachment of microorganisms sometimes relies on the materials of bionic plants. The distinguished mechanism of bionic plants is magnifying microbial biomass or biofilm tens or hundreds of times, so these improved and abundant biofilms directly remove contaminants or pollutants like ammonia from wastewater. With this technology, artificial aquatic plants are mainly used as biological carriers where microbes are attached. The main target of these techniques and mechanisms is to enable the desired organisms to flourish, thus an increased number of microbes could participate in the unceasing degradation of pollutants. The principle of this technology is generating a larger surface and concomitantly creating an environment congenial for biofilm to burgeon. It is vital to have a substantial number of biofilms on the carrier filler with a view to strengthening and utilizing the metabolism of microorganisms for removing pollutants from water (Zhang *et al.*, 2011). Actually, the cardinal rule for degrading of water pollutants is played by the biofilm associated with the surface area–enhancing bionic plants. As the used bionic plant can magnify the surface area up to hundreds fold, it could be the epitome to be utilized in restoring estuaries and marine environments – and for forming the biofilms on those extended surface areas, marine microorganisms can be easily introduced.

12.9   CONCLUSION

Biofilms are aggregated colonies of microorganisms that are attached with natural and artificial surfaces, being highly explored for research and commercial prospects. Thereby, marine biofilms are of great importance and are being viably studied for commercial applications in some major sectors to address existing problems. Growing concern implies that marine biofilms are optimistically very appreciated in aquaculture sector – for instance, aquaculture food sources, larval settlement, disease control, aquaculture wastewater management and developing vaccinations. Moreover, it is being prominently considered in extracting antifouling metabolites and compounds, developing antifouling coatings and thus controlling marine biofouling, though commercial applications are yet to become sophisticated. Besides these applications, marine biofilms are also gaining potentiality in shipping and industrial wastewater treatment, bioremediation of marine organic pollutants, and environmental adaptation and restoration to different adverse situations and their consequent survival. Thus, biofilm-derived treatment strategies could offer an efficient approach for the decontamination and restoration of polluted environments, showing the versatile uses of microorganisms. Still, the multiple uses of marine biofilms and biofilm-based technologies are not highly applied in commercial perspectives and there is a long way to go to achieve large-scale applications. Therefore, a molecular insight of marine biofilms and their action strategies and technological sophistication is required for multiple and commercial exploration of marine biofilms.

REFERENCES

Abdel-Raouf, N., Al-Homaidan, A.A. and Ibraheem, I.B.M., 2012. Microalgae and wastewater treatment. *Saudi Journal of Biological Sciences*, *19*(3), pp.257–275. DOI: https://doi.org/10.1016/j.sjbs.2012.04.005.

Abrahamsson, K., Ekdahl, A., Collen, J. and Pedersen, M., 1995. Marine algae-a source of trichloroethylene and perchloroethylene. *Limnology and Oceanography*, *40*(7), pp.1321–1326. DOI: https://doi.org/10.4319/lo.1995.40.7.1321

Abreu, P.C., Ballester, E.L., Odebrecht, C., Wasielesky Jr, W., Cavalli, R.O., Granéli, W. and Anesio, A.M., 2007. Importance of biofilm as food source for shrimp (*Farfantepenaeus paulensis*) evaluated by stable isotopes (δ13C and δ15N) *Journal of Experimental Marine Biology and Ecology*, *347*(1–2), pp.88–96. DOI: https://doi.org/10.1016/j.jembe.2007.03.012

Adnan, M., Alshammari, E., Patel, M., Ashraf, S.A., Khan, S. and Hadi, S., 2018. Significance and potential of marine microbial natural bioactive compounds against biofilms/biofouling: necessity for green chemistry. *PeerJ*, 6, p.e5049. DOI: https://doi.org/10.7717/peerj.5049.

Aeckersberg, F., Bak, F. and Widdel, F., 1991. Anaerobic oxidation of saturated hydrocarbons to CO_2 by a new type of sulfate-reducing bacterium. *Archives of Microbiol*, *156*(1), pp.5–14. DOI: https://doi.org/10.1007/BF00418180

Ali, M., Shaw, D.R. and Saikaly, P.E., 2020. Application of an enrichment culture of the marine anammox bacterium "Ca. Scalindua sp. AMX11" for nitrogen removal under moderate salinity and in the presence of organic carbon. *Water research*, *170*, p.115345. DOI: https://doi.org/10.1016/j.watres.2019.115345

Arai, M., Niikawa, H. and Kobayashi, M., 2013. Marine-derived fungal sesterterpenes, ophiobolins, inhibit biofilm formation of *Mycobacterium* species. *Journal of Natural Medicines*, *67*(2), pp.271–275. DOI: https://doi.org/10.1007/s11418-012-0676-5.

Asaduzzaman, M., Wahab, M.A., Verdegem, M.C.J., Benerjee, S., Akter, T., Hasan, M.M. and Azim, M.E., 2009. Effects of addition of tilapia *Oreochromis niloticus* and substrates for periphyton developments on pond ecology and production in C/N-controlled freshwater prawn *Macrobrachium rosenbergii* farming systems. *Aquaculture*, *287*(3–4), pp.371–380. DOI: https://doi.org/10.1016/j.aquaculture.2008.11.011

Atlas, R.M. and Hazen, T.C., 2011. Oil biodegradation and bioremediation: a tale of the two worst spills in US history. *Environmental Science and Technology*, *45*(16), pp.6709–6715. DOI: https://doi.org/10.1021/es2013227

Azad, I.S., Shankar, K.M., Mohan, C.V. and Kalita, B., 1999. Biofilm vaccine of *Aeromonas hydrophila* – standardization of dose and duration for oral vaccination of carps. *Fish and Shellfish Immunology*, *9*(7), pp.519–528. DOI: https://doi.org/10.1006/fsim.1998.0206

Azad, I.S., Shankar, K.M., Mohan, C.V. and Kalita, B., 2000. Uptake and processing of biofilm and free-cell vaccines of *Aeromonas hydrophila* in Indian major carps and common carp following oral vaccination antigen localization by a monoclonal antibody. *Diseases of Aquatic Organisms*, 43(2), pp.103–108. DOI:10.3354/dao043103

Azim ME, Wahab MA (2005) Periphyton-based pond polyculture. In Azim MEJ, Verdegem, van Dam AA, Beveridge MCM (Eds.), *Periphyton: Ecology, exploitation and management*. CABI Publishing, vol. 18, 207.

Bagi, A., Pampanin, D.M., Lanzén, A., Bilstad, T. and Kommedal, R., 2014. Naphthalene biodegradation in temperate and arctic marine microcosms. *Biodegradation*, 25(1), pp.111–125. DOI: https://doi.org/10.1007/s10532-013-9644-3

Balaban, N., Gelman, F., Taylor, A.A., Walker, S.L., Bernstein, A., Ronen, Z. 2021. Degradation of brominated organic compounds (Flame Retardants) by a four-strain consortium isolated from contaminated groundwater. *Applied Sciences*, 11(14), pp.6263. DOI: https://doi.org/10.3390/app11146263

Bao, W.Y., Lee, O.O., Chung, H.C., Li, M. and Qian, P.Y., 2010. Copper affects biofilm inductiveness to larval settlement of the serpulid polychaete *Hydroides elegans* (Haswell) *Biofouling*, 26(1), pp.119–128. DOI: https://doi.org/10.1080/0892701 0903329680

Bartrons, M., Grimalt, J.O. and Catalan, J., 2011. Altitudinal distributions of BDE-209 and other polybromodiphenyl ethers in high mountain lakes. *Environmental Pollution*, 159(7), pp.1816–1822. DOI: https://doi.org/10.1016/j.envpol.2011.03.027

Bertino A (2011) *Study on one-stage partial nitritation-anammox process in moving bed biofilm reactors: a sustainable nitrogen removal* (Master Thesis). Royal Institute, DiVA, diva2:530273.

Bhattarai, H.D., Ganti, V.S., Paudel, B., Lee, Y.K., Lee, H.K., Hong, Y.K. and Shin, H.W., 2007. Isolation of antifouling compounds from the marine bacterium, *Shewanella oneidensis* SCH0402. *World Journal of Microbiol and Biotechnology*, 23(2), pp.243–249. DOI: https://doi.org/10.1007/s11274-006-9220-7.

Brown, G.D., Wong, H.F., Hutchinson, N., Lee, S.C., Chan, B.K. and Williams, G.A., 2004. Chemistry and biology of maculalactone A from the marine cyanobacterium *Kyrtuthrix maculans*. *Phytochemistry Reviews*, 3(3), pp.381–400. DOI: https://doi.org/10.1007/s11101-004-6552-5.

Burford, M.A., Sellars, M.J., Arnold, S.J., Keys, S.J., Crocos, P.J. and Preston, N.P., 2004. Contribution of the natural biota associated with substrates to the nutritional requirements of the post-larval shrimp, *Penaeus esculentus* (Haswell), in high-density rearing systems. *Aquaculture Research*, 35(5), pp.508–515. DOI: https://doi.org/10.1111/j.1365-2109.2004.01052.x

Busetti, A., Thompson, T.P., Tegazzini, D., Megaw, J., Maggs, C.A. and Gilmore, B.F., 2015. Antibiofilm activity of the brown alga *Halidrys siliquosa* against clinically relevant human pathogens. *Marine Drugs*, 13(6), pp.3581–3605. DOI: https://doi.org/10.3390/md13063581.

Callow, J.A. and Callow, M.E., 2011. Trends in the development of environmentally friendly fouling-resistant marine coatings. *Nature Communications*, 2(1), pp.1–10. DOI: https://doi.org/10.1038/ncomms1251.

Campbell, B.J., Engel, A.S., Porter, M.L. and Takai, K., 2006. The versatile ε-proteobacteria: key players in sulphidic habitats. *Nat Rev Microbiol*, 4(6), pp.458–468. DOI: https://doi.org/10.1038/nrmicro1414

Carneiro, R.F., Viana, J.T., Torres, R.C.F., da Silva, L.T., Andrade, A.L., de Vasconcelos, M.A., Pinheiro, U., Teixeira, E.H., Nagano, C.S. and Sampaio, A.H., 2019. A new mucin-binding lectin from the marine sponge *Aplysina fulva* (AFL) exhibits antibiofilm effects. *Archives of Biochemistry and Biophysics*, 662, pp.169–176. DOI: https://doi.org/10.1016/j.abb.2018.12.014.

Cerniglia, C.E., White, G.L. and Heflich, R.H., 1985. Fungal metabolism and detoxification of polycyclic aromatic hydrocarbons. *Archives of Microbiol*, *143*(2), pp.105–110. DOI: https://doi.org/10.1007/BF00411031

Chang, Y.C., Okeke, B.C., Hatsu, M. and Takamizawa, K., 2001. In vitro dehalogenation of tetrachloroethylene (PCE) by cell-free extracts of Clostridium bifermentans DPH-1. *Bioresource Technology*, *78*(2), pp.141–147. DOI: https://doi.org/10.1016/S0960-8524(01)00005-0

Chen, M., Shao, C.L., Wang, K.L., Xu, Y., She, Z.G. and Wang, C.Y., 2014. Dihydroisocoumarin derivatives with antifouling activities from a gorgonian-derived *Eurotium* sp. fungus. *Tetrahedron*, 70(47), pp.9132–9138. DOI: https://doi.org/10.1016/j.tet.2014.08.055.

Cho, J.Y. and Kim, M.S., 2012. Induction of antifouling diterpene production by *Streptomyces cinnabarinus* PK209 in co-culture with marine-derived *Alteromonas* sp. KNS-16. *Bioscience, Biotechnology, and Biochemistry*, 76(10), pp.1849–1854. DOI: https://doi.org/10.1271/bbb.120221.

Christenson, L. and Sims, R., 2011. Production and harvesting of microalgae for wastewater treatment, biofuels, and bioproducts. *Biotechnology Advances*, *29*(6), pp.686–702. DOI: https://doi.org/10.1016/j.biotechadv.2011.05.015

Chung, H.C., Lee, O.O., Huang, Y.L., Mok, S.Y., Kolter, R. and Qian, P.Y., 2010. Bacterial community succession and chemical profiles of subtidal biofilms in relation to larval settlement of the polychaete *Hydroides elegans*. *The ISME Journal*, *4*(6), pp.817–828. DOI: https://doi.org/10.1038/ismej.2009.157

Dabrock, B., Riedel, J., Bertram, J. and Gottschalk, G., 1992. Isopropylbenzene (cumene) – a new substrate for the isolation of trichloroethene-degrading bacteria. *Archives of Microbiol*, *158*(1), pp.9–13. DOI: https://doi.org/10.1007/BF00249058

Dafforn, K.A., Lewis, J.A. and Johnston, E.L., 2011. Antifouling strategies: history and regulation, ecological impacts and mitigation. *Marine Pollution Bulletin*, *62*(3), pp.453–465. DOI: https://doi.org/10.1016/j.marpolbul.2011.01.012.

Dahms, H.U. and Dobretsov, S., 2017. Antifouling compounds from marine macroalgae. *Marine Drugs*, 15(9), p.265. DOI: https://doi.org/10.3390/md15090265.

Daigger, G.T., 2014. Oxygen and carbon requirements for biological nitrogen removal processes accomplishing nitrification, nitritation, and anammox. *Water Environment Research*, *86*(3), pp.204–209. DOI: https://doi.org/10.2175/106143013X13807328849459

Dalmo, R.A., Leifson, R.M. and Bøgwald, J., 1995. Microspheres as antigen carriers: studies on intestinal absorption and tissue localization of polystyrene microspheres in Atlantic salmon, *Salmo salar* L. *Journal of Fish Diseases*, *18*(1), pp.87–91. DOI: https://doi.org/10.1111/j.1365-2761.1995.tb01270.x

Dang, H. and Lovell, C.R., 2016. Microbial surface colonization and biofilm development in marine environments. *Microbiol and Molecular Biology Reviews*, 80(1), pp.91–138. DOI: https://doi.org/10.1128/MMBR.00037-15.

Danko, A.S., Luo, M., Bagwell, C.E., Brigmon, R.L. and Freedman, D.L., 2004. Involvement of linear plasmids in aerobic biodegradation of vinyl chloride. *Applied and Environmental Microbiol*, *70*(10), pp.6092–6097. DOI: https://doi.org/10.1128/AEM.70.10.6092-6097.2004

Daume, S., Brand, S. and Woelkerling, J., 1997. Effects of post-larval abalone (*Haliotis rubra*) grazing on the epiphytic diatom assemblage of coralline red algae. *Molluscan Research*, *18*(2), pp.119–130. DOI: https://doi.org/10.1080/13235818.1997.10673686

Daume, S., Brand-Gardner, S. and Woelkerling, W.J., 1999. Settlement of abalone larvae (*Haliotis laevigata* Donovan) in response to non-geniculate coralline red algae (Corallinales, Rhodophyta) *Journal of Experimental Marine Biology and Ecology*, 234(1), pp.125–143. DOI: https://doi.org/10.1016/S0022-0981(98)00143-9

De Carvalho, C.C. and Caramujo, M.J., 2012. Lipids of prokaryotic origin at the base of marine food webs. *Marine Drugs*, *10*(12), pp.2698–2714. DOI: https://doi.org/10.3390/md10122698

De Carvalho, C.C. and Caramujo, M.J., 2017. Carotenoids in aquatic ecosystems and aquaculture: a colorful business with implications for human health. *Frontiers in Marine Science*, *4*, p.93. DOI: https://doi.org/10.3389/fmars.2017.00093

Decho, A.W., 2000. Microbial biofilms in intertidal systems: an overview. *Continental Shelf Research*, *20*(10–11), pp.1257–1273. DOI: https://doi.org/10.1016/S0278-4343(00)00022-4

DeWeerd, K.A., Mandelco, L., Tanner, R.S., Woese, C.R. and Suflita, J.M., 1990. *Desulfomonile tiedjei* gen. nov. and sp. nov., a novel anaerobic, dehalogenating, sulfate-reducing bacterium. *Archives of Microbiol*, *154*(1), pp.23–30. DOI: https://doi.org/10.1007/BF00249173

Díaz, Y.M., Laverde, G.V., Gamba, L.R., Wandurraga, H.M., Arévalo-Ferro, C., Rodríguez, F.R., Beltrán, C.D. and Hernández, L.C., 2015. Biofilm inhibition activity of compounds isolated from two Eunicea species collected at the Caribbean Sea. *Revista Brasileira de Farmacognosia*, 25, pp.605–611. DOI: https://doi.org/10.1016/j.bjp.2015.08.007.

Dolinová, I., Štrojsová, M., Černík, M., NěmeTimes New Romanek, J., Macháčková, J. and Ševců, A., 2017. Microbial degradation of chloroethenes: a review. *Environmental Science and Pollution Research*, *24*(15), pp.13262–13283. DOI: https://doi.org/10.1007/s11356-017-8867-y

Duff, D.C.B. 1942. The oral immunization of trout against *Bacterium salmonicida*. *Journal of Immunology*, 44:87–94.

Dutta, T.K. and Harayama, S., 2001. Biodegradation of n-alkylcycloalkanes and n-alkylbenzenes via new pathways in *Alcanivorax* sp. strain MBIC 4326. *Applied and Environmental Microbiol*, *67*(4), pp.1970–1974. DOI: https://doi.org/10.1128/AEM.67.4.1970-1974.2001

Dyksterhouse, S.E., Gray, J.P., Herwig, R.P., Lara, J.C. and Staley, J.T., 1995. *Cycloclasticus pugetii* gen. nov., sp. nov., an aromatic hydrocarbon-degrading bacterium from marine sediments. *International Journal of Systematic and Evolutionary Microbiol*, *45*(1), pp.116–123. DOI: https://doi.org/10.1099/00207713-45-1-116

Egan, S., Thomas, T. and Kjelleberg, S., 2008. Unlocking the diversity and biotechnological potential of marine surface associated microbial communities. *Current Opinion in Microbiol*, *11*(3), pp.219–225. DOI: https://doi.org/10.1016/j.mib.2008.04.001

Falas, P., Longrée, P., la Cour Jansen, J., Siegrist, H., Hollender, J. and Joss, A., 2013. Micropollutant removal by attached and suspended growth in a hybrid biofilm-activated sludge process. *Water Research*, *47*(13), pp.4498–4506. DOI: https://doi.org/10.1016/j.watres.2013.05.010

Fathepure, B.Z. and Boyd, S.A., 1988. Reductive dechlorination of perchloroethylene and the role of methanogens. *FEMS Microbiol Letters*, *49*(2), pp.149–156. DOI: https://doi.org/10.1111/j.1574-6968.1988.tb02706.x

Freese, E., Rütters, H., Köster, J., Rullkötter, J. and Sass, H., 2009. Gammaproteobacteria as a possible source of eicosapentaenoic acid in anoxic intertidal sediments. *Microbial Ecology*, *57*(3), pp.444–454. DOI: https://doi.org/10.1007/s00248-008-9443-2

Gallego, S., Vila, J., Tauler, M., Nieto, J.M., Breugelmans, P., Springael, D. and Grifoll, M., 2014. Community structure and PAH ring-hydroxylating dioxygenase genes of a marine pyrene-degrading microbial consortium. *Biodegradation*, *25*(4), pp.543–556. DOI: https://doi.org/10.1007/s10532-013-9680-z

Gao, F., Li, C., Yang, Z.H., Zeng, G.M., Feng, L.J., Liu, J.Z., Liu, M. and Cai, H.W., 2016. Continuous microalgae cultivation in aquaculture wastewater by a membrane photobioreactor for biomass production and nutrients removal. *Ecological Engineering*, *92*, pp.55–61. DOI: https://doi.org/10.1016/j.ecoleng.2016.03.046

Gil-Turnes, MS, Fenical, W, 1992. Embryos of *Homarus americanus* are protected by epibiotic bacterial. *Biological Bulletin, 182*(1), pp.105–108.

Golyshin, P.N., Chernikova, T.N., Abraham, W.R., Lünsdorf, H., Timmis, K.N. and Yakimov, M.M., 2002. *Oleiphilaceae* fam. nov., to include *Oleiphilus messinensis* gen. nov., sp. nov., a novel marine bacterium that obligately utilizes hydrocarbons. *International Journal of Systematic and Evolutionary Microbiol*, *52*(3), pp.901–911. DOI: https://doi.org/10.1099/00207713-52-3-901

Guérin, T., Sirot, V., Volatier, J.L. and Leblanc, J.C., 2007. Organotin levels in seafood and its implications for health risk in high-seafood consumers. *Science of the Total Environment*, 388(1–3), pp.66–77. DOI: https://doi.org/10.1016/j.scitotenv.2007.08.027.

Hadfield, M.G., 2011. Biofilms and marine invertebrate larvae: what bacteria produce that larvae use to choose settlement sites. *Annual Review of Marine Science*, *3*, pp.453–470. DOI: https://doi.org/10.1146/annurev-marine-120709-142753

Hall-Stoodley, L., Costerton, J.W. and Stoodley, P., 2004. Bacterial biofilms: from the natural environment to infectious diseases. *Nat Rev Microbiol*, *2*(2), pp.95–108. DOI: https://doi.org/10.1038/nrmicro821

Han, W., Mao, Y., Wei, Y., Shang, P. and Zhou, X., 2020. Bioremediation of aquaculture wastewater with algal-bacterial biofilm combined with the production of selenium rich biofertilizer. *Water*, *12*(7), p.2071. DOI: https://doi.org/10.3390/w12072071

Harayama, S., Kasai, Y. and Hara, A., 2004. Microbial communities in oil-contaminated seawater. *Current Opinion in Biotechnology*, *15*(3), pp.205–214. DOI: https://doi.org/10.1016/j.copbio.2004.04.002

Harris BC, Dimitriou-Christidis P, Sterling MC, Autenrieth RL, Bonner JS, McDonald TJ, Fuller CB, Page CA (2002, July) *Bioavailability of chemically-dispersed crude oil*. Arctic and Marine Oilspill Program Program Technical Seminar. 2:895–906. Environment Canada.

Head, I.M., Jones, D.M. and Röling, W.F., 2006. Marine microorganisms make a meal of oil. *Nat Rev Microbiol*, *4*(3), pp.173–182. DOI: https://doi.org/10.1038/nrmicro1348

Hong, Y.K. and Cho, J.Y., 2013. Effect of seaweed epibiotic bacterium Streptomyces violaceoruber SCH-09 on marine fouling organisms. *Fisheries Science*, 79(3), pp.469–475. DOI: https://doi.org/10.1007/s12562-013-0604-y.

Iwabuchi, N., Sunairi, M., Urai, M., Itoh, C., Anzai, H., Nakajima, M. and Harayama, S., 2002. Extracellular polysaccharides of *Rhodococcus rhodochrous* S-2 stimulate the degradation of aromatic components in crude oil by indigenous marine bacteria. *Applied and Environmental Microbiol*, *68*(5), pp.2337–2343. DOI: https://doi.org/10.1128/AEM.68.5.2337-2343.2002

Iyer, A., Mody, K. and Jha, B., 2005. Biosorption of heavy metals by a marine bacterium. *Marine Pollution Bulletin*, *50*(3), pp.340–343. DOI: https://doi.org/10.1016/j.marpolbul.2004.11.012

Jha, R.K. and Zi-Rong, X., 2004. Biomedical compounds from marine organisms. *Marine Drugs*, 2(3), pp.123–146. DOI: https://doi.org/10.3390/md203123

Johnson, K.A. and Amend, D.F., 1983. Efficacy of *Vibrio anguillarum* and *Yersinia ruckeri* bacterins applied by oral and anal intubation of salmonids. *Journal of Fish Diseases*, 6(5), pp.473–476. DOI: https://doi.org/10.1111/j.1365-2761.1983.tb00101.x

Ju, Z.Y., Forster, I., Conquest, L. and Dominy, W., 2009. Enhanced growth effects on shrimp (*Litopenaeus vannamei*) from inclusion of whole shrimp floc or floc fractions to a formulated diet. *Aquaculture Nutrition*, *14*(6), pp.533–543. DOI: https://doi.org/10.1111/j.1365-2095.2007.00559.x

Jun, J.Y., Jung, M.J., Jeong, I.H., Yamazaki, K., Kawai, Y. and Kim, B.M., 2018. Antimicrobial and antibiofilm activities of sulfated polysaccharides from marine algae against dental plaque bacteria. *Marine Drugs*, 16(9), p.301. DOI: https://doi.org/10.3390/md16090301.

Kanaly, R.A. and Harayama, S., 2010. Advances in the field of high-molecular-weight polycyclic aromatic hydrocarbon biodegradation by bacteria. *Microbial Biotechnology*, *3*(2), pp.136–164. DOI: https://doi.org/10.1111/j.1751-7915.2009.00130.x

Kao, C.M., Chen, S.C. and Liu, J.K., 2001. Development of a biobarrier for the remediation of PCE-contaminated aquifer. *Chemosphere*, *43*(8), pp.1071–1078. DOI: https://doi.org/10.1016/S0045-6535(00)00190-9

Kaplan, D., Christiaen, D. and Arad, S., 1987. Chelating properties of extracellular polysaccharides from *Chlorella* spp. *Applied and Environmental Microbiol*, *53*(12), pp.2953–2956. DOI: https://doi.org/10.1128/aem.53.12.2953-2956.1987

Kartal, B., de Almeida, N.M., Maalcke, W.J., Op den Camp, H.J., Jetten, M.S. and Keltjens, J.T., 2013. How to make a living from anaerobic ammonium oxidation. *FEMS Microbiol Reviews*, *37*(3), pp.428–461. DOI: 10.1111/1574-6976.12014.

Kasai, Y., Shindo, K., Harayama, S. and Misawa, N., 2003. Molecular characterization and substrate preference of a polycyclic aromatic hydrocarbon dioxygenase from *Cycloclasticus* sp. strain A5. *Applied and Environmental Microbiol*, *69*(11), pp.6688–6697. DOI: https://doi.org/10.1128/AEM.69.11.6688-6697.2003

Kim, E.S., Nomura, I., Hasegawa, Y. and Takamizawa, K., 2006. Characterization of a newly isolatedcis-1, 2-dichloroethylene and aliphatic compound-degrading bacterium, *Clostridium* sp. strain KYT-1. *Biotechnology and Bioprocess Engineering*, *11*(6), pp.553–556. DOI: https://doi.org/10.1007/BF02932083

Kim, S.J., Kweon, O., Sutherland, J.B., Kim, H.L., Jones, R.C., Burback, B.L., Graves, S.W., Psurny, E. and Cerniglia, C.E., 2015. Dynamic response of *Mycobacterium vanbaalenii* PYR-1 to BP Deepwater Horizon crude oil. *Applied and Environmental Microbiol*, *81*(13), pp.4263–4276. DOI: https://doi.org/10.1128/AEM.00730-15

Kirchman D, Mitchell R (1983) Biochemical interactions between microorganisms and marine fouling invertebrates. In Oxley TA, Barry S (Eds.), *Biodeterioration*. John Wiley and Sons, vol. 5, 281–290.

Kong, J.Y., Lee, H.W., Hong, J.W., Kang, Y.S., Kim, J.D., Chang, M.W. and Bae, S.K., 1998. Utilization of a cell-bound polysaccharide produced by the marine bacterium *Zoogloea* sp.: New biomaterial for metal adsorption and enzyme immobilization. *Journal of Marine Biotechnology*, *6*(2), pp.0099–0103.

Krumholz, L.R., 1997. *Desulfuromonas chloroethenica* sp. nov. uses tetrachloroethylene and trichloroethylene as electron acceptors. *International Journal of Systematic and Evolutionary Microbiol*, *47*(4), pp.1262–1263. DOI: https://doi.org/10.1099/00207713-47-4-1262

Kumar, J., Lind, P.M., Salihovic, S., van Bavel, B., Ekdahl, K.N., Nilsson, B., Lind, L. and Ingelsson, E., 2014. Influence of persistent organic pollutants on the complement system in a population-based human sample. *Environment International*, *71*, pp.94–100. DOI: https://doi.org/10.1016/j.envint.2014.06.009

Kurane Ryuichiro, 1997. Microbial degradation and treatment of polycyclic aromatic hydrocarbons and plasticizers. *Annals of the New York Academy of Sciences*, *829*, pp.118–134. DOI: https://doi.org/10.1111/j.1749-6632.1997.tb48570.x

Kwong, T.F.N., Miao, L., Li, X. and Qian, P.Y., 2006. Novel antifouling and antimicrobial compound from a marine-derived fungus *Ampelomyces* sp. *Marine Biotechnology*, *8*(6), pp.634–640. DOI: 10.1007/s10126-005-6146-2.

Langford, K., Scrimshaw, M. and Lester, J., 2007. The impact of process variables on the removal of PBDEs and NPEOs during simulated activated sludge treatment. *Archives of Environmental Contamination and Toxicology*, *53*(1), pp.1–7. DOI: https://doi.org/10.1007/s00244-006-0052-0

Lee, J.W., Nam, J.H., Kim, Y.H., Lee, K.H. and Lee, D.H., 2008. Bacterial communities in the initial stage of marine biofilm formation on artificial surfaces. *The Journal of Microbiol*, *46*(2), pp.174–182. DOI: https://doi.org/10.1007/s12275-008-0032-3

Lee, O.O., Chung, H.C., Yang, J., Wang, Y., Dash, S., Wang, H. and Qian, P.Y., 2014. Molecular techniques revealed highly diverse microbial communities in natural marine biofilms on polystyrene dishes for invertebrate larval settlement. *Microbial Ecology*, 68(1), pp.81–93. DOI: https://doi.org/10.1007/s00248-013-0348-3

Li, D., Xu, Y., Shao, C.L., Yang, R.Y., Zheng, C.J., Chen, Y.Y., Fu, X.M., Qian, P.Y., She, Z.G., Voogd, N.J.D. and Wang, C.Y., 2012. Antibacterial bisabolane-type sesquiterpenoids from the sponge-derived fungus *Aspergillus* sp. *Marine Drugs*, 10(1), pp.234–241. DOI: https://doi.org/10.3390/md10010234.

Li, S., Zhang, S., Ye, C., Lin, W., Zhang, M., Chen, L., Li, J. and Yu, X., 2017. Biofilm processes in treating mariculture wastewater may be a reservoir of antibiotic resistance genes. *Marine Pollution Bulletin*, 118(1–2), pp.289–296. DOI: https://doi.org/10.1016/j.marpolbul.2017.03.003

Linan-Cabello, M.A., Paniagua-Michel, J. and Hopkins, P.M., 2002. Bioactive roles of carotenoids and retinoids in crustaceans. *Aquaculture Nutrition*, 8(4), pp.299–309. DOI: https://doi.org/10.1046/j.1365-2095.2002.00221.x

Liu, Y., Lv, J., Feng, J., Liu, Q., Nan, F. and Xie, S., 2019. Treatment of real aquaculture wastewater from a fishery utilizing phytoremediation with microalgae. *Journal of Chemical Technology and Biotechnology*, 94(3), pp.900–910. DOI: https://doi.org/10.1002/jctb.5837

Lohner, S.T. and Spormann, A.M., 2013. Identification of a reductive tetrachloroethene dehalogenase in *Shewanella sediminis*. *Philosophical Transactions of the Royal Society B: Biological Sciences*, 368(1616), p.20120326. DOI: https://doi.org/10.1098/rstb.2012.0326

Lolas, I.B., Chen, X., Bester, K. and Nielsen, J.L., 2012. Identification of triclosan-degrading bacteria using stable isotope probing, fluorescence in situ hybridization and micro-autoradiography. *Microbiol*, 158(11), pp.2796–2804. DOI: https://doi.org/10.1099/mic.0.061077-0

Ma, Y., Liu, P., Zhang, Y., Cao, S., Li, D. and Chen, W., 2010. Inhibition of spore germination of *Ulva pertusa* by the marine bacterium *Pseudoalteromonas haloplanktis* CI4. *Acta Oceanologica Sinica*, 29(1), pp.69–78. DOI: https://doi.org/10.1007/s13131-010-0009-z.

Martins, M., Costa, P.M., Raimundo, J., Vale, C., Ferreira, A.M. and Costa, M.H., 2012. Impact of remobilized contaminants in *Mytilus edulis* during dredging operations in a harbour area: Bioaccumulation and biomarker responses. *Ecotoxicology and Environmental Safety*, 85, pp.96–103. DOI: https://doi.org/10.1016/j.ecoenv.2012.08.008

Matsuda, M., Worawattanamateekul, W., Okutani, K., 1992. Simultaneous production of muco- and sulfated polysaccharides by marine *Pseudomonas*. *Nippon Susian Gakkaishi* 58, 1735–1741. DOI: https://doi.org/10.2331/suisan.58.1735

Metz, J.G., Roessler, P., Facciotti, D., Levering, C., Dittrich, F., Lassner, M., Valentine, R., Lardizabal, K., Domergue, F., Yamada, A. and Yazawa, K., 2001. Production of polyunsaturated fatty acids by polyketide synthases in both prokaryotes and eukaryotes. *Science*, 293(5528), pp.290–293. DOI: 10.1126/science.1059593

Michaud, L., Di Marco, G., Bruni, V. and Giudice, A.L., 2007. Biodegradative potential and characterization of psychrotolerant polychlorinated biphenyl-degrading marine bacteria isolated from a coastal station in the Terra Nova Bay (Ross Sea, Antarctica) *Marine Pollution Bulletin*, 54(11), pp.1754–1761. DOI: https://doi.org/10.1016/j.marpolbul.2007.07.011

Mohapatra, B.R. and Bapuji, M., 1998. Characterization of acetylcholinesterase from *Arthrobacter ilicis* associated with the marine sponge (*Spirastrella* sp.). *Journal of Applied Microbiol*, 84(3), pp.393–398. DOI: https://doi.org/10.1046/j.1365-2672.1998.00360.x

Narro, M.L., Cerniglia, C.E., Van Baalen, C. and Gibson, D.T., 1992a. Evidence for an NIH shift in oxidation of naphthalene by the marine cyanobacterium *Oscillatoria* sp. strain JCM. *Applied and Environmental Microbiol*, 58(4), pp.1360–1363. DOI: https://doi.org/10.1128/aem.58.4.1360-1363.1992

Narro, M.L., Cerniglia, C.E., Van Baalen, C.H.A.S.E. and Gibson, D.T., 1992b. Metabolism of phenanthrene by the marine cyanobacterium *Agmenellum quadruplicatum* PR-6. *Applied and Environmental Microbiol*, *58*(4), pp.1351–1359. DOI: https://doi.org/10.1128/aem.58.4.1351-1359.1992

Natrah, F.M., Bossier, P., Sorgeloos, P., Yusoff, F.M. and Defoirdt, T., 2014. Significance of microalgal – bacterial interactions for aquaculture. *Reviews in Aquaculture*, *6*(1), pp.48–61. DOI: https://doi.org/10.1111/raq.12024

Nayak, D.K., Asha, A., Shankar, K.M. and Mohan, C.V., 2004. Evaluation of biofilm of *Aeromonas hydrophila* for oral vaccination of *Clarias batrachus* – a carnivore model. *Fish and Shellfish Immunology*, *16*(5), pp.613–619. DOI: https://doi.org/10.1016/j.fsi.2003.09.012

Nielsen, S.J., Harder, T. and Steinberg, P.D., 2015. Sea urchin larvae decipher the epiphytic bacterial community composition when selecting sites for attachment and metamorphosis. *FEMS Microbiol Ecology*, *91*(1), pp.1–9. DOI: 10.1093/femsec/fiu011

Ortega-Morales, B.O., Chan-Bacab, M.J., De la Rosa, S.D.C. and Camacho-Chab, J.C., 2010. Valuable processes and products from marine intertidal microbial communities. *Current Opinion in Biotechnology*, *21*(3), pp.346–352. DOI: https://doi.org/10.1016/j.copbio.2010.02.007

Ortiz-Estrada, Á.M., Gollas-Galván, T., Martínez-Córdova, L.R., Burgos-Hernández, A., Scheuren-Acevedo, S.M., Emerenciano, M. and Martínez-Porchas, M., 2019. Diversity and bacterial succession of a phototrophic biofilm used as complementary food for shrimp raised in a super-intensive culture. *Aquaculture International*, *27*(2), pp.581–596. DOI: https://doi.org/10.1007/s10499-019-00345-x

Page, C.A., Bonner, J.S., Sumner, P.L., McDonald, T.J., Autenrieth, R.L. and Fuller, C.B., 2000. Behavior of a chemically-dispersed oil and a whole oil on a near-shore environment. *Water Research*, *34*(9), pp.2507–2516. DOI: https://doi.org/10.1016/S0043-1354(99)00398-X

Pissetti TL (2004) *Efeitos da densidade de estocagem e do substrato artificial no cultivo do camarão-rosa Farfantepenaeus paulensis (Pérez-Farfante, 1967) em cercados* (M.Sc. Thesis). Federal University of Rio Grande, 48.

Prabhakaran, S., Rajaram, R., Balasubramanian, V. and Mathivanan, K., 2012. Antifouling potentials of extracts from seaweeds, seagrasses and mangroves against primary biofilm forming bacteria. *Asian Pacific Journal of Tropical Biomedicine*, *2*(1), pp.S316-S322. DOI: https://doi.org/10.1016/S2221-1691(12)60181-6.

Qi, S.H. and Ma, X., 2017. Antifouling compounds from marine invertebrates. *Marine Drugs*, *15*(9), p.263. DOI: https://doi.org/10.3390/md15090263.

Qian, P.Y., Xu, Y. and Fusetani, N., 2009. Natural products as antifouling compounds: recent progress and future perspectives. *Biofouling*, *26*(2), pp.223–234. DOI: https://doi.org/10.1080/08927010903470815.

Rajivgandhi, G., Vijayan, R., Maruthupandy, M., Vaseeharan, B. and Manoharan, N., 2018. Antibiofilm effect of *Nocardiopsis* sp. GRG 1 (KT235640) compound against biofilm forming Gram negative bacteria on UTIs. *Microbial Pathogenesis*, 118, pp.190–198. DOI: https://doi.org/10.1016/j.micpath.2018.03.011.

Ram, M.K., Kumar, B.N., Poojary, S.R., Abhiman, P.B., Patil, P., Ramesh, K.S. and Shankar, K.M., 2019. Evaluation of biofilm of Vibrio anguillarum for oral vaccination of Asian seabass, *Lates calcarifer* (BLOCH, 1790). *Fish and Shellfish Immunology*, 94, pp.746–751. DOI: https://doi.org/10.1016/j.fsi.2019.09.053

Riquelme, C., Hayashida, G., Araya, R., Uchida, A., Satomi, M. and Ishida, Y., 1996. Isolation of a native bacterial strain from the scallop *Argopecten purpuratus* with inhibitory effects against pathogenic *Vibrios*. *Journal of Shellfish research*, *15*(2), pp.369–374.

Rodriguez, S. and Bishop, P.L., 2008. Enhancing the biodegradation of polycyclic aromatic hydrocarbons: Effects of nonionic surfactant addition on biofilm function and structure. *Journal of Environmental Engineering*, *134*(7), pp.505–512. DOI: https://doi.org/10.1061/(ASCE)0733-9372(2008)134:7(505)

Rombout, J.H.W.M., Lamers, C.H.J., Helfrich, M.H., Dekker, A. and Taverne-Thiele, J.J., 1985. Uptake and transport of intact macromolecules in the intestinal epithelium of carp (*Cyprinus carpio* L.) and the possible immunological implications. *Cell and Tissue Research*, *239*(3), pp.519–530. DOI: https://doi.org/10.1007/BF00219230

Rombout, J.H.W.M., Blok, L.J., Lamers, C.H. and Egberts, E., 1986. Immunization of carp (*Cyprinus carpio*) with a *Vibrio anguillarum* bacterin: indications for a common mucosal immune system. *Developmental and Comparative Immunology*, *10*(3), pp.341–351. DOI: https://doi.org/10.1016/0145-305X(86)90024-8

Roques, J.A.C., Micolucci, F., Hosokawa, S., Sundell, K., Kindaichi, T. 2021. Effects of recirculating aquaculture system wastewater on anammox performance and community structure. *Processes*, 9, 1183. DOI: https://doi.org/ 10.3390/pr9071183

Rubini, D., Banu, S.F., Nisha, P., Murugan, R., Thamotharan, S., Percino, M.J., Subramani, P. and Nithyanand, P., 2018. Essential oils from unexplored aromatic plants quench biofilm formation and virulence of Methicillin resistant *Staphylococcus aureus*. *Microbial Pathogenesis*, 122, pp.162–173. DOI: https://doi.org/10.1016/j.micpath.2018.06.028.

Ruiz, J., Olivieri, G., De Vree, J., Bosma, R., Willems, P., Reith, J.H., Eppink, M.H., Kleinegris, D.M., Wijffels, R.H. and Barbosa, M.J., 2016. Towards industrial products from microalgae. *Energy and Environmental Science*, *9*(10), pp.3036–3043. DOI: https://doi.org/10.1039/C6EE01493C

Rusten, B., Eikebrokk, B., Ulgenes, Y. and Lygren, E., 2006. Design and operations of the kaldnes moving bed biofilm reactors. *Aquacultural Engineering*, *34*(3), pp.322–331. DOI: https://doi.org/10.1016/j.aquaeng.2005.04.002

Rusten, B., Hem, L.J. and Ødegaard, H., 1995. Nitrification of municipal wastewater in moving-bed biofilm reactors. *Water Environment Research*, *67*(1), pp.75–86. DOI: https://doi.org/10.2175/106143095X131213

Rusten, B., Siljudalen, J.G. and Nordeidet, B., 1994. Upgrading to nitrogen removal with the KMT moving bed biofilm process. *Water Science and Technology*, *29*(12), p.185.

Santos, A.L.S.D., Galdino, A.C.M., Mello, T.P.D., Ramos, L.D.S., Branquinha, M.H., Bolognese, A.M., Columbano Neto, J. and Roudbary, M., 2018. What are the advantages of living in a community? A microbial biofilm perspective!. *Memórias do Instituto Oswaldo Cruz*, 113(9). DOI: https://doi.org/10.1590/0074-02760180212.

Satheesh, S., Soniamby, A.R., Shankar, C.S. and Punitha, S.M.J., 2012. Antifouling activities of marine bacteria associated with sponge (*Sigmadocia* sp.). *Journal of Ocean University of China*, 11(3), pp.354–360. DOI: 10.1007/s11802-012-1927-5.

Schultz, M.P., Bendick, J.A., Holm, E.R. and Hertel, W.M., 2011. Economic impact of biofouling on a naval surface ship. *Biofouling*, 27(1), pp.87–98. DOI: 10.1080/08927014.2010.542809.

Schwartz, N., Dobretsov, S., Rohde, S. and Schupp, P.J., 2017. Comparison of antifouling properties of native and invasive *Sargassum* (Fucales, Phaeophyceae) species. *European Journal of Phycology*, 52(1), pp.116–131. DOI: https://doi.org/10.1080/09670262.2016.1231345.

Seo, Y., Lee, W.H., Sorial, G. and Bishop, P.L., 2009. The application of a mulch biofilm barrier for surfactant enhanced polycyclic aromatic hydrocarbon bioremediation. *Environmental Pollution*, 157(1), pp.95–101. DOI: https://doi.org/10.1016/j.envpol.2008.07.022

Shao, C.L., Wang, C.Y., Wei, M.Y., Gu, Y.C., She, Z.G., Qian, P.Y. and Lin, Y.C., 2011a. Aspergilones A and B, two benzylazaphilones with an unprecedented carbon skeleton from the gorgonian-derived fungus *Aspergillus* sp. *Bioorganic and Medicinal Chemistry Letters*, 21(2), pp.690–693. DOI: https://doi.org/10.1016/j.bmcl.2010.12.005.

Shao, C.L., Wu, H.X., Wang, C.Y., Liu, Q.A., Xu, Y., Wei, M.Y., Qian, P.Y., Gu, Y.C., Zheng, C.J., She, Z.G. and Lin, Y.C., 2011b. Potent antifouling resorcylic acid lactones from the gorgonian-derived fungus *Cochliobolus lunatus*. *Journal of Natural Products*, 74(4), pp.629–633. DOI: https://doi.org/10.1021/np100641b.

Shao, C.L., Xu, R.F., Wang, C.Y., Qian, P.Y., Wang, K.L. and Wei, M.Y., 2015. Potent antifouling marine dihydroquinolin-2 (1 H)-one-containing alkaloids from the gorgonian coral-derived fungus Scopulariopsis sp. *Marine Biotechnology*, 17(4), pp.408–415. https://doi.org/10.1007/s10126-015-9628-x.

Sharma, P.K. and McCarty, P.L., 1996. Isolation and characterization of a facultatively aerobic bacterium that reductively dehalogenates tetrachloroethene to cis-1, 2-dichloroethene. *Applied and Environmental Microbiol*, 62(3), pp.761–765. DOI: https://doi.org/10.1128/aem.62.3.761-765.1996

Shivaraman, N. and Shivaraman, G., 2003. Anammox-A novel microbial process for ammonium removal. *Current Science-Bangalore*, 84(12), pp.1507–1508.

Silkina, A., Bazes, A., Mouget, J.L. and Bourgougnon, N., 2012. Comparative efficiency of macroalgal extracts and booster biocides as antifouling agents to control growth of three diatom species. *Marine Pollution Bulletin*, 64(10), pp.2039–2046. DOI: https://doi.org/10.1016/j.marpolbul.2012.06.028.

Stroband, H.W.J. and Kroon, A.G., 1981. The development of the stomach in *Clarias lazera* and the intestinal absorption of protein macromolecules. *Cell and Tissue Research*, 215(2), pp.397–415. DOI: http://dx.doi.org/10.1007/BF00239123

Suttinun, O., Müller, R. and Luepromchai, E., 2009. Trichloroethylene cometabolic degradation by *Rhodococcus* sp. L4 induced with plant essential oils. *Biodegradation*, 20(2), pp.281–291. DOI: https://doi.org/10.1007/s10532-008-9220-4

Swannell, R.P.J., Mitchell, D., Lethbridge, G., Jones, D., Heath, D., Hagley, M., Jones, M., Petch, S., Milne, R., Croxford, R. and Lee, K., 1999. A field demonstration of the efficacy of bioremediation to treat oiled shorelines following the sea empress incident. *Environmental Technology*, 20(8), pp.863–873. DOI: https://doi.org/10.1080/09593332008616881

Tello, E., Castellanos, L., Arevalo-Ferro, C., Rodríguez, J., Jiménez, C. and Duque, C., 2011. Absolute stereochemistry of antifouling cembranoid epimers at C-8 from the Caribbean octocoral *Pseudoplexaura flagellosa*. Revised structures of plexaurolones. *Tetrahedron*, 67(47), pp.9112–9121. DOI: https://doi.org/10.1016/j.tet.2011.09.094.

Thompson, F.L., Abreu, P.C. and Wasielesky, W., 2002. Importance of biofilm for water quality and nourishment in intensive shrimp culture. *Aquaculture*, 203(3–4), pp.263–278. DOI: https://doi.org/10.1016/S0044-8486(01)00642-1

Timmons, M.B., Holder, J.L. and Ebeling, J.M., 2006. Application of microbead biological filters. *Aquacultural Engineering*, 34(3), pp.332–343. DOI: https://doi.org/10.1016/j.aquaeng.2005.07.003

Tokarz Iii, J.A., Ahn, M.Y., Leng, J., Filley, T.R. and Nies, L., 2008. Reductive debromination of polybrominated diphenyl ethers in anaerobic sediment and a biomimetic system. *Environmental Science and Technology*, 42(4), pp.1157–1164. DOI: https://doi.org/10.1021/es071989t

Tyagi, M., da Fonseca, M.M.R. and de Carvalho, C.C., 2011. Bioaugmentation and biostimulation strategies to improve the effectiveness of bioremediation processes. *Biodegradation*, 22(2), pp.231–241. DOI: https://doi.org/10.1007/s10532-010-9394-4

Unabia, C.R.C. and Hadfield, M.G., 1999. Role of bacteria in larval settlement and metamorphosis of the polychaete *Hydroides elegans*. *Marine Biology*, 133(1), pp.55–64. DOI: https://doi.org/10.1007/s002270050442

Uyttebroek, M., Spoden, A., Ortega-Calvo, J.J., Wouters, K., Wattiau, P., Bastiaens, L. and Springael, D., 2007. Differential Responses of Eubacterial, *Mycobacterium*, and *Sphingomonas* communities in polycyclic aromatic hydrocarbon (PAH)-contaminated soil to artificially induced changes in PAH profile. *Journal of Environmental Quality*, 36(5), pp.1403–1411. DOI: https://doi.org/10.2134/jeq2006.0471

Vaikundamoorthy, R., Rajendran, R., Selvaraju, A., Moorthy, K. and Perumal, S., 2018. Development of thermostable amylase enzyme from *Bacillus cereus* for potential anti-biofilm activity. *Bioorganic Chemistry*, 77, pp.494–506. DOI: https://doi.org/10.1016/j.bioorg.2018.02.014.

Vinay, T.N., Girisha, S.K., D'souza, R., Jung, M.H., Choudhury, T.G. and Patil, S.S., 2016. Bacterial biofilms as oral vaccine candidates in aquaculture. *Indian Journal of Comparative Microbiol, Immunology and Infectious Diseases*, 37(2), pp.57–62. DOI: http://dx.doi.org/10.5958/0974-0147.2016.00011.8

Wackett, L.P. and Gibson, D.T., 1988. Degradation of trichloroethylene by toluene dioxygenase in whole-cell studies with *Pseudomonas putida* F1. *Applied and Environmental Microbiol*, 54(7), pp.1703–1708. DOI: https://doi.org/10.1128/aem.54.7.1703-1708.1988

Walter, U., Beyer, M., Klein, J. and Rehm, H.J., 1991. Degradation of pyrene by *Rhodococcus* sp. UW1. *Applied Microbiol and Biotechnology*, 34(5), pp.671–676. DOI: https://doi.org/10.1007/BF00167921

Wang, C., Bao, W.Y., Gu, Z.Q., Li, Y.F., Liang, X., Ling, Y., Cai, S.L., Shen, H.D. and Yang, J.L., 2012. Larval settlement and metamorphosis of the mussel *Mytilus coruscus* in response to natural biofilms. *Biofouling*, 28(3), pp.249–256. DOI: https://doi.org/10.1080/08927014.2012.671303

Wang, H., Qi, J., Dong, Y., Li, Y., Xu, X. and Zhou, G., 2017b. Characterization of attachment and biofilm formation by meat-borne Enterobacteriaceae strains associated with spoilage. *LWT*, 86, pp.399–407. DOI: https://doi.org/10.1016/j.lwt.2017.08.025.

Wang, H., Wang, H., Xing, T., Wu, N., Xu, X. and Zhou, G., 2016. Removal of *Salmonella* biofilm formed under meat processing environment by surfactant in combination with bio-enzyme. *LWT-Food Science and Technology*, 66, pp.298–304. DOI: https://doi.org/10.1016/j.lwt.2015.10.049.

Wang, K.L., Wu, Z.H., Wang, Y., Wang, C.Y. and Xu, Y., 2017a. Mini-review: antifouling natural products from marine microorganisms and their synthetic analogs. *Marine Drugs*, 15(9), p.266. DOI: https://doi.org/10.3390/md15090266.

Wang, W., Wang, L. and Shao, Z., 2018. Polycyclic aromatic hydrocarbon (PAH) degradation pathways of the obligate marine PAH degrader *Cycloclasticus* sp. strain P1. *Applied and Environmental Microbiol*, 84(21), pp. e01261–18. DOI: https://doi.org/10.1128/AEM.01261-18

Wang, Y., Lau, P.C. and Button, D.K., 1996. A marine oligobacterium harboring genes known to be part of aromatic hydrocarbon degradation pathways of soil pseudomonads. *Applied and Environmental Microbiol*, 62(6), pp.2169–2173. DOI: https://doi.org/10.1128/aem.62.6.2169-2173.1996

Weiland-Bräuer, N., Fischer, M.A., Schramm, K.W. and Schmitz, R.A., 2017. Polychlorinated biphenyl (PCB)-degrading potential of microbes present in a cryoconite of Jamtalferner glacier. *Front Microbiol.*, 8, p.1105. DOI: https://doi.org/10.3389/fmicb.2017.01105

Weiner R (1993) Periphytic bacteria cue oyster larvae to set on fertile benthic biofil larvae by microbial biofilm cues. *Biofouling.* 12:81–93.

Whalan, S. and Webster, N.S., 2014. Sponge larval settlement cues: the role of microbial biofilms in a warming ocean. *Scientific Reports*, 4(1), pp.1–5. DOI: https://doi.org/10.1038/srep04072

Woollacott RM, Hadfield MG. (1996). Induction of metamorphosis in larvae of a sponge. *Invertebr Biol.* 257–262.

Xing, Q., Gan, L.S., Mou, X.F., Wang, W., Wang, C.Y., Wei, M.Y. and Shao, C.L., 2016. Isolation, resolution and biological evaluation of pestalachlorides E and F containing both point and axial chirality. *RSC Advances*, 6(27), pp.22653–22658. DOI: https://doi.org/10.1039/C6RA00374E.

Xiong, J.Q., Kurade, M.B. and Jeon, B.H., 2018. Can microalgae remove pharmaceutical contaminants from water? *Trends in biotechnology*, 36(1), pp.30–44. DOI: https://doi.org/10.1016/j.tibtech.2017.09.003

Xu, Y., Li, H., Li, X., Xiao, X. and Qian, P.Y., 2009. Inhibitory effects of a branched-chain fatty acid on larval settlement of the polychaete *Hydroides elegans*. *Marine Biotechnology*, 11(4), pp.495–504. DOI: https://doi.org/10.1007/s10126-008-9161-2.

Yakimov, M.M., Timmis, K.N. and Golyshin, P.N., 2007. Obligate oil-degrading marine bacteria. *Current Opinion in Biotechnology*, *18*(3), pp.257–266. DOI: https://doi.org/10.1016/j.copbio.2007.04.006

Yebra, D.M., Kiil, S. and Dam-Johansen, K., 2004. Antifouling technology – past, present and future steps towards efficient and environmentally friendly antifouling coatings. *Progress in Organic Coatings*, 50(2), pp.75–104. DOI: https://doi.org/10.1016/j.porgcoat.2003.06.001

Yoshida, K., Yamamoto, T., Kuroki, T. and Okubo, M., 2009. Pilot-scale experiment for simultaneous dioxin and NOx removal from garbage incinerator emissions using the pulse corona induced plasma chemical process. *Plasma chemistry and plasma processing*, *29*(5), pp.373–386. DOI: https://doi.org/10.1007/s11090-009-9184-0

Yuan, H.L., Yang, J.S., Wang, Z.S., Li, B.Z., Zhang, L. and Lin, R.Z., 2003. Microorganism screening for petroleum degradation and its degrading characteristics. *China Environmental Science*, *23*(2), pp.157–161.

Yufen, L., Jingyuan, Y. and Chuangxin, L., 2009. Biomimetic medium purifies rural micro-polluted water sources. *Environmental Science and Management*, *34*(7), pp.95–97.

Zhang, H., Wang, H., Jie, M., Zhang, K., Qian, Y. and Ma, J., 2020. Performance and microbial communities of different biofilm membrane bioreactors with pre-anoxic tanks treating mariculture wastewater. *Bioresource Technology*, 295, p.122302. DOI: https://doi.org/10.1016/j.biortech.2019.122302

Zhang, R., Han, Z.Y., Chen, Z.J., Shi, D.Z., Huang, X.X. and Wu, W.X., 2011. Microstructure and microbial ecology of biofilm in the bioreactor for nitrogen removing from wastewater: a review. *Chinese journal of ecology*, 30, pp.2628–2636.

Zhao, B., Zhang, S. and Qian, P.Y., 2003. Larval settlement of the silver-or goldlip pearl oyster *Pinctada maxima* (Jameson) in response to natural biofilms and chemical cues. *Aquaculture*, *220*(1–4), pp.883–901. DOI: https://doi.org/10.1016/S0044-8486(02)00567-7

Zheng, C.J., Shao, C.L., Wu, L.Y., Chen, M., Wang, K.L., Zhao, D.L., Sun, X.P., Chen, G.Y. and Wang, C.Y., 2013. Bioactive phenylalanine derivatives and cytochalasins from the soft coral-derived fungus, *Aspergillus elegans*. *Marine Drugs*, 11(6), pp.2054–2068. DOI: https://doi.org/10.3390/md11062054

Zhu, L., Li, Z. and Ketola, T., 2011. Biomass accumulations and nutrient uptake of plants cultivated on artificial floating beds in China's rural area. *Ecological Engineering*, *37*(10), pp.1460–1466. DOI: https://doi.org/10.1016/j.ecoleng.2011.03.010

Part III

Control of Microbial Biofilms

Emerging Methods

13 Use of Nanotechnology for Biofilm Mitigation

Pabbati Ranjit, Sai Manisha, Kondakindi Venkateswar Reddy, and Palakeerti Srinivas Kumar

CONTENTS

13.1 INTRODUCTION

The majority of human infections prevailing at higher rates are due to the formation of biofilms, particularly bacterial species. Biofilms formed by the involvement of bacteria are known as bacterial biofilm, eventually leading to chronic manifestations in humans (Xin *et al.*, 2020). Biofilms are basically described as the community of the organisms that are encapsulated within the matrix consisting of the essential moieties such as DNA, proteins and lipids. These molecules aid in protecting the bacteria internally from the adverse effects that are caused by environmental factors (Permana *et al.*, 2020).

DOI: 10.1201/9781003184942-16

13.2 FORMATION OF BIOFILM

Biofilm formation is mediated by series of steps that eventually facilitate attachment of the species to the surface: i) organisms either reversibly or irreversibly attach to the surface; ii) cells are attached together, leading to the formation of monolayer and matrix; iii) cells further continue to increase their colonies massively, resulting in the establishment of a matured biofilm – the energy required for all these processes is mediated by the bacterial adenosine triphosphate (ATP) (Sazzad *et al.*, 2019). From the mature biofilm, cells disperse and reverts to planktonic growth, which initiates a new cell cycle (Figure 13.1). It is found that bacteria have the potential capacity to distribute rapidly along the surfaces, most commonly known as the extracellular polymeric substances (EPS) (Zhuang *et al.*, 2020). Associations such as intra and inter specific are found among the bacterial species are of monomeric or mixed types of populations (Sazzad *et al.*, 2019).

Bacterial biofilms are a huge threat to humankind in terms of their growth in water distribution pipelines. These pipes are mostly made of iron stainless steel and galvanized steel or copper-based materials, so biofilms growing on these metals corrode the equipment used in the industry – and thus deteriorate the quality of water, leading to infectious diseases (Pabbati *et al.*, 2021). Bacterial biofilms such as *E. coli*, *Salmonella*, and *Lactobacillus* are only a few examples of bacterial bio-films that develop on surfaces (Giaouris *et al.*, 2020). Biofilm infections are typi-cally persistent, including chronic lung infections of cystic fibrosis patients, chronic

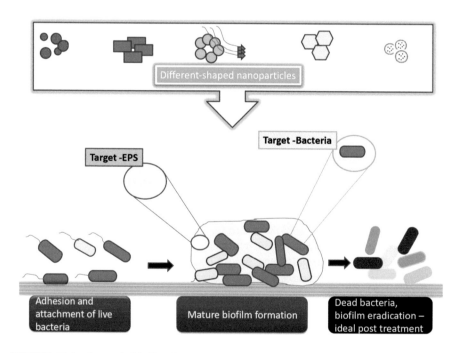

FIGURE 13.1   Stages in biofilm development (modified after Vasudevan, 2014).

osteomyelitis, chronic prostatitis, chronic rhinosinusitis, chronicotitis media, chronic wounds, recurrent urinary tract infection, endocarditis, periodontitis and dental caries. Biofilms produce a variety of infections, including dimorphism, which is mostly caused by the presence of *Candida* sp., one of the most distressing species (Jacqueline *et al.*, 2019).

13.2.1 Need to Mitigate Biofilms

A major disadvantage of biofilms is biofouling, which is caused on various surfaces and leads to negative impact on industries, though various methods have been developed to control the impact of biofouling such as membrane surface modification and usage of biocides and metabolic un-couplers to control the biofouling mechanism; these tactics were not very successful because of the complex mediation of the biofilm structure (Xiaochi *et al.*, 2020). In order to screen the presence of biofilms, various methods were developed. One of the methods is usage of staining dyes such as ethidium bromide, which stains the DNA present in the biofilm – through which the biofilm can be detected. Other techniques for detecting the existence of biofilms include in vitro scanning acoustic microscopy and the use of Indiocyan green to see the bacterial species in the biofilm (Zhuang *et al.*, 2020).

Aqueous systems deal with fluidic types of environments, and is one of the major sectors concerned with biofilm formation. Water is the main source for the supply to various industries and other divisions, if there is occurrence of biofilm. It leads to deterioration of the water, resulting in corrosion of the equipment, which can form a major obstacle to the reverse osmosis process; it can also affect the process of heat transfer efficiency (Jiaping *et al.*, 2020).

Regions where the deposits of the proteins and the emulsions retain form the basis for the growth and development of biofilms, which is mainly seen in industrial processes (Guruprakash *et al.*, 2020). Biofilm formation is adversely affected the food industries leads to contaminate and poisoning of the foods (Giaouris *et al.*, 2020). The progression of biofilm depends mainly on various factors such as location and area of formation (Sazzad *et al.*, 2019).

13.2.1.1 Constraints in Biofilm Mitigation

Different treatments were developed and are being used to remove biofilms, but most of them failed, as these biofilms are highly resistant toward the various antibiotic treatments (Permana *et al.*, 2020). The matrix around the biofilm prevents the entry of anti-microbial agents inside the biofilms, acting as a protective barrier against various kinds of agents. Various forms of resistance are being inculcated by bacteria by modifying itself to adjust according to the environment. One of its modifications lead toward involvement in the quorum sensing mechanism. A few modifications also facilitate in the transfer of genetic materials from one species to the other species naturally.

Due to its various strategies on adjusting the conditions toward prevailing, it has become a difficult task to remove or detach biofilms from their surfaces. Further inactivation of the microorganisms was tried using various compounds and was found to be moderately effective with the usage of copper and its derivative forms (Hye-jin *et al.*, 2016).

In contrast to this, the biofilms are used in treatments involving bioremediation and wastewater management (Guruprakash *et al.*, 2020). Considering all the characteristics of the biofilms, they are also used in various reactor designing mechanisms such as sequencing batch biofilm and moving-bed biofilm reactors (Fuzhong *et al.*, 2020).

13.2.2.2 Strategies to Mitigate Biofilm

Progression of an effective and an efficient strategy or a treatment process for reducing, eliminating the biofilms in nature is still a complicated issue in today's era (Jinshan *et al.*, 2016). It is found that bacteria are subjected to dispersal from its biofilm (bacteria gets removed from the matrix due to numerous factors such as degradation of components, alternation in the signaling molecules, depletion of oxygen, presence of inhibitors, and various other physicochemical properties leading to reduction in the function of the matrix that normally adheres to the biofilm); this becomes susceptible to the action of various immune responses and wide range of anti-microbial actions (Premana *et al.*, 2020).

Nanoparticles are gaining wide range of importance in various sectors such as pharmacy, medicine, textiles, energy, electrical, and cosmetics. Iron, zinc oxide, silicon dioxide, and titanium dioxide are few of the widely used nanoparticles. Many morphological and physiological properties like size, shape, and charge have their impact on the efficiency and performance of the nanoparticles. These drawbacks can be minimized by significant characterization of the nanoparticle which facilitates to analyze and access its toxicity levels. Various strategies to minimize the growth of biofilms are developed in the field of nanotechnology. In recent years, nanoantimicrobials have been utilized to inhibit biofilm development. Where antimicrobials are encapsulated in pH-responsive and target-specific matrices, responsive nanocarriers possess additional modifying properties such as pH, which enables them to have a strong affinity toward the negatively charged bacterial species (Kecheng *et al.*, 2020).

13.3 NANOPARTICLES FOR MITIGATION OF PURE/ MULTISPECIES BACTERIAL BIOFILMS

The science that deals with the nano-sized materials is collectively known as nanotechnology; particles that are usually ranging from 1–10 nm are known as nanoparticles or nanomaterials (Sazzad *et al.*, 2019). The physiochemical properties that are displayed by nanoparticles vary from others due to decrease and reductions in their sizes (Xin Yi *et al.*, 2020).

Nanoparticles now have been widely used to analyze anti-microbial properties using various mechanisms, such as the following.

- Investigating and analyzing the microbial activity using nanoparticles.
- Nanomaterials are used as drug delivery compounds against specific targeted particles.
- Biofilm formation was reduced using the surfaces that are modified with the coating of the nanomaterials (Holban *et al.*, 2016).

The nanomaterials having anti-microbial activity are characterized into various forms such as nanotubes, hierarchal structures nanoparticles, and unit and functional nanoparticles. Dispersal is commonly known as the important mechanism by which biofilms can be inactivated or eliminated partially or completely, depending upon presence of the bacterial species in the biofilms (Figure 13.2). Using this phenomenal evidence, the nanomaterials were used as the carrier molecules which serve as the nanocarriers. The ability of the nanomaterials to facilitate as nanocarriers was mainly used for the generation of the reactive oxygen species (Shuang *et al.*, 2021).

Recent developments and advancements have demonstrated that uses of nanomaterials in combination with anti-microbial agents were used to kill bacterial species effectively. Nanoparticles are generally used in medical fields and are characterized into various classes such as inorganic, organic, and mixed-type particles. While all the three types are extensively used in anti-microbial therapies (Jeevanandam *et al.*, 2018), different types of nanoparticles are currently used against mitigation of biofilms. The most commonly and widely used nanoparticles are zinc, silver, polyamine, and chitosan nanoparticles. Anti-fouling surfaces are synthesized mostly using polydopamine nanomaterials (Ding *et al.*, 2012). Photo-thermal therapy is a therapy that is actively obtained from the polydopamine nanoparticles and is used for the bacterial biofilm inactivation; it is done by transforming surfaces with polydopamine nanoparticle coats. This phenomenon was widely used to eliminate the biofilms caused by *Staphylococcus aureus* species (Yuan *et al.*, 2019).

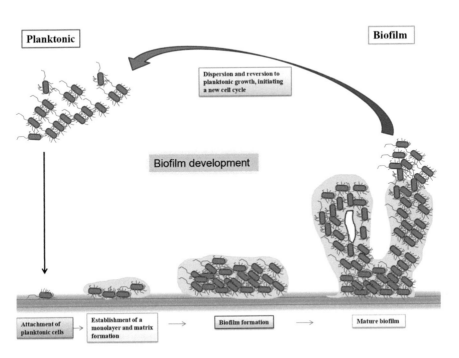

FIGURE 13.2 Treatment of established biofilms using nanoparticles (modified after Tran *et al.*, 2020).

13.3.1 Interactions of Nanoparticles

Chitosan is an effective nanoparticle for the eradication of biofilms. It is a cationic particle with the capability to pierce into biofilm layers and start initiation activity by interacting with negatively charged molecules in bacterial membranes, thereby leading to the disruption and causing death of the species (Carlson *et al.*, 2019). Drug-resistant bacteria are effectively destroyed by combinational therapies. One of the most dangerous infections is caused by gram-positive bacteria that are present in the biofilms known as the *Staphylococcus aureus* (Zoubos *et al.*, 2012). These infectious diseases can be minimized to an extent by adapting the target oriented drug delivery mechanism using nanoparticles (Zhu *et al.*, 2014). Desperate properties are crucial during dispersal of nanoparticles, which are found naturally demonstrated by pH-responsive nanomaterials. Several nanoparticles such as citrate capped gold nanospheres, iron oxide, Ferumoxitol, and PEG (polyethylene glycol)-coated gold nanomaterials, and mixed-shell polymer combinations, are some of those nanoparticles that are widely used as dispersants to destroy and terminate the organisms in the biofilms (Table 13.1). Citrate-capped gold nanospheres were found to cause modifications in the *Legionella pneumophilia* structural components of the biofilm, ultimately leading to its dispersal.

The total biovolume of *Legionella pneumophila* species was shown to be reduced by PEG-coated gold nanomaterials, facilitating biofilm dispersion (Raftery *et al.*, 2014). Iron oxide nanomaterials interact with the bacterial species, leading to the generation of the reactive oxygen species in *Escherichia coli* (Al Shabib *et al.*, 2018). Matrix degradation was found to occur by production of the reactive oxygen species using the ferumoxitol nanomaterials in the *S. mutans* species (Liu *et al.*, 2018). Further, the mixed-shell polymer combinations were also found to interact with the matrix complexes, leading to the biofilm dispersal in the *S. aureus* species (Tian *et al.*, 2020). Several other monometallic nanoparticles have also being found to be actively involved in the mitigation of the biofilms. Anti-microbial activity was observed against the *Bacillus subtilis* species using the copper nanoparticles at 30 nm of size (Lotha *et al.*, 2019), whereas crystalline copper oxide nanoparticles have an inhibitory effect toward *Escherichia coli* species (Govindaswamy *et al.*, 2018). Spherical crystalline gold nanoparticles at a range of 48 nm were found to be effective against inhibition of the *Aspergillus niger* species (Jayaseelan *et al.*, 2013). Crystalline rod-shaped iron nanoparticles have effective roles in inhibiting the *S. aureus* species at range of 11 nm (Radini *et al.*, 2018). Various studies have also found that one of the infectious species that is primarily responsible for the biofilm formation is *Staphylococcus aureus*. *Staphylococcus aureus* interacts using its intracellular polysaccharide membranes. These protein-mediated biofilms are found to be more virulent than the other species. AMG (alpha-mangostin) is one of the compounds that is found to inhibit the biofilms that are caused due to the *Staphylococcus aureus* species (Ibrahim *et al.*, 2016). Nanoparticles in combination with AMG are used to act as anti-biofilm agents. Gold nanoparticles have anti-microbial activities toward both gram-positive and gram-negative species. When gold nanoparticles are coupled with photosensitizers, photo-thermal radiation is produced, which aids in the reduction of biofilm development (Liakos *et al.*, 2014; Gebru *et al.*, 2013). Photo-thermal

TABLE 13.1
An Overview of Nanoparticles' Effects on Biofilms

S. No.	Nanoparticles	Biofilm	Details	Reference
1	Iron oxide nanoparticles (NPs)	*E.coli*	It interacts with bacteria and produces reactive oxygen species	Al-Shabib *et al.*, 2018
2	Ferumoxytol NPs	*S. mutans*	It interacts with bacteria, producing reactive oxygen species from H_2O_2	Liu *et al.*, 2018
3	Graphene	*S. aureus*	It interacts with bacteria and degrades the amyloid fibrils	Wang *et al.*, 2019
4	2,3-dimethylmaleic-anhydride modified carbon	*S.epidermidis*	It interacts with bacteria and minimizes the biofilm density	Wu *et al.*, 2020
5	Cationic dextran-block copolymer NPs	*Enterococcus faecalis*	It acts as a barrier between the bacteria and the biofilm	Li *et al.*, 2018
6	Cationic, poly(2-[dimethylamino] ethyl methacrylate)	*P. aeruginosa*	It interacts with bacteria and disrupts the extracellular matrix	Borisova *et al.*, 2018
7	Cationic polymer silver ions	*P. aeruginosa*	It interacts with bacteria and acidic ions of the extracellular matrix	Paunova-Krasteva *et al.*, 2020
8	Zwitter ionic, mixed shell	*S. aureus*	It interacts with bacteria and reacts with the extracellular matrix	Tian *et al.*, 2020
9	Silver-binding peptide fused to Dispersin B	*S. epidermidis*	It interacts and distrupt the extracellular matrix pnag	Chen *et al.*, 2018
10	Nanocapsules	*S. aureus*	It interacts with and degrades the eDNA	Liu *et al.*, 2020
11	Hybrid micelles	*P. aeruginosa*	It interacts with and degrades the amyloid fibrils	Chen *et al.*, 2019
12	Chitin-based nanocomposite containing D-amino acids and iron oxide	*S. aureus*	It interacts with and reacts to the amino acids	Abenojar *et al.*, 2018
13	Polydopamine-coated iron oxide	*P. aeruginosa*	It interacts with bacterial components	Adnan *et al.*, 2018
14	PLGA nanoparticles	Methicillin-resistant *S. aureus*	It interacts with the resistant components	Hasan *et al.*, 2019
15	Micelles	*P. aeruginosa*	It interacts with the bacteria and triggers the visible light	Shen *et al.*, 2019

(Continued)

TABLE 13.1   (*Continued*)

S. No.	Nanoparticles	Biofilm	Details	Reference
16	Liposomes	*S. aureus*	It triggers release of the biosurfactants	Giordani *et al.*, 2019
17	Chitosan nanoparticles	*P. aeruginosa*	Interacts with bacteria and degrades the eDNA	Patel *et al.*, 2020
18	Gold nanoparticles	*S. epidermidis*	It interacts with bacteria and degrades the eDNA	Xie *et al.*, 2020
19	Gold nanoparticles	*P. fluorescens*	It interacts with bacteria and disrupts the essential proteins	Habimana *et al.*, 2018
20	Silver nanoparticles	*S. aureus*	It interacts with bacteria and degrades the biofilm	Huang *et al.*, 2020

nanoparticles are being efficiently used to inhibit biofilm formation. This effect is seen using the polydopamine nanoparticles, whereby the growth of the biofilm was inhibited (Park *et al.*, 2019). This polydopamine is generally used as a coating on the structures of nanoparticles. Nanoparticles derived from polydopamines are bio-degradable and biocompatible (Bettinger *et al.*, 2009). *Staphylococcus aureus* and many other species, when brought in contact with the polydopamine nanoparticles, have shown reduction patterns in their growth (Reddy *et al.*, 2017). Several other methods are also being used to mitigate biofilm that are formed by *Staphylococcus* species, such as use of the alpha-mangostin, a regular xanthan compound which has anti-fungal, anti-inflammatory, and anti-bacterial properties (Ibrahmin *et al.*, 2016). *Bacillus subtilis* forms biofilms on various surfaces. Usage of tetrachlorosalicylani-lide (TCS) in low-grade concentrations inhibits and blocks the expression of the EPS genes, by which the biofilm formation can be suppressed (Feng *et al.*, 2018). Spherical crystalline manganese nanoparticles isolated from extracts of cinnamon were found to inhibit the biofilm of *S. aureus* and *E.coli* species (Kamran *et al.*, 2019).

Hexagonal crystalline nickel oxide nanoparticles isolated from extracts of the Neem leaf were proved to be effective against biofilms of the *Staphylococcus aureus* species (Helan *et al.*, 2016). Spherical crystalline nanoparticles of titanium dioxide isolated from *Morindacitrifolia* leaf extracts is effective against various strains of *Bacillus subtilis* and *Aspergillus niger* (Sundrarajan *et al.*, 2017). Cubic-structured zinc oxide and spherical-shaped palladium nanoparticles were isolated from the leaf extracts of olive, and *Filicium decipiens* inhibits the biofilms of *Xanthomonas orzyae*, *E.coli*, and *S. aureus*, respectively (Oguyemi *et al.*, 2019; Govindaswamy *et al.*, 2017). *S.typi*, *E.coli*, and *S.mutans* species were shown to be mitigated by crystalline spherical silver nanoparticles derived from honey (Sreelakshmi *et al.*, 2011). *Tropaeolummajus* spherical crystalline nanoparticles were active against the *Penicillin notatam* species (Valsalam *et al.*, 2018). Crystalline silver nanopar-ticles derived from *Aspergillus tamari* were discovered to exhibit anti-fungal and anti-bacterial effects when tested against a various strains of *C. albicans* forming biofilms (Nanda *et al.*, 2018). Turmeric powder extracts provide flower-shaped crystalline nanoparticles that are efficient against food-borne infections (Alsammarraie *et al.*, 2018). In the treatment of *S. mutans*, spherical selenium nanoparticles derived from berry extracts are employed (Yazhiniprabha *et al.*, 2019). *S. aureus, Bacillus subtilis*, and *Candida* species are all inhibited by rectangular-shaped platinum nanoparticles derived from the *Xanthium strumariam*. The *Bacillus subtilis* and *S. aureus* species may be harmed by crystalline zinc oxide microparticles derived from *Rhamnusvirgata* extracts (Iqbal *et al.*, 2019) Quasi-spherical lead oxide nanoparticles taken from the *Sagateria* are used to obstruct the *P. arugenosa* species (Khalil *et al.*, 2020). Spherical crystalline zirconium nanoparticles extracted from the *Sargassumwightii* were found to show positive results in mitigating the *S. typhi* and the *E. coli* species (Kumaresan *et al.*, 2018).

13.4 APPLICATIONS OF NANOTECHNOLOGY IN MEDICINE, INDUSTRIES, AND WATER AND WASTEWATER FOR BIOFILMS MITIGATION

When the biofilms are present in the body, then the time that is taken for healing is usually very long. More than 80% of the wounds that are formed are results of the biofilm accumulations at the infected sites (Duckworth *et al.*, 2018). Most frequently, organisms found in the wounds forms biofilms are *Pseudomonas aeruginosa* and *Staphylococcus aureus* (Clinmicrorevs *et al.*, 2001). If these wounds are removed surgically, there will be high chances where the biofilms tends to form again after 1–2 days (Wolcott *et al.*, 2010). Doxycycline, which is an anti-microbial compound, is active toward inhibiting the *Pseudomonas aeruginosa* and *Staphylococcus aureus* species (Adhirajan *et al.*, 2009).

Poly(lactic-co-glycolic) acid and Ɛ-poly caprolactone nanoparticles belonging to the class of the biodegradable materials have been proved to have favorable results toward handing of infection (Guo *et al.*, 2016). Poly(lactic-co-glycolic) acid and Ɛ-poly caprolactone are polymers used specifically to deliver the nanomaterials to the site of the infection (Xiong *et al.*, 2012). Positively charged chitosan has the ability to actively interact with the site of the infection (Rabea *et al.*, 2003). Thus, this is an approach of nanoparticles whereby mainly poly(lactic-co-glycolic) acid and Ɛ-poly caprolactone in the presence of chitosan are coupled with the doxycycline micro needles. This methodology therefore provides an efficient way of transferring the doxycycline to the site of infection. Anti-microbial activity was demonstrated after 48 hours in the ex vivo infected skin.

Photo-thermal nanoparticles such as polydopamine are used for the anti-microbial treatment to inhibit the infections that are caused due to the presence of the biofilms. The main purpose of using these photo-thermal nanomaterials is that they have the capability of transforming the near infrared rays into the heat; this heat is generally above normal temperatures (Meng *et al.*, 2016). These photo-thermal nanoparticles are thus widely used in cancer treatments and prophylactic methods (Nam *et al.*, 2018); mainly polydopamine nanoparticles are extensively used in the study of tumor-causing microorganisms and their inhibition.

Indocyanin green is a photo-sensitive compound that is mainly generated by the reactive oxygen species, when this is taken in combination with polydopamine nanoparticles. It demonstrated the ability to mitigate the biofilms that are formed by the *Staphylococcus aureus* species (Yuan *et al.*, 2019). Ethylene glycol, when taken with polydopamine nanoparticles, has shown positive results after exposing them to near-infrared radiation against the anti-bacterial–resistant *Staphylococcus aureus*, by which the species was destroyed (Gao *et al.*, 2016).This mechanism of treatment was also found to be effective on various other microorganisms causing biofilms, such as *Klebsiella, Enterococcus, Enterobacter*, and *Pseudomonas*.

Methicillin-resistant *Staphylococcus aureus* (MRSA) is one of the deadly micro-organisms that is involved in various diseases causing infections. Numerous strategies have been developed for treating this species. *Staphylococcus* species are those that tend to grow and propagate on the surfaces. Long-term complications of this species include endocarditis, pneumonia, and sepsis (Gomez and Prince, 2007). This deadly species arises due to various factors such as long-term antibiotic usage, overdose, or heavy doses of antibiotics. This eventually leads to the formation of the resistant species (Zhu *et al.*, 2014). One of these approaches to mitigate this species includes targeting the silver nanoparticles that are coupled with benzodioxane and piperazine with chitosan onto the species. Chitosan has anti-biofilm and anti-microbial properties, which play critical roles in the mitigation of the biofilms (Orgaz *et al.*, 2011). Chitosans are cationic species which have the ability to interact with organisms' negatively charged membranes; this interaction facilitates in disrupting the membrane surface of the species, which eventually leads to the death of the organism (Carlson *et al.*, 2019).

Another emerging sector in the field of nanomaterials toward treating the deadly infections that cause biofilms are magnetic nanoparticles. These magnetic nanoparticles have the ability to generate heat at the site of the infections, by which the organisms that are present in those sites are subjected to destruction leading to the death of the species. Magnetic nanoparticles can also be taken in combination with the antimicrobials which serve as the delivering vehicles for targeted actions toward the infections. Magnetic nanoparticles' effects were initially demonstrated in tumor therapy and imagining processes (Ulbrich *et al.*, 2016).The diameter scale of the infections makes it possible to develop accurately sized magnetic nanoparticles accordingly in order to effectively target the site of the infection. Various magnetic nanoparticles such as iron, cobalt, and nickel are most widely used in the therapeutic mechanism of mitigating the biofilms (Bjarnsholt *et al.*, 2013). Using one of the simplest methodologies, such as co precipitation, the iron magnetic nanoparticles can be easily prepared (Wu *et al.*, 2008). Several other methods through which the magnetic nanoparticles can be prepared are hydrothermal synthesis, sol gel synthesis, thermal decomposition, electrochemical reactions, gas phase decomposition, and microbial synthesis. When the methicillin-resistant species are encapsulated with the iron magnetic nanoparticles, this leads to the destruction of most resistant species from the site of the infection (Ulbrich *et al.*, 2016). Magnetic nanoparticles are known to induce hyperthermia, a condition whereby heat at a given location increases, and this mechanism can be utilized in destroying the biofilms when it is specifically and accurately targeted toward the site of infection. Inaccurate or non-specific targeting may lead to destruction of healthy cells and tissues (Chao *et al.*, 2019). *Pseudomonas fluorescens, Escherichia coli, P. aeruginosa,* and *S. aureus* species are selectively and efficiently destroyed by this process.

Green technology is known as the technology that involves in the synthesis of the nanoparticles through natural components such as human cells, plant cells, and fungal cells, and this mechanism is now being widely being used in the mitigation of biofilms (Pabbati *et al.*, 2021). Various species that arise due to exposure toward multi-drug resistance can be inhibited by treating them with green nanoparticles. Green nanoparticles also reduce the risk of contamination of the particles. Taking the biomedical sector into consideration, the gold and silver nanoparticles that are produced through green technology play a vital role (Wong *et al.*, 2010).

Staphylococcus aureus is one of the most dangerous strains which causes adverse effects. In order to inhibit this strain, various mechanisms have been developed; one among those is the use of the alpha-mangostin–based nanoparticles. Alpha-mangostin being a xanthan compound, when it is combined with the nanoparticles, it is found to inhibit the growth of the *Staphylococcus aureus* species. This was observed due to its minimal water solubility patterns that were created due to the interaction with the alpha-mangostin (Sundrarajan *et al.*, 2017).

Pseudomonas aeruginosa is the species that is mainly responsible for the severe and deadly infections that are caused due to the incorporation of the catheters in the human body; these catheters are introduced in various organs to facilitate their performance when their natural performance is inhibited. One among these catheter-associated diseases is urinary tract infection, which is mainly caused due to the *Pseudomonas aeruginosa* species. Immune-compromised patients are most readily affected by this species (Cole *et al.*, 2014). This, to a certain extent, can be prevented by coating a layer of alpha-mangostin on the surfaces of the catheters, as it prevents the accumulations of these strains on their surfaces. Gold nanoparticles, when taken in combination with chitosan-streptomycin, are used in targeting the biofilms that are caused due to the gram-positive and gram-negative species. This strategy was found to inhibit the most deadly antibiotic-resistant species.

Based on the synthesis mechanism, these nanoparticles can be divided into various groups such as mono metallic nanoparticles, bi-metallic nanoparticles, and tri-metallic nanoparticles (Gebru *et al.*, 2013). A few important mono-metallic nanoparticles are copper, copper oxide, iron, gold, iron oxide, manganese, lead oxide, nickel oxide, nickel, zinc oxide, tellurium, and silver nanoparticles; all of these are widely used in mitigating the biofilms caused by various species such as *E. coli, K. pneumonia, S. aureus, S. pyogenes, S. epidermidis, Bacillus,* and *E. aerogenes.*

13.4.1 INDUSTRIES

Silver nanoparticles are widely applied in various sectors due to their effectiveness and efficiency toward the targeting sites. It is proved that these silver nanoparticles have the ability to interact with the enzymes thiol groups, eventually leading to the inactiveness of the enzyme. These nanoparticles also have the capability to interact with the essential compounds that are present in the DNA, which leads to death of the species (Prabhu *et al.*, 2012).

One of the most affected causes in the industrial sector is accumulation of the biofilms in industrial equipment, which can lead to corrosion buildup, eventually disrupting the mechanism of reverse osmosis in process called biofouling. Biofouling is when the organisms present in the biofilm grow in bulk colonies on the membranes of the reverse osmosis chambers. One among the few species that was found on the reverse osmosis membranes is *P. aeruginosa*; in order to control and mitigate this species, hydroxylamine and hydrogen peroxide are taken in combination with the cupric ions, by treating the feed solution with the combinations of the hydroxylamine, hydrogen peroxide, and cupric ions, prevented the growth and proliferation of the *P. aeruginosa* on the membranes of the reverse osmosis columns (Kim *et al.*, 2015).

13.4.2 WATER AND WASTEWATER TREATMENTS

Biofilms are widely used in wastewater treatment applications. Biofilms are growing in the liquid environments generally possess the following components.

- Surfaces.
- Microorganisms.
- Nutrients.
- Gas zones.

In order to optimize the biofilm activity, biofilm reactors are generally used. The rate of the growth-limiting substrate is used as biofilm activity. The amount at which the biofilm accumulates should be examined to see if it helps or hinders the formation of the biofilm. A few methods widely used to optimize the biofilm activity are measuring the intensity of light reflected from the colonies through optical microscopy, imaging the biofilm colonies, and piezoelectric models using surface sensors (Kantawanichkui *et al.*, 2009).

In dispersed growth systems, biofilms form freely in the liquid media without any attachment to the surface. The microorganisms present in these systems absorb nutrients, organic matter, and nitrogen, which allow them to grow and reproduce to form microcolonies. These micro colonies settle as sludge, which is either treated or removed in sludge treatment process (Kantawanichkui *et al.*, 2009).

Organisms that require support for their growth and propagation are known as attached growth systems. In attached growth systems, biofilms are widely used to break down the various nutrients, such as carbon, nitrogen, and phosphorus compounds, as well as to capture the pathogens from the wastewater (Kantawanichkui *et al.*, 2009).

Elimination of various pollutants such as hydrocarbons, toxic metals, nitrogen, phosphorus, and organic compounds from wastewater are eliminated from the biofilm on the filter media. Once pollutants are eliminated, treated water is either released to the environment or used for agriculture and other recreational purposes.

Biofilms in wastewater treatment systems have several advantages, including flexibility in operation, low space requirements, hydraulic retention time reduction, resilience to changes in the environment, increased biomass residence time, giant active biomass concentration, and increased ability to break down intractable compounds; additionally, moderate microbial growth rate and evolution in lower sludge production (Kantawanichkui *et al.*, 2009).

13.5   CHALLENGES AND FUTURE DIRECTIONS

Infections that are seen mainly due to antibiotic-resistant species are of serious concern. These resistances are mostly due to their long-term or prolonged exposure to antibiotics (Gandra *et al.*, 2014). Thus, it is now essential to develop and formulize a new technology in order to tackle these highly resistant organisms. Most of the bacterial species follow dispersal mechanisms; this is a phenomenon whereby the species gets detached from the surface for a period of time, which is a normal process in its growth phase cycle. Combinations of antibiotics with multi-scale nanoparticles can be formulated in order to suppress its effects.

As a part of handling the challenges, it is important to progress toward new strategies in developing the anti-biofilms Dispersals are not anti-biofilm compounds, but these dispersals promote loosening the extracellular polymeric substances' ability to adhere.

A few nanoparticles, such as the pH-responsive nanomaterials, mimic the dispersals' behavior (Tian *et al.*, 2020). There is a variety of molecular desperants existing; a few of them are DNase, peptidase, nitric oxide, and reactive oxygen species. Currently, molecular dispersals are in verification stages. *E coli*, *Pseudomonas*, and *Staphylococcus* are a few among the species that are effectively targeted by the DNase (Kaplan *et al.*, 2012).

Extracellular polymeric substances were found to get disrupted by the action of the reactive oxygen species on the covalent bond breakage; this methodology is effective against *Streptococcus mutans* and *E.coli* (Gao *et al.*, 2016). Treatment of *Staphylococcus aureus* and *Streptococcus mutants* with monomeric alpha sheet peptidases has shown inhibitory effects.

In order to protect against removal by the host responsive systems, the nanocarriers are generally used. Metal-based nanocarriers are known to produce reactive oxygen species, where this acts as the key property in stating the nanoparticles as dispersal compounds. There are few among these nanoparticles that possess dispersal properties. Changes are reported in the biofilms of *Legionella pneumophila* using the citrate-capped gold nanospheres (Bleem *et al.*, 2017).

P. aeruginosa biofilms are generally formed in the lungs; in order to treat these, Pulmoyzme has been certified for treatment (Hoiby *et al.*, 2010). Pulmoyzme, in combination with antibiotics, is effective in treating the respiratory disorders (Frederiksen *et al.*, 2006). Furthermore, biofilms are caused due to *P. aeruginosa* are targeted with Levofloxacin to reduce the concentrations of the biofilm populations (Slan *et al.*, 2016).

Thus encapsulating the dispersals in the nanomaterials enhaces its transportation into the bloodstream. Unexpected rise in the concentrations of biofilms in the blood flow is challenging issue as it can be a threat for increase in infection to neighboring organs and may lead to sepsis. The hurdle of occurrence of sepsis can be minimized by including antibiotics in combination with the dispersals. Another challenging part is maintaining the required concentration of the antibiotics; in order to overcome these issues, anti-microbial facilitated dispersals should be developed.

Nanoparticles as the dispersals is not completely studied. There are still few elements that need to be analyzed to get a better conclusion; hence, complete research on how nanoparticles facilitate the dispersal mechanism – and the possible factors that need to be considered in order to avoid the risk of sepsis – should be focused.

Many different types of biofilms are formed on the marine layers, where they get settled on the barnacle larvae. Various species that are usually found in these systems are *Bacillus cereus* and *Alteromonas* sp., taking the dispersal mechanism into the consideration. It is where the loosely bound extracellular polymeric substances are found. The components of the extracellular polymeric substances such as carbohydrates are usually involved, due to their infitment in the matrix. Thus, future studies should mainly focus on the internal larval interactions, evaluating the levels of toxicity, and – most importantly – they need to focus on new strategies to develop anti-fouling mechanisms (Schroeder *et al.*, 2001).

Traditional medicines – such as antibiotic treatments, targeting monoclonal antibody therapy, and quorum sensing inhibition – are only a few of the traditional techniques available to prevent biofilm formation. Furthermore, in order to successfully reduce biofilms, more study and analysis of diverse natural treatments is required.

Various methodologies have being investigated, but they have certain challenges which need to be overcome. To confirm the reliability of the natural therapies, quality control plays a critical role. As natural components are composed of complex structures and arrangements, evaluating them under quality control protocols is of a major challenge (Zhao *et al.*, 2018).

Various challenges arise in conducting the pharmacokinetic studies, due to confinement properties of the molecules. It is further important to perform a comparative study focusing on the pharmacokinetic and pharmacodynamics properties (Yan *et al.*, 2018).

Materials which have the ability to naturally act as anti-biofilms should also be investigated. Strategies for combining the natural modes with the existing antibiotic compounds should also be formulated. Analysis should utilize ex vivo models to determine their efficiency and accuracy.

In lettuce cultures, biofilms can exist in mono or mixed forms; one viable method for controlling or inhibiting multi-cultures of biofilms developing in lettuce cultures is to use atmospheric cold plasma. Atmospheric cold plasma has been shown to minimize contamination in the food sector, which is mostly caused by species of *E. coli* and *Salmonella*. It gave effective results in inhibiting the growth of organisms such as *E. coli* and *Salmonella* on food particles. Further research is necessary to know the mechanisms of how atmosphere cold plasma interacts with the internal components and validation of the food process, and control studies are also needed (Ziuzina *et al.*, 2015).

The presence of multispecies biofilms is of great concern. It is important to develop biological controls that are specific toward the substrate. This forms a major challenge toward mitigating the biofilms (Nahar *et al.*, 2018). Thus, it may be sufficient to use a single strategy toward battling the biofilms; combinations of two or more methodologies in effectively targeting the species have to be developed. Utmost care has to be taken while analyzing and combining the combinations, as improper mixture can lead to worsening of the target (Lim *et al.*, 2019).

Enzyme-responsive coatings can be applied on the enzymes that are usually targeted toward the critical environments; all these formulations are quite effective in the small-scale medium, but might vary when taken into the larger-scale mediums. Bioinformatics tools can be used in order to investigate and analyze the characterization of different organisms; updating the methodologies accordingly can eliminate the risk of worsening the conditions.

13.6 CONCLUSION

A wide variety of deadly diseases such as cystic fibrosis, pneumonia, and periodontitis are being associated with the growth of the biofilms in different organs of a human body, making it a serious concern that is to be immediately addressed. These biofilms reduce the human in-built natural immune mechanisms to fight infection; further, these species become resistant to wide range of antibiotics. Thus, it is very important to inhibit these biofilms. It is important to prevent the occurrence of such infections, and this can be done by minimal or less dosage exposure to antimicrobials. A few of the bacterial species involved in the biofilms formations are *Streptococcus mutants, Enterococcus foecalis, Klebsiella pneumonia, Pseudomonas aeuginosa, Staphylococcus aureus, Staphylococcus epidermidis*, and *Escherichia coli*. Intercellular biofilms are formed by the resistant species, which leads to the failure of antibiotic therapies. To facilitate mitigation of these biofilms, various methodologies have been proposed and developed. One among the emerging fields to investigate the possible approaches and to mitigate biofilms is nanotechnology. It utilizes nano-sized particles in targeting the site of infection. Nanoparticles are small, self-responsive, and target specific molecules. These nanoparticles can be either applied individually in a single-approach mechanism, or even combined with existing antimicrobials in a multi-approach mechanism to mitigate the biofilms. Some of these nanomaterials contain anti-biofilm characteristics that are naturally present; thus, utilization of such nanomaterials can eliminate the risk of hazardous infections. Most of these nanoparticles have been proved to demonstrate promising results toward mitigating the existing biofilms and also in inhibiting their further growth.

REFERENCES

Abenojar, E.C., S. Wickramasinghe, M. Ju, S. Uppaluri, A. Klika, J. George, W.Barsoum, S.J. Frangiamore, C.A. Higuera-Rueda, A.C.S. Samia, (2018) Magnetic glycol chitin-based hydrogel nanocomposite for combined thermal and D-amino-acid-assisted biofilm disruption ACS Infect. Dis.4:1246–1256.

Adhirajan, N., Shanmugasundaram, N., Shanmuganathan, S., Babu, M., (2009) Collagen based wound dressing for doxycycline delivery: in-vivo evaluation in an infected excisional wound model in rats. J. Pharm. Pharmacol. 61: 1617–1623.

Adnan NNM, Sadrearhami Z, Bagheri A, Nguyen TK, Wong EHH, Ho KKK, Lim M, Kumar N, Boyer C (2018) Exploiting the versatility of polydopamine-coated nanoparticles to deliver nitric oxide and combat bacterial biofilm macromol. *Rapid Commun.* 39:e1800159.

Alsammarraie, F.K., Wang, W., Zhou, P., Mustapha, A., Lin, M., (2018) Green synthesis of silver nanoparticles using turmeric extracts and investigation of their antibacterial activities. Coll. Surf. B 171: 398–405.

Al-Shabib NA, Husain FM, Ahmed F, Khan RA, Khan MS, Ansari FA, Alam MZ, Ahmed MA, Khan MS, Baig MH, Khan JM, Shahzad SA, Arshad M, Alyousef A, Ahmad I (2018) Low temperature synthesis of superparamagnetic iron oxide (Fe3O4) nanoparticles and their ROS mediated inhibition of biofilm formed by food-associated bacteria *Front Microbiol.* 9:2567.

Bettinger CJ, Bruggeman JP, Misra A, Borenstein JT, Langer R. (2009) Biocompatibility of biodegradable semiconducting melanin films for nerve tissue engineering, Biomaterials 30:3050–7. doi: 10.1016/j.biomaterials.2009.02.018

Bjarnsholt, T, M. Alhede, M. A lhede, S.R. Eickhardt-Sørensen, C. Moser, M.Kühl, P.Ø. Jensen, N. Høiby, (2013) The in vivo biofilm, Trends Microbiol. 21(9): 466–474

Bleem A, Francisco R, Bryers JD, Daggett V.(2017) Designed α-sheet peptides suppress amyloid formation in *Staphylococcus aureus* biofilms. NPJ Biofilms Microbiomes, 3:16. doi: 10.1038/s41522-017-0025-2.

Borisova, D., E. Haladjova, M. Kyulavska, P. Petrov, S. Pispas, S. Stoitsova, T.Paunova-Krasteva, (2018) Application of cationic polymer micelles for the dispersal of bacterial biofilms Eng. Life Sci. 18:943–948.

Carlson RP, Taffs R, Davison WM, Stewart PS. (2019) Anti-biofilm properties of chitosan-coated surfaces. J BiomaterSciPolym Ed.19(8):1035–46. doi: 10.1163/156856208784909372. PMID: 18644229

Chao, Y., G. Chen, C. Liang, J. Xu, Z. Dong, X. Han, C. Wang, Z. Liu, (2019) Iron nanoparticles for low-power local magnetic hyperthermia in combination with immune checkpoint blockade for systemic antitumor therapy. *Nano letters*, 19(7), 4287–4296.

Chen, K.J., C.K. Lee, (2018) Twofold enhanced dispersin B activity by N-terminal fusion to silver-binding peptide for biofilm eradication. Int. J. Biol. Macromol. 118: 419–426

Chen, M, J. Wei, S. Xie, X. Tao, Z. Zhang, P. Ran, X. Li, (2019) Bacterial biofilm destruction by size/surface charge-adaptive micelles Nanoscale 11(3):1410–1422

Clinmicrorevs, M., Bowler, P., Armstrong, D.G., (2001) Wound microbiology and associated approaches to wound. Clin. Microbiol. Rev. 14: 244–269.

Cole SJ, Records AR, Orr MW, Linden SB, Lee VT. (2014) Catheter-associated urinary tract infection by Pseudomonas aeruginosa is mediated by exopolysaccharide-independent biofilms. Infect Immun, 82(5):2048–58. doi: 10.1128/IAI.01652-14.

Ding X, Yang C, Lim TP, Hsu LY, Engler AC, Hedrick JL, Yang YY. (2012) Antibacterial and antifouling catheter coatings using surface grafted PEG-b-cationic polycarbonate diblockcopolymers. Biomaterials, 33(28):6593–603. doi: 10.1016/j.biomaterials.2012.06.001. Epub 2012 Jun 27. PMID: 22748920.

Duckworth, P.F., Rowlands, R.S., Barbour, M.E., Maddocks, S.E., (2018) A novel flowsystem to establish experimental biofilms for modelling chronic wound infection and testing the efficacy of wound dressings. Microbiol. Res. 215:141–147.

Feng, X., *et al.*, (2018) Inhibition of biofilm formation by chemical uncoupler, 3, 3', 4', 5-tetrachlorosalicylanilide (TCS): from the perspective of quorum sensing and biofilm related genes. Biochem. Eng. J. 137: 95–99.

Frederiksen B, Pressler T, Hansen A, Koch C, Høiby N.(2006) Effect of aerosolized rhDNase (Pulmozyme) on pulmonary colonization in patients with cystic fibrosis. Acta Paediatr. 95(9):1070–4. doi: 10.1080/08035250600752466.

FuzhongXiong, Xiaoxi Zhao, Donghui Wen, QilinLi, (2020) Effects of N-acyl-homoserine lactones-based quorum sensing on biofilm formation, sludge characteristics, and bacterial community during the start-up of bioaugmented reactors, Science of The Total Environment,735: 139449. doi: 10.1016/j.scitotenv.2020.139449.

Gandra S, Barter DM, Laxminarayan R.(2014) Economic burden of antibiotic resistance: how much do we really know? ClinMicrobiol Infect. 20(10):973–80. doi: 10.1111/1469-0691.12798.

Gao L, Liu Y, Kim D, Li Y, Hwang G, Naha PC, Cormode DP, Koo H. Nanocatalysts promote Streptococcus mutans biofilm matrix degradation and enhance bacterial killing to suppress dental caries in vivo. Biomaterials, 101:272–84. doi: 10.1016/j.biomaterials.2016.05.051

Gebru, H., Taddesse, A., Kaushal, J., Yadav, O.P.J., (2013) Green synthesis of silver nanoparticles and their antibacterial activity. Surf. Sci. Technol. 29: 47–66.

Giaouris E, Heir E, Hébraud M, Chorianopoulos N, Langsrud S, Moretro T, Habimana O, Desvaux M, Renier S, Nychas GJ (2020) Attachment and biofilm formation by foodborne bacteria in meat processing environments: causes, implications, role of bacterial interactions and control by alternative novel methods. Meat Sci, 97(3):298–309. doi: 10.1016/j.meatsci.2013.05.023.

Giordani, B, P.E. Costantini, S. Fedi, M. Cappelletti, A. Abruzzo, C. Parolin, C.Foschi, G. Frisco, N. Calonghi, T. Cerchiara, F. Bigucci, B. Luppi, B. Vitali (2019), Liposomes containing biosurfactants isolated from Lactobacillus gasseri exert antibiofilm activity against methicillin resistant Staphylococcus aureus strains Eur. J.Pharm. Biopharm. 139:246–252

Gomez MI, Prince A. (2007) Opportunistic infections in lung disease: Pseudomonas infections in cystic fibrosis. Curr Opin Pharmacol. 7(3):244–51. doi: 10.1016/j.coph.2006.12.005.

Govindasamy, S., Fathima, M.F., Haries, S., Geetha, S., Kumar, N.M., Muthukumaran, C.,(2017) Biogenic using green synthesis, characterization and antibacterial efficacy of palladium nanoparticles synthesized using *Filiciumdecipiens*leaf extract. J. Mol. Struct. 1138: 35–40.

Govindasamy, S., Sandiya, R.S.P.K., Santhiya, S., Muthukumaran, C., Jeyanthi, J., Kumar, N.M., Thirumarimurugan, M., (2018) Biogenic synthesis of CuO nanoparticles using *Bauhinia tomentosa* leaves: characterization and its antibacterial application. *Journal of Molecular Structure* 1165: 288–292.

Guo, J., Wang, W., Hu, J., Xie, D., Gerhard, E., Nisic, M., Shan, D., Qian, G., Zheng, S., Yang, J., (2016) Synthesis and characterization of anti-bacterial and anti-fungal citrate- based mussel-inspired bioadhesives. Biomaterials 85: 204–217.

Guruprakash, Subbiahdoss, Erik Reimhult, (2020) Biofilm formation at oil-water interfaces is not a simple function of bacterial hydrophobicity, Colloids and Surfaces B: Biointerfaces,194:111163. doi: 10.1016/j.colsurfb.2020.111163

Habimana, O., M. Zanoni, S. Vitale, T. O'Neill, D. Scholz, B. Xu, E. Casey, (2018) One particle, two targets: a combined action of functionalised gold nanoparticles, against Pseudomonas fluorescens biofilms J.Colloid Interface Sci. 526: 419–428.

Hasan N, Cao J, Lee J, Naeem M, Hlaing SP, Kim J, Jung Y, Lee BL, Yoo JW (2019) PEI/NONOates-doped PLGA nanoparticles for eradicating methicillin-resistant Staphylococcus aureus biofilm in diabetic wounds via binding to the biofilm matrix. *Mater Sci Eng C* 103:109741.

Helan, V., Prince, J.J., Al-Dhabi, N.A., Arasu, M.V., Ayeshamariam, A., Madhumitha, G., Roopan, S.M., Jayachandran, M.,(2016) Neem leaves mediated preparation of NiO nanoparticles and its magnetization coercivity and antibacterial analysis. Results in Phys. 6:712–718.

Hoiby N, Ciofu O, Bjarnsholt T. (2010) Pseudomonas aeruginosa biofilms in cystic fibrosis. Future Microbiol, 5(11):1663–74. doi: 10.2217/fmb.10.125. PMID: 21133688.

Holban AM, Gestal MC, Grumezescu AM.(2016) Control of biofilm-associated infections by signaling molecules and nanoparticles. Int J Pharm, 510(2):409–18. doi: 10.1016/j.ijpharm.2016.02.044. Epub 2016 Mar 2. PMID: 26945736

Huang L, Lou Y, Zhang D, Ma L, Qian H, Hu Y, Ju P, Xu D, Li DX (2020) Cysteine functionalised silver nanoparticles surface with a "disperse-then-kill" antibacterial synergy. *Chem Eng J.* 381:122662.

Hye-Jin Lee, Hyung-Eun Kim, ChanghaLee,(2017) Combination of cupric ion with hydroxylamine and hydrogen peroxide for the control of bacterial biofilms on RO membranes, Water Research: 83–90, https://doi.org/10.1016/j.watres.2016.12.014.

Ibrahim, M.Y., Hashim, N.M., Mariod, A.A., Mohan, S., Abdulla, M.A., Abdelwahab, S. I., Arbab, I.A., (2016) a-Mangostin from Garciniamangostana Linn: An updated review of its pharmacologicalproperties.Arab.J.Chem.9(3):317–329. https://doi.org/10.1016/j.arabjc.2014.02.011.

Iqbal, J., Abbasi, B.A., Mahmood, T., Kanwal, S., Ahmad, R., Ashraf, M.,(2019) Plantextractmediated green approach for the synthesis of ZnO-NPs: characterization andevaluation of cytotoxic, antimicrobial and antioxidant potentials. J. Mol. Struct.1189: 315–327.

Jacqueline Cosmo Andrade, Ana Raquel Pereira da Silva, Maria AudileneFreitas, Bárbara de Azevedo Ramos, ThiagoSampaioFreitas, Franz de Assis G. dos Santos, Melyna C. Leite-Andrade, MichellângeloNunes, SauloRelisonTintino, MárciaVanusa da Silva, Maria Tereza dos Santos Correia, ReginaldoGonçalves de Lima-Neto, Rejane P. Neves, Henrique Douglas MeloCoutinho, (2019) Control of bacterial and fungal biofilms by natural products of Ziziphusjoazeiro Mart. (Rhamnaceae), Comparative Immunology, Microbiology and Infectious Diseases 65: 226–233. doi: 10.1016/j.cimid.2019.06.006

Jayaseelan, C., Ramkumar, R., Rahuman, A.A., Perumal, P., (2013) Green synthesis of gold nanoparticles using seed aqueous extract of *Abelmoschusesculentus*and its antifungal activity. Ind. Crop. Prod. 45: 423–429.

Jeevanandam J, Barhoum A, Chan YS, Dufresne A, Danquah MK.(2018) Review on nanoparticles and nanostructured materials: history, sources, toxicity and regulations. *Beilstein J Nanotechnol.* 9:1050–1074. doi:10.3762/bjnano.9.98

Jiaping Wang, Guiying Li, Hongliang Yin, Taicheng (2020) Bacterial response mechanism during biofilm growth on different metal material substrates: EPS characteristics, oxidative stress and molecularregulatory network analysis, Environmental Research, 185:109451, https://doi.org/10.1016/j.envres.2020.109451.

JinshanGuo, Wei Wang, Jianqing Hu, DenghuiXie, Ethan Gerhard, MerisaNisic, Dingying Shan, GuoyingQian, SiyangZheng, JianYang, (2016) Synthesis and characterization of anti-bacterial and anti-fungal citrate-based mussel-inspired bioadhesives, Biomaterials, 85:204–217, https://doi.org/10.1016/j.biomaterials.2016.01.069.

Kamran, U., Bhatti, H.N., Iqbal, M., Jamil, S., Zahid, M., (2019) Biogenic synthesischaracterizationand investigation of photocatalytic and antimicrobial activity of manganese nanoparticles synthesized from *Cinnamomumverum* bark extract. J. Mol. Struct. 1179:532–539.

Kantawanichkul S, Kladprasert S, Brix H. (2009) Treatment of high-strength wastewater in tropical vertical flow constructed wetlands planted with Typhaangustifolia and Cyperusinvolucratus. Ecological Engineering.35(2): 238–247

Kaplan JB, LoVetri K, Cardona ST, Madhyastha S, Sadovskaya I, Jabbouri S, Izano EA.(2012) Recombinant human DNase I decreases biofilm and increases antimicrobial susceptibility in staphylococci. J Antibiot (Tokyo) 65(2):73–7. doi: 10.1038/ja.2011.113.

Kecheng Quan, Zexin Zhang, YijinRen, Henk J. Busscher, Henny C. van der Mei, Brandon W. Peterson,(2021) Possibilities and impossibilities of magnetic nanoparticle use in the control of infectious biofilms, Journal of Materials Science & Technology, 69:69–78, https://doi.org/10.1016/j.jmst.2020.08.031.

Khalil, A. T., Ovais, M., Ullah, I., Ali, M., Jan, S. A., Shinwari, Z. K., Maaza, M. (2020) Bioinspired synthesis of pure massicot phase lead oxide nanoparticles and assessment of their biocompatibility, cytotoxicity and in-vitro biological properties. Arabian Journal of Chemistry, *13*(1): 916–931.

Kim, H.E., Nguyen, T.T.M., Lee, H., Lee, C., (2015) Enhanced inactivation of Escherichia coli and MS2 coliphage by cupric ion in the presence of hydroxylamine: dual microbicidal effects. Environ. Sci. Technol. 49 (24), 14416e14423.

Kumaresan, M., Anand, K.V., Govindaraju, K., Tamilselvan, S., Kumar, V.G., (2018) Seaweed *Sargassumwightii*mediated preparation of zirconia (ZrO2) nanoparticles and their antibacterial activity against Gram positive and Gram negative bacteria. Microb.Pathog. 124: 311–315.

Li, J, K. Zhang, L. Ruan, S.F. Chin, N. Wickramasinghe, H. Liu, V. Ravikumar, J.Ren, H. L. Yang, M.B. Chan-Park, (2018) Block copolymer nanoparticles remove biofilms of drug-resistant gram-positive bacteria by nanoscale bacterial debridement.Nano Lett. 18 4180–4187.

Liakos, I., Grumezescu, A.M., Holban, A.M., (2014) Magnetite nanostructures as novel strategies for anti-Infectious therapy. Molecules 19 (8): 12710–12726.

Lim, E.S., Koo, O.K., Kim, MJ. *et al.* (2019) Bio-enzymes for inhibition and elimination of Escherichia coli O157:H7 biofilm and their synergistic effect with sodium hypochlorite. Sci Rep 9(1): 1–10. https://doi.org/10.1038/s41598-019-46363

Liu, C, Y. Zhao, W. Su, J. Chai, L. Xu, J. Cao, Y. Liu (2020) Encapsulated DNase improving the killing efficiency of antibiotics in staphylococcal biofilms. *Journal of Materials Chemistry.* B 8: 4395–4401.

Liu., Y, Naha, P. C., Hwang, G., Kim, D., Huang, Y., Simon-Soro, A., Koo, H. (2018) Topical ferumoxytol nanoparticles disrupt biofilms and prevent tooth decay in vivo via intrinsic catalytic activity. *Nature communications*, *9*(1), 1–12.

Lotha, R., Shamprasad, B.R., Sundaramoorthy, N.S., Nagarajan, S., Sivasubramanian, A., (2019) Biogenic phytochemicals (cassinopin and isoquercetin) capped copper nanoparticles (ISQ/CAS@CuNPs) inhibits MRSA biofilms. Microb. Pathog. 132: 178–187.

Meng Y, Wang S, Li C, Qian M, Yan X, Yao S, *et al.* (2016) Photothermal combined gene therapy achieved by polyethyleneimine-grafted oxidized mesoporous carbon nanospheres, Biomaterials, 100:134–42.

Nahar, S., Mizan, M. F. R., Ha, A. J.-W., Ha, S.-D. (2018) Advances and future prospects of enzyme-based biofilm prevention approaches in the food industry. ComprehensiveReviews in Food Science and Food Safety, *17*:1484–1502.

Nam J, Son S, Ochyl LJ, Kuai U, Schwendeman A, Moon JJ. (2018) Printing of small molecular medicines from the vapor phase Nat Commun, 8(1):1–9

Nanda, A., Nayak, B.K., Krishnamoorthy, M., (2018) Antimicrobial properties of biogenic silver nanoparticles synthesized from phylloplane fungus*, Aspergillustamarii.* Biocat. Agric. Biotechnol. 16: 225–228.

Ogunyemi, S. O., Abdallah, Y., Zhang, M., Fouad, H., Hong, X., Ibrahim, E., Li, B. (2019) Green synthesis of zinc oxide nanoparticles using different plant extracts and their antibacterial activity against Xanthomonas oryzae pv. oryzae. *Artificial cells, nanomedicine, and biotechnology*, *47*(1), 341–352.

Orgaz, B, M.M. Lobete, C.H. Puga, C.S. Jose, (2011) Effectiveness of chitosan against mature biofilms formed by food related bacteria, Int J MolSci 12: 817–828.

Pabbati R., Aerupula M., Shaik F., Kondakindi V.R. (2021) Nanoparticles for Biofilm Control. In: Maddela N.R., Chakraborty S., Prasad R. (eds) Nanotechnology for Advances in Medical Microbiology. Environmental and Microbial Biotechnology. Springer, Singapore. https://doi.org/10.1007/978-981-15-9916-3_9

Park, D, Yeo, J., Lee, Y. M., Lee, J., Kim, K., Kim, J., . . ., Kim, W. J. (2019) Nitric oxide-scavenging nanogel for treating rheumatoid arthritis. *Nano letters*, *19*(10), 6716–6724.

Patel, K.K., A.K. Agrawal, M.M. Anjum, M. Tripathi, N. Pandey, S.Bhattacharya, R. Tilak, S. Singh, (2020) DNase-I functionalization of ciprofloxacin-loaded chitosan nanoparticles overcomes the biofilm-mediated resistance of Pseudomonas aeruginosa Appl. Nanosci. 10: 563–575

Paunova-Krasteva, T., E. Haladjova, P. Petrov, A. Forys, B. Trzebicka, T.Topouzova-Hristova, S.R. Stoitsova, (2020) Destruction of Pseudomonas aeruginosa pre-formed biofilms by cationic polymer micelles bearing silver nanoparticles, Biofouling 36: 679–695.

Permana Andi Dian, Maria Mir, Emilia Utomo, Ryan F. Donnelly (2020) Bacterially sensitive nanoparticle-based dissolving microneedles of doxycycline for enhanced treatment of bacterial biofilm skin infection: A proof of concept study, International Journal of Pharmaceutics, X2: 100047 https://doi.org/10.1016/j.ijpx.2020.100047

Prabhu, S., Poulose, E.K., (2012) Silver nanoparticles: mechanism of antimicrobial action, synthesis, medical applications, and toxicity effects. Int Nano Lett, 2: 32 https://doi.org/10.1186/2228-5326-2-32

Rabea, E.I., Badawy, M.E.T., Stevens, C.V., Smagghe, G., Steurbaut, W., (2003) Chitosan as antimicrobial agent: applications and mode of action. Bio macromolecules 4: 1457–1465.

Radini, I.A., Hasan, N., Malik, M.A., Khan, Z., (2018) Biosynthesis of iron nanoparticles using *Trigonellafoenum-graecum*seed extract for photocatalytic methyl orange dye degradation and antibacterial applications. J. Photochem. Photobiol. B 183: 154–163.

Raftery TD, Kerscher P, Hart AE, Saville SL, Qi B, Kitchens CL, Mefford OT, McNealy TL., (2014) Discrete nanoparticles induce loss of Legionella pneumophila biofilms from surfaces. Nanotoxicology, 8(5) 477–484. doi:10.3109/17435390.2013.796537.

Reddy PN, Srirama K, Dirisala VR (2017) An update on clinical burden, diagnostic tools, and therapeutic options of staphylococcus aureus. *Infect Dis Res Treat*. 10:1179916117703999.

Sazzad HossenToushik, Md. FurkanurRahamanMizan, Md. Iqbal Hossain, Sang-Do Ha, (2019) Fighting with old foes: The pledge of microbe-derived biological agents to defeat mono- and mixed-bacterial biofilms concerning food industries, Trends in Food Science &Technology,99: 413–425, https://doi.org/10.1016/j.tifs.2020.03.019.

Schroeder TH, Reiniger N, Meluleni G, Grout M, Coleman FT, Pier GB. (2001) Transgenic cystic fibrosis mice exhibit reduced early clearance of Pseudomonas aeruginosa from the respiratory tract. J Immunol. 166(12):7410–8. doi: 10.4049/jimmunol.166.12.7410

Shen, Z, He, K., Ding, Z., Zhang, M., Yu, Y., Hu, J. (2019) Visible-light-triggered self-reporting release of nitric oxide (NO) for bacterial biofilm dispersal. *Macromolecules*, 52(20):7668–7677.

Shuang Tian, Henny C. van der Mei, YijinRen, Henk J. Busscher, LinqiShi, (2021) Recent advances and future challenges in the use of nanoparticles for the dispersal of infectious biofilms, Journal of Materials Science &Technology,84: 208–218, https://doi.org/10.1016/j.jmst.2021.02.007.

Slan GA, Tornello PC, Abraham GA, Duran N, Castro GR. (2016) Smart lipid nanoparticles containing levofloxacin and DNase for lung delivery. Design and characterization. Colloids Surf B Biointerfaces. 143:168–176. doi: 10.1016/j.colsurfb.2016.03.040.

Sreelakshmi, C., Datta, K.K., Yadav, J.S., Reddy, B.V., (2011) Honey derivatized Au and Ag nanoparticles and evaluation of its antimicrobial activity. J. Nanosci. Nanotechnol.11:6995–7000.

Sundrarajan, M., Bama, K., Bhavani, M., Jegatheeswaran, S., Ambika, S., Sangili, A., Nithya, P., Sumathi, R., (2017) Obtaining titanium dioxide nanoparticles with spherical shape and antimicrobial properties using *M. Citrifolia* leaves extract by hydrothermal method. J. Photochem. Photobiol. B 171:117–124.

Tian, S.L. Su, Y. Liu, J. Cao, G. Yang, Y. Ren, F. Huang, J. Liu, Y. An, H.C. van derMei, H.J. Busscher, L. Shi,(2020) Self-targeting, zwitterionic micellar dispersants enhance antibiotic killing of infectious biofilms – An intravital imaging study in mice. *Science advances*, 6(33): p.eabb1112.

Tran, H. M., Tran, H., Booth, M. A., Fox, K. E., Nguyen, T. H., Tran, N., Tran, P. A. (2020) Nanomaterials for treating bacterial biofilms on implantable medical devices. Nanomaterials, *10*(11): 2253.

Ulbrich K, Holá K, Šubr V, Bakandritsos A, Tuček J, Zbořil R.(2016) Targeted Drug Delivery with Polymers and Magnetic Nanoparticles: Covalent and Noncovalent Approaches, Release Control, and Clinical Studies. Chem Rev. 116(9):5338–431. doi: 10.1021/acs. chemrev.5b00589.

Valsalam, S., Agastian, P., Arasu, M. V., Al-Dhabi, N. A., Ghilan, A. K. M., Kaviyarasu, K.,. . ., Arokiyaraj, S. (2018) Rapid biosynthesis and characterization of silver nanoparticles from the leaf extract of Tropaeolum majus L. and its enhanced in-vitro antibacterial, antifungal, antioxidant and anticancer properties. Journal of Photochemistry and Photobiology B: Biology, *191*: 65–74.

Vasudevan R (2014) Biofilms: microbial cities of scientific significance. *J Microbiol Exp.* 1(3):00014.

Wang, Y, Kadiyala, U., Qu, Z., Elvati, P., Altheim, C., Kotov, N. A., VanEpps, J. S. (2019) Anti-biofilm activity of graphene quantum dots via self-assembly with bacterial amyloid proteins. *ACS nano*, 13(4), 4278–4289.

Wolcott, R.D., Rum baugh, K.P., James, G., Schultz, G., Phillips, P., Yang, Q., Waiters, C., Stewart, P.S., Dowd, S.E., (2010) Biofilm maturity studies indicate sharp debridement opens a time dependent therapeutic window. J. Wound Care 19: 320–328.

Wong, K.K., Liu, X., (2010) Silver nanoparticles – the real "silver bullet" in clinical medicine? Med. Chem. Commun. 1: 125–131.

Wu W, He Q, Jiang C. Magnetic iron oxide nanoparticles: synthesis and surface functionalization strategies. Nanoscale Res Lett. (2008) 3(11): 397–415. doi: 10.1007/s11671-008-9174-9.

Wu Y, van der Mei HC, Busscher HJ, Ren Y (2020) Enhanced bacterial killing by vancomycin in staphylococcal biofilms disrupted by novel, DMMA-modified carbon dots depends on EPS production. *Colloids Surf B Biointerfaces*. 193:111114.

Xiaochi Feng, Qinglian Wu, Lin Che, Nanqi Ren, (2020) Analyzing the inhibitory effect of metabolic uncoupler on bacterial initial attachment and biofilm development and the underlying mechanism, EnvironmentalResearch,185:109390, https://doi.org/10.1016/j. envres.2020.109390.

Xie., Y., Zheng, W., Jiang, X. (2020) Near-infrared light-activated phototherapy by gold nanoclusters for dispersing biofilms. *ACS applied materials & interfaces*, 12(8), 9041–9049.

Xin Yi, Chengyong Wang, Xiao Yu, ZhishanYuan, A novel bacterial biofilms eradication strategy based on the microneedles with antibacterial properties, ProcediaCIRP,89:159–163, https://doi.org/10.1016/j.procir.2020.05.136.

Xiong, M.H., Bao, Y., Yang, X.Z., Wang, Y.C., Sun, B., Wang, J., (2012) Lipase-sensitive polymeric triple-layered nanogel for "on-demand" drug delivery. J. Am. Chem. Soc.134: 4355–4362.

Yan R, Yang Y, Chen Y. (2018) Pharmacokinetics of Chinese medicines: strategies and perspectives. Chin Med. 13:24. doi: 10.1186/s13020-018-0183-z.

Yazhiniprabha M, Vaseeharan B (2019) In vitro and in vivo toxicity assessment of selenium nanoparticles with significant larvicidal and bacteriostatic properties. *Mater Sci Eng C*. 103:109763.

Yuan Z, Tao B, He Y, Mu C, Liu G, Zhang J, Liao Q, Liu P, Cai K.(2019) Remote eradi-
    cation of biofilm on titanium implant via near-infrared light triggered photother-
    mal/photodynamic therapy strategy. Biomaterials. 223:119479. doi: 10.1016/j.
    biomaterials.2019.119479

Zhao J, Ma SC, Li SP. (2018) Advanced strategies for quality control of Chinese medicines. J
    Pharm Biomed Anal.147:473–478. doi: 10.1016/j.jpba.2017.06.048.

Zhu X, Radovic-Moreno AF, Wu J, Langer R, Shi J.(2014) Nanomedicine in the Management
    of Microbial Infection – Overview and Perspectives. Nano Today. 9(4):478–498. doi:
    10.1016/j.nantod.2014.06.003.

Zhuang Ma, Jie Li, YayunBai, Yufei Zhang, Haonan Sun, XingeZhang (2020) A bacterial
    infection-microenvironment activated nanoplatform based on spiropyran-conjugated
    glycoclusters for imaging and eliminating of the biofilm, Chemical Engineering
    Journal,399:125787, https://doi.org/10.1016/j.cej.2020.125787.

Ziuzina D, Han L, Cullen PJ, Bourke P. (2015) Cold plasma inactivation of internalised
    bacteria and biofilms for Salmonella entericaserovar Typhimurium, Listeria mono-
    cytogenes and Escherichia coli. Int JFoodMicrobiol. 210:53–61. doi: 10.1016/j.
    ijfoodmicro.2015.05.019.

Zoubos, A. B., Galanakos, S. P., Soucacos, P. N. (2012) Orthopedics and biofilm – what do we
    know? A review. *Medical science monitor: international medical journal of experimen-
    tal and clinical research*, *18*(6), RA89 – RA96. https://doi.org/10.12659/msm.882893

14 Postbiotics Produced by Lactic Acid Bacteria

Current and Recent Advanced Strategies for Combating Pathogenic Biofilms

Lucimeire Fernandes Correia, Maria Vitoria Minzoni De Souza Iacia, Letícia Franco Gervasoni, Karolinny Cristiny de Oliveira Vieira, Erika Kushikawa Saeki, and Lizziane Kretli Winkelströter

CONTENTS

DOI: 10.1201/9781003184942-17

14.1 INTRODUCTION

A novel concept is gaining relevance in science and microbiology studies. Postbiotics are characterized as a new class of biologically active microbial agents responsible for modifying biological responses (Barros *et al.*, 2019; Shenderov and Gabrichevsky, 2017; Oleskin and Shenderov, 2019).

Also known as metabiotics, biogenics, PCFs (probiotic cell fragments), or simply metabolites/CFS (cell free supernatants), they are soluble factors (metabolic products or byproducts) secreted by living bacteria (probiotics or non-probiotics) or released after bacterial lysis (Barros *et al.*, 2019; Cuevas-González *et al.*, 2020).

Postbiotics can have important applications in areas such as medicine, veterinary medicine, pharmaceuticals, and food. They demonstrate advantageous characteristics compared to probiotics such as: greater safety, longer shelf life, standardized safety dosage, and defined chemical structure. These particularities make them attractive from economic, biotechnological, and clinical perspectives (Mantziari *et al.*, 2020; Rad *et al.*, 2020; Teame *et al.*, 2020; Tomar *et al.*, 2015). Also, due to their immunomodulatory and antimicrobial effects, postbiotics have been suggested as a preventive strategy for infectious diseases (Cuevas-González *et al.*, 2020; Rossoni *et al.*, 2020).

In this context, the potential of postbiotics as promising tools in combating pathogenic biofilms is highlighted (Ishikawa *et al.*, 2021; Rossoni *et al.*, 2020). This fact may be associated with the mechanism of action by which probiotic bacteria antagonize pathogenic microorganisms and thus could lead to possible impacts on the formation of biofilms (Kareem *et al.*, 2014; Oleskin and Shenderov, 2019; Ishikawa *et al.*, 2021; Rossoni *et al.*, 2020).

Data in the literature arouse interest in the development of new discoveries in the field of research and also boost the application and inclusion of several beneficial strains for combating pathogenic biofilms (Shenderov and Gabrichevsky, 2017). However, further clarification is needed about the mechanisms of action, effects, dosages, and interactions of the diversity of existing postbiotic compounds.

14.2   LACTIC ACID BACTERIA

The term lactic acid bacteria (LABs) describes an important group of very diverse and well-established microorganisms that share common physiological and metabolic characteristics, such as the production of lactic acid as the main metabolic product of glucose (Papadimitriou *et al.*, 2016).

LABs were one of the first groups of bacteria to be widely studied, mainly due to the facts that they are responsible for the processes of food fermentation, they present beneficial effects on human health because of the probiotic properties of some strains, and also because they play important roles in biotechnological processes (De Filippis *et al.*, 2020; Papadimitriou *et al.*, 2016). This group includes microorganisms from the environmental microbiota, human and animal commensals, opportunistic microorganisms, and obligatory pathogens mainly associated with the genera *Lactococcus, Enterococcus, Oenococcus, Pediococcus, Streptococcus, Leuconostoc*, and *Lactobacillus* (Ganzle, 2015; Makarova *et al.*, 2006).

Advances in molecular techniques during the 1990s allowed the use of more refined criteria to define LABs. This group is currently described as gram-positive, non-motile, non–spore-forming bacteria, usually cocci or rods, microaerophilic or facultative anaerobic, typically fastidious, aerotolerant, acid-tolerant, catalase, and cytochrome negative, with a base molar DNA composition less than 50% of G + C (Papadimitriou *et al.*, 2016). In addition to the production of lactic acid, this group is also recognized for the production of antimicrobial substances such as bacteriocins, hydrogen peroxide, and diacyl, among others (Mokoena, 2017).

LABs can be classified based on the final products of carbohydrate metabolism that it is divided into: homofermentative and heterofermentative, then subdivided into facultative and obligate fermentative species (Ganzle, 2015).

LABs considered obligate homofermentative produce mainly lactic acid from the fermentation of hexoses via Emden-Meyerhoff. The genera included in this group are *Pediococcus, Lactococcus*, and *Streptococcus*. LABs considered facultative heterofermentative, such as *Leuconostoc* and certain *Lactobacillus* are capable of degrading hexoses; however, they also metabolize pentoses and often gluconate due to the production of the enzyme aldolase and phosphoketolase. Finally, the obligate heterofermentative produces lactic acid, along with significant amounts of ethanol or acetic acid and carbon dioxide (Ganzle, 2015; Buron-Moles *et al.*, 2019).

Currently, LABs are known to produce a variety of antimicrobial compounds that may be known as postbiotics (Yordshahi *et al.*, 2020). It is noted that LABs are a multifunctional group, their characteristics are advantageous for a variety of functions, and the postbiotics produced by them have the potential to affect and to improve various aspects associated with problems in health and food production areas.

14.3   POTENTIAL USES OF LACTIC ACID BACTERIA

14.3.1   Starter Cultures in the Food Industry

LABs play a significant role in food development due to their fermentation capacity in different food matrices such as dairy products, meat, fruits, vegetables, and cereals. LABs influence the flavor, aroma, texture, and nutritional value of food due to the metabolic compounds generated, for example, in glycolysis (sugar fermentation, with lactic acid as the main final product), lipolysis (degradation of fat), and proteolysis (degradation of proteins, having great importance for the development of flavor) (Papadimitriou *et al.*, 2016; Wu *et al.*, 2017).

The traditional method for manufacturing fermented foods consists of a series of steps such as inoculation, whereby a sample of the fermented matrix from a previous process is used as a starter in a new fermentation process, thus, environmental cultures composed of various species and strains of LABs are naturally selected by manufacturing (raw material, environmental conditions, type of fermentation, etc.) in a process called back-slopping (Bintsis, 2018; De Filippis *et al.*, 2020).

The variety of microorganisms combinations and raw material originated from this process allows the enormous diversity of fermented products with different organoleptic properties. Nowadays, the production of fermented foods is performed on an industrial scale in automated processes, and for this reason, there was a need to use selected starter LABs with defined strains and known physiological characteristics to ensure products with good quality (Bintsis, 2018; De Filippis *et al.*, 2020).

The main area for LABs application is still in the food industry as a starter culture, but due to their versatile metabolisms and abilities to synthesize a wide range of beneficial metabolites, these bacteria are highly attractive for use in biotechnology and therapeutic products (Mora-Villalobos *et al.*, 2020).

14.3.2 PROBIOTICS

Many species of LABs are considered potential probiotics and are marketed in functional foods and/or probiotic preparations, such as some belonging to the genera *Lactobacillus* and *Bifidobacterium* (De Filippis *et al.*, 2020; Evivie *et al.*, 2017). Probiotics are defined as "live microorganisms that, when administered in adequate amounts, confer a benefit to the health of the host" (Hill *et al.*, 2014). In general, consumers are already familiar with the concept of probiotics due to their intense promotion in the pharmaceutical and nutraceutical industries as therapeutic foods and health promoters. Probiotics have been applied in the treatment of various health conditions such as lactose intolerance, food allergies, and Crohn's disease, shortening the duration of acute infectious diarrhea in infants and children, rheumatoid arthritis, and colon cancer (Păcularu-Burada *et al.*, 2020; Evivie *et al.*, 2017).

Despite the many health benefits, the use of probiotics has limitations, and some data show that not all probiotic bacteria are completely safe and may – rarely – be associated with human opportunistic infections such as endocarditis, sepsis, pneumonia, urological infections, abdominal abscesses, and others, especially in immunologically compromised patients. In addition, the facts that there are unknown molecular mechanisms, strain-specific behavior, development of antibiotic resistance, transfer of virulence genes, and questions about viability in the production process are also highlighted (Evivie *et al.*, 2017; Nataraj *et al.*, 2020; Shenderov, 2013).

14.3.3 POSTBIOTICS

The term postbiotic was quickly adopted in several areas of study, and the definition widely used for it is "soluble metabolic products secreted by probiotics that have physiological benefits to the host" (Bermudez-Brito *et al.*, 2012). These postbiotics have drawn attention because of their known chemical structures, safe dose parameters, easy production and storage, and longer shelf lives (Nataraj *et al.*, 2020; Teame *et al.*, 2020).

In vitro and in vivo studies suggest that they have similar effects to probiotics. It comes down to health benefits, although the mechanisms are not fully elucidated. Their functional properties include antimicrobial, antioxidant, and immunomodulatory capacities that will positively influence specific immunological, physiological, neuro-hormonal, and metabolic reactions. With this, postbiotics have become a valid and safer alternative to live probiotic bacteria. In addition, the benefits of these metabolites go beyond health, proving to be important participants in different bioprocesses (Aguilar-Toalá *et al.*, 2018; Nataraj *et al.*, 2020; Teame *et al.*, 2020).

The various postbiotics products of LABs include vitamins, organic acids, short-chain fatty acids, proteins/peptides, bacteriocins, neurotransmitters, biosurfactants, amino acids, and more (Nataraj *et al.*, 2020).

14.3.4 PRODUCTION OF SUBSTANCES OF INDUSTRIAL INTEREST

Lactic acid bacteria have attributes that favor their general application in industry, such as their GRAS (generally regarded as safe) status, stress tolerance, simple metabolism, and ability to metabolize various carbon sources. Their genetic characteristics and metabolism control are already well elucidated, making LABs potential candidates for the production of important metabolites in different industrial sectors, as shown in Figure 14.1 (Hatti-Kaul *et al.*, 2018; Wu *et al.*, 2017).

Recent advances in metabolic engineering strategies, genetic manipulation, and cell metabolism adaptation are already used in LABs. They focus mainly on redirecting pyruvate metabolism in order to increase the production of commercially important secondary metabolites such as organic acids, sweeteners, aromatic compounds, vitamins, and biosynthetic complexes, making the group promising candidates for cell manufacturing (Wu *et al.*, 2017).

LABs are of great importance in the fermentative production of lactic acid (LA), a product with great application in the cosmetic, pharmaceutical, chemical, and agricultural industries (Eiteman and Ramalingam, 2015; Mora-Villalobos *et al.*, 2020). The demand for LA in recent decades has mainly been driven by the growing market for polylactic acid (PLA), a biodegradable, thermostable, and biocompatible polyester that can be used in packaging, textiles, pharmaceuticals, and prosthetic devices (Hatti-Kaul *et al.*, 2018). The rapid growth of the lactic acid market relies on metabolic engineering as a powerful tool to manipulate the characteristics of LA produced by LABs, in order to increase the yield, productivity, and purity of the final product (Upadhyaya *et al.*, 2014).

Several important polyols used as sweeteners in dietary foods are produced by LABs – for example L-alanine, mannitol, sorbitol, and xylitol – that can be added to

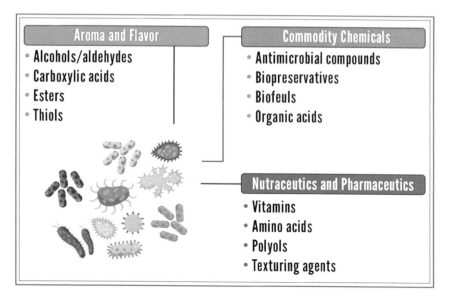

Aroma and Flavor
- Alcohols/aldehydes
- Carboxylic acids
- Esters
- Thiols

Commodity Chemicals
- Antimicrobial compounds
- Biopreservatives
- Biofeuls
- Organic acids

Nutraceutics and Pharmaceutics
- Vitamins
- Amino acids
- Polyols
- Texturing agents

FIGURE 14.1   Metabolites produced by lactic acid bacteria with potential use in industry.

food or generated in situ during fermentation. In addition, some LABs also demonstrate the ability to biosynthesize soluble vitamins belonging to the B group – such as riboflavin (B2), folate (B11), thiamine (B1), and cobalamin (B12) – which are essential cofactors for metabolic activities. Metabolites such as acetaldehyde, bacteriocins, benzaldehyde, diacetyl, formaldehyde, propionic acid, and acetoin contribute positively to the flavor, aroma, stability, and texture of fermented foods. The use of metabolites goes even further with the production of biofuels such as ethanol and butanol, chemical compounds for the production of solvents, and polymers such as 2,3 butanediol, succinic acid, and 1,3-propanediol (LeBlanc *et al.*, 2020; Sauer *et al.*, 2017; Sharma *et al.*, 2020).

14.3.5   BIOPRESERVATION

Biopreservation is a new approach applied to food. It is based on the use of specific microorganisms and their metabolites to extend the shelf life of products and prevent their deterioration. LABs and their postbiotics are strongly associated with biopreservation processes due to the production of organic acids and a decrease in pH during fermentation. However, LABs still synthesize other products with antimicrobial action such as bacteriocins, enzymes, alcohols, and low molecular weight substances (Singh, 2018; Mora-Villalobos *et al.*, 2020).

Bacteriocins are peptides with antimicrobial activity and have the advantage of not being toxic and immunogenic, and they are also heat stable. Due to these properties, they are considered good bio-preservative agents. To date, only the bacteriocins nisin and pediocin are marketed as food additives. However, other LABs bacteriocins offer promising prospects (Castellano *et al.*, 2017; Settanni and Corsetti, 2008).

Bacteriocins can be added in pure form, or from inoculation of the producing strain. The most recent application of this postbiotic product involves its connection to polymeric packaging, a technology known as active packaging; in this way, they are gradually released and meet consumer demand for extended shelf life of food (Castellano *et al.*, 2017; Settanni and Corsetti, 2008; Alvarez-Sieiro *et al.*, 2016).

14.3.6 BIOREFINERIES

Our main source of energy is obtained from a non-renewable resource oil. In this scenario, there is currently a great demand to find alternative and sustainable raw materials. One alternative is the production of fuels and other chemical products from microbial fermentation of sugars in different types of biomass, a process known as biorefinery. LABs are among the most promising microorganisms to be applied in biorefineries, as they naturally produce many metabolites (bacteriocins, vitamins, and products for the manufacture of plastic polymers, among others) and are microorganisms easily adapted to industrial processes (Galbe and Wallberg, 2019; Mazzoli *et al.*, 2014; Sabater *et al.*, 2020).

We can highlight the development of biorefineries with the use of *L. lactis*. This microorganism was modified to produce ethanol from lactose present in whey, a residue resulting from the manufacture of cheese. We also emphasize the production of commercially valuable chemical products such as ethanol produced from sugarcane bagasse (Liu *et al.*, 2016; Mora-Villalobos *et al.*, 2020).

14.3.7 BIOFERTILIZERS, BIOSTIMULANTS, AND BIOCONTROL

The constant increase of the population has generated a huge need for food production, which has consequently resulted in the excessive use of synthetic agrochemicals with serious environmental impacts. Thus, society is facing the challenge of increasing agricultural production through innovative technologies that avoid environmental degradation. LABs isolated from different sources are being studied to improve agricultural production, and have shown promising properties as biofertilizers, biocontrol agents, and biostimulants (Gabra *et al.*, 2019; Garcia-Gonzalez and Sommerfeld, 2016).

As biofertilizers, LABs improved the availability of nutrients from the decomposition of organic compounds present in the soil, as they are capable of fixing atmospheric nitrogen, an essential element for the structure and function of plant cells (Gabra *et al.*, 2019). LABs have also shown good results in the biocontrol of a wide variety of fungal and bacterial phytopathogens due to the production of antimicrobial compounds, reactive oxygen species, and bacteriocins (Lamont *et al.*, 2017; Mącik *et al.*, 2020). In addition, LABs can promote biostimulation through plant growth or seed germination, as well as alleviating abiotic stress (Lamont *et al.*, 2017).

14.3.8 BIOREMEDIATION

The increasing use of heavy metals, synthetic dyes, and pesticides poses a threat to environmental safety and serious concerns for human health. In this context, the expansion of new technologies has stimulated bioremediation, a type of biological recycling mechanism whereby microorganisms are used in an attempt to degrade toxic waste (Abatenh et al., 2017). The ability of LABs to bioadsorb heavy metals has been reported. It is suggested that oral ingestion of these microorganisms in the form of probiotics could significantly avoid the absorption of cadmium, for example (Kumar et al., 2018). It is also believed that probiotic strains decrease the absorption and toxicity of pesticides in the body (Sarlak et al., 2021).

14.4 POSTBIOTICS AND POTENTIAL MECHANISMS OF ACTION

The human microbiota is made up of several microorganisms that are present in body tissues and fluids. The gastrointestinal tract contains microbiota that directly affect the health of individuals. It is a stable environment that provides nutrients and favors the installation of microorganisms that generate benefits for the immune system, control the emergence of pathogenic strains, and ensure intestinal protection. In this context, prebiotics, probiotics, or postbiotics favor intestinal modulation, promoting the balance of the microbiota (Żółkiewicz, et al., 2020).

In general, ingested carbohydrates are acted upon by several intestinal enzymes and then they can be absorbed. In addition to digestible carbohydrates, non-digestible fibers are also present, and have great importance as a prebiotic and in the passage of food through the gastrointestinal tract (Bhat et al., 2019). Prebiotics are used as food by probiotic bacteria in a beneficial way for the health of the individual, and in this context, several postbiotics are produced such as short-chain fatty acids (SCFAs), enzymes, vitamins, and others (Figure 14.2) (O'Toole et al., 2017).

Postbiotics are bioactive substances produced by bacteria or fungi that aid the individual's health in several aspects. The mechanism of action of postbiotics is mainly based on the production of anti-inflammatory substances such as IL-4, IL-6, and IL-10 cytokines and the reduction in reactive oxygen species (ROS), generating an anti-inflammatory action and induction of the immune system through the differentiation of T lymphocytes into Th1 and Th2 lymphocytes, which favors the immunoregulation process (Malagón-Rojas et al., 2020).

Postbiotics can act through different mechanisms of action and can be distributed into a variety of classes, as shown in Table 14.1.

14.4.1 EXTRACELLULAR POLYMERIC SUBSTANCES

Extracellular polymeric substances (EPSs) are products formed from biopolymers located in the bacterial cell wall and have been widely used by the food and pharmaceutical industries (Singh and Saini, 2017). EPSs can modulate the immune system, especially in some cells such as macrophages, dendritic cells, and T and NK lymphocytes. EPSs can also affect lipids and reduce cholesterol absorption. In addition, some EPSs generated from *Lactobacillus* isolated from durian fruits showed antimicrobial and antioxidant effects (Khalil et al., 2018).

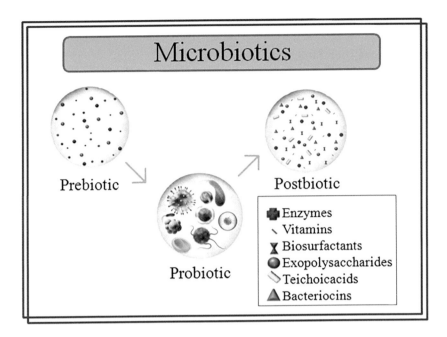

FIGURE 14.2   The connection between prebiotic, probiotic and postbiotic.

A type of EPS known as β-glucans can influence the action of probiotics, such as providing adhesion of *Lactobacilli* to the intestinal tissue, in addition to boosting the cellular immune response against microorganisms and cancer cells (Żółkiewicz *et al.*, 2020). These β-glucans can also enhance the bioavailability and absorption of substances with antioxidant and anti-inflammatory capacities in the intestinal tissue (Masashi *et al.*, 2020).

14.4.2   ENZYMES

ROSs are unstable chemical compounds formed during oxidative stress. Some bacteria have the ability to fight ROSs through antioxidant enzymes, especially catalase, glutathione peroxidase, NADH oxidase, and peroxide dismutase (Żółkiewicz, *et al.*, 2020). High concentrations of glutathione peroxidase were identified in *Lactobacillus fermentum*, suggesting its antioxidant potential (Tomusiak-Plebanek *et al.*, 2018).

14.4.3   SHORT-CHAIN FATTY ACIDS

Short-chain fatty acids are organic acids composed of carboxylic acids, such as propionate, acetate, and butyrate, produced from fermentation processes. *Lactobacillus* uses fermentation via the glycolytic pathway; however, heterofermentative bacteria use the phosphoketolase pathway to obtain them (LeBlanc *et al.*, 2017). The metabolization of acetate takes place at the level of muscle tissue and other tissues, while

TABLE 14.1

Postbiotics and Their Functions

Postbiotic	Microorganism	Mechanism of Action	Application	Reference
Exopolysaccharides	*Lactobacillus casei*	Modulation of the immune response	Increasing the effectiveness of the foot-and-mouth disease vaccine	Lei *et al.*, 2018
	Lactobacillus kefiranofaciens	Inhibition of cholesterol absorption	Prevention of cardiovascular diseases	Khalil *et al.*, 2018
Enzymes	*Lactobacillus fermentun*	Antioxidant action from glutathione peroxidase	Decrease in oxidative stress	Żółkiewicz *et al.*, 2020
Short-chain fatty acids	*Roseburia intestinalis*	Inhibition of the atherogenesis process	Avoid formation of atheromatous plaques	Kasahara *et al.*, 2018
Biosurfactants	*Rhodococcus erythropolis*	Antibacterial and antifungal action	Therapeutic applications	Hu *et al.*, 2020
Vitamins	*Lactobacillus breeis, L. fermentum, L. reuteri, L. salivarius*	Riboflavin synthesis	Improved absorption of nutrients	Nataraj *et al.*, 2020
Bacteriocins	*Lactobacillus lactis*	Inhibitory action against food pathogens	Intestinal protection	Nataraj *et al.*, 2020
Teichoic acids	*Lactobacillus*	Antibiofilm in the oral cavity and gastrointestinal tract	Protection against biofilm formation	Jung *et al.*, 2019
	Lactobacillus rhamnosus	Production of pro-inflammatory cytokines	Participation in modulation of the inflammatory process	Kim *et al.*, 2017

liver cells carry out the absorption of propionate. Butyrate has a primary energetic action and regulates the microbiota, programmed cell death, and multiplication and differentiation of intestinal cells. Also noteworthy is the participation in the modulation of the immune response and in cell signaling (Zeng *et al.*, 2019).

14.4.4 BIOSURFACTANTS

Biosurfactants (BSs) are polymeric substances produced and secreted during bacterial growth and are directly involved in obtaining nutrients, cell defense, and substrate metabolism (Kaczorek *et al.*, 2018). The amphiphilic characteristics of BSs contribute to controlling the formation of biofilms of pathogenic bacteria. In addition, BSs present wetting and emulsifying action, which hinders bacterial adhesion and also the formation of biofilms (Nataraj *et al.*, 2020).

The chemical structure of BS has a hydrophilic and a hydrophobic part, capable of reducing surface tension and interfacial tension (immiscible phases) generating microemulsions. This fact favors the use of BS in the cleaning, hygiene, and cosmetics industries (Varjani & Upasani, 2017).

Biosurfactants also present antimicrobial action, such as rhamnolipids from *Pseudomonas aeruginosa* and surfactin from *Bacillus subtilis*. In this way, the BS-producing bacteria increase their chances of survival and competition for nutrients (Emmanuel *et al.*, 2019).

14.4.5 VITAMINS

Vitamins are organic components acquired mainly through food (Giori *et al.*, 2018). Some extra vitamins are needed through specific supplementation such as folic acid, cobalamin, and riboflavin (Nataraj *et al.*, 2020). These substances are important in different cellular processes such as energy metabolism (B-complex) and catalytic action (vitamin K). The use of vitamins produced by microorganisms has been an advantageous option in the pharmaceutical industry, such as the vitamin cobalamin obtained from the fermentation process (Deptula *et al.*, 2017). Other vitamins have also been associated with biosynthesis from bacteria in the gastrointestinal tract, such as biotin, cobalamin, thiamine, folate, niacin, pyridoxine, and riboflavin. In this way, lactic acid bacteria such as *Lactococcus* and *Lactobacillus* can express genes for the production of several vitamins such as riboflavin (Takamura *et al.*, 2021).

14.4.6 BACTERIOCINS

Bacteriocins are ribosomal peptides produced by gram-positive and gram-negative bacteria that have antimicrobial action (Field *et al.*, 2018). Bacteriocins have been used as natural biopreservatives in the processing and fermentation of foods such as cheese, meat, and vegetables. In addition, their clinical use has been evaluated due to their ability to inhibit several pathogens (Hasan *et al.*, 2019; Nataraj *et al.*, 2020).

14.4.7 TEICHOIC ACIDS

Teichoic acids are glycopolymers that can appear as lipoteichoic acids (linked to glycolipids) and wall teichoic acids (linked to peptidoglycans). In gram-positive bacteria, they contribute to antimicrobial resistance (Sumrall *et al.*, 2019). Wall teichoic acids play an important role in defining cell shape, cell division, and other metabolic processes essential to cell physiology (Nataraj *et al.*, 2020).

It is recognized that lipoteichoic acid (LTA) from *Lactobacillus* acts by preventing the formation of biofilms in the oral cavity and gastrointestinal tract (Jung *et al.*, 2019). The antibiofilm effect of LTA against *Staphylococcus aureus* demonstrates action against ica operon, responsible for producing N-acetylglucosamine, the main metabolite in the production of biofilms of *S. aureus*. However, studies using

LTA from *Lactobacillus rhamnosus* GG produced pro-inflammatory cytokines (Kim *et al.*, 2017).

14.5 BIOFILM FORMATION AND TREATMENT STRATEGIES FOR COMBATING BACTERIAL BIOFILM

Biofilms are complex biological systems formed by a structured community of microorganisms that adhere to a surface (biotic or abiotic) and are surrounded by a rigid structural matrix secreted by the microorganisms themselves. The matrix is composed of EPSs, proteins, lipids, and extracellular DNA (eDNA), among others (Barzegari *et al.*, 2020; Brindhadevi *et al.*, 2020). Biofilm formation is also influenced by bacterial structures and environmental factors such as temperature and pH (Zhao *et al.*, 2017).

Bacterial biofilm development can be grouped into five main steps. (I) The process begins with the formation of a reversible adhesion composed of exopolysaccharides, flagellar movement and type IV pili. (II) Subsequently, there is an intercellular interaction with the quorum sensing (QS) system that favors sessile growth. (III) First stage of maturation, in which there is the initial development of the biofilm architecture, called microcolonies. (IV) Second stage of biofilm maturation, with the formation of three-dimensional mushroom-shaped structures. (V) Dispersion stage, in which mobile cells disperse from microcolonies, which allows the cycle to start again (Saxena *et al.*, 2019; Thi *et al.*, 2020).

Biofilm formation depends on signaling molecules to coordinate their group activities through the QS system. This inter- and intra-species behavior is essential for the adaptation and survival of bacterial communities. The QS system involves the synthesis and release of signaling molecules resulting in the expression of specific genes that promote the regulation of secondary metabolites and several virulence factors (Barzegari *et al.*, 2020; Brindhadevi *et al.*, 2020; Thi *et al.*, 2020).

The implications of biofilms can vary in nature and magnitude, being considered harmful or beneficial, depending on the applied field (Lianou *et al.* 2020). Biofilm formation can cause serious problems in the food industry, water quality, hospital environments, and human health.

In the food industry, biofilms are largely responsible for spoilage and outbreaks of foodborne illnesses, as well as damage to food processing equipment. *Salmonella enterica*, *Listeria monocytogenes*, *Campylobacter* spp., *Escherichia coli*, *Bacillus cereus*, and *Staphylococcus aureus* are some pathogens that affect the quality and safety of food (Carrascosa *et al.*, 2021). The formation of biofilms in the piping of water supply systems contributes to the presence of heterotrophic bacteria, which can indicate hygiene problems (Riley *et al.*, 2011).

In the hospital environment, the formation of biofilms on medical devices is responsible for several chronic infections in hospitalized patients (Stewart and Bjarnsholt, 2020). Among the various negative impacts that microbial biofilms cause on human health, we highlight persistent infections by biofilm-forming bacteria, especially *P. aeruginosa* in immunocompromised patients (patients with burns and/ or cystic fibrosis) (Høiby *et al.*, 2017; Das *et al.*, 2020).

In contrast, biofilms can have beneficial effects when applied to the environment for bioremediation, such as for the treatment of domestic sewage and bioremediation of toxic compounds (Catania *et al.*, 2020; Han *et al.*, 2020).

Currently, the treatment of infections caused by bacteria that form biofilms is considered a complex challenge for medicine, since antibiotic therapy has become insufficient due to the spread of multi-resistant bacteria. Therefore, understanding the nature of biofilms helps us to support efforts to prevent and fight bacterial infections (Barzegari *et al.*, 2020).

The best way to prevent the formation of biofilms is to inhibit the growth of bacteria (Zhao *et al.*, 2017). In this sense, different prevention and control methods are currently described, such as the use of antibacterials, physical and chemical treatments, inhibitors of the quorum sensing system, nanotechnology, phagotherapy, and the use of postbiotics.

14.5.1   ANTIBACTERIAL AGENTS

The use of antibacterials is one of the ways to prevent the formation of biofilms, as they act to inhibit the growth of bacteria. In the food industry, food sterilization methods or the use of food additives (surfactants, flavorings, and dyes) that have antibacterial activities against pathogenic microorganisms are used (Zhao *et al.*, 2017).

14.5.2   PHYSICAL TREATMENTS

Physical treatments to control biofilm formation in the food industry can be carried out with the application of ultraviolet (UV) radiation, thermal shock, and ultrasound (Gayán *et al.*, 2012; Bhargava *et al.*, 2021). Usually, a single treatment is not able to eliminate the formation of biofilms, requiring a combination of different methods (Mizan *et al.*, 2020; Shao *et al.*, 2020).

Ultraviolet light (UV) technology is based on the emission of radiation within the ultraviolet (100–400 nm) region, being categorized as UV-A (320–400 nm), UV-B (280–320 nm), and UV-C (200–280 nm). UV-C radiation is considered the lethal germicidal wavelength for most microorganisms (Gayán *et al.*, 2012).

The application of thermal shock at temperatures at or higher than 80°C can be considered an easy method for inhibiting biofilm. This type of treatment is commonly used to inactivate bacteria (Chang *et al.*, 2017).

Ultrasound technology is environmentally sustainable and is a form of vibrational energy produced by an ultrasound transducer that converts electrical energy into acoustic energy. This process directly damages the cell wall or cell membrane of bacteria, promotes the release of free radicals, breaks down DNA, and inactivates enzymes (Chen *et al.*, 2020).

14.5.3   CHEMICAL TREATMENTS

When the formation of mature biofilm occurs, physical treatment alone may not be enough for its removal. Some studies have shown that chemical treatments play an important role in removing bacterial cells (Zhao *et al.*, 2017). Disinfection using quaternary ammonium compounds, aldehydes, alcohols, and halogens can be used, especially in hospital environments (Abreu *et al.*, 2013).

14.5.4 QUORUM SENSING INHIBITORS

Some inhibitors of the quorum sensing (QS) system, employed in a process also called quorum quenching (QQ), are being described as an alternative and promising therapy for controlling the spread of pathogenic bacteria, and therefore providing a new possibility to try to solve the problem of microbial multi-resistance (Carradori *et al.*, 2020). QQ consists of the enzymatic degradation of signal molecules from the quorum sensing system in order to prevent their accumulation in the environment and inhibit the expression of genes that regulate virulence and biofilm formation (Paluch *et al.*, 2020). Among the compounds already described with this activity, there are examples of natural compounds, synthetic compounds, and even metallic nanoparticles (Garcia-Lara *et al.*, 2015; Zhang *et al.* 2018; Cáceres *et al.*, 2020).

Nanotechnology applied in the medical field has become widely recognized for the treatment of various bacterial infections and proposed as one of the current strategies for controlling biofilm. Metals are widely used in the synthesis of nanoparticles due to their antibacterial potential; the most commonly used are silver, gold, zinc, copper, and titanium nanometals (Chaudhary *et al.*, 2020).

For the control of bacterial biofilm, the size, shape, and properties of the nanoparticles must be considered, as well as the type of target microorganism and site of infection (Liu *et al.*, 2019; Malaekeh-Nikouei *et al.*, 2020).

14.5.5 PHAGOTHERAPY

Bacteriophages are viruses that can infect and kill bacteria without any negative effect on human or animal cells. For this reason, they can be used alone or in combination with antibiotics to treat bacterial infections. They can be found in soil and seawater, ocean and land surfaces, and extreme environments with very high or very low temperatures (Principi *et al.* 2019).

The use of phages as therapeutic agents is called phage therapy and is also considered an alternative and potential strategy, in which the use of two or more bacteriophage mixtures is generally more effective in inhibiting bacterial infection and biofilm formation (Gu *et al.*, 2012; Pires *et al.*, 2017).

Another promising method is the use of postbiotics that have been used to improve food safety due to their antimicrobial action in food preservation and packaging, and in the control and elimination of bacterial biofilms (Sarikhani *et al.*, 2018; Cui *et al.*, 2020, Rad *et al.*, 2021).

Table 14.2 presents some examples of methods to combat the formation of bacterial biofilms.

14.6 APPLICATION AND CHALLENGES OF POSTBIOTICS USE AGAINST PATHOGENIC BIOFILM

Bacterial biofilm can be considered a protective mechanism developed by certain microorganisms to ensure their growth and survival in adverse conditions (Deng

TABLE 14.2
Description of Methods for Combating Biofilm

Methods	Biofilm-Forming Microorganism	Activity	Reference
Antibacterials			
Ascorbic acid additive	*Staphylococcus aureus*, *Escherichia coli*, and *Listeria monocytogenes*	Reduction in biofilm formation.	Przekwas *et al.* (2020)
Azorubine additive	*Chromobacterium violaceum* 12472 *Pseudomonas aeruginosa* PAO1 *E. coli* O157: H7, *Serratia marcescens*, and *Listeria monocytogenes*.	Reduction in EPS production and swarming motility in *P. aeruginosa* Reduction in biofilm formation in *C. violaceum*, *P. aeruginosa*, *E. coli* O157: H7, *Serratia marcescens*, and *Listeria monocytogenes*.	Al-Shabib *et al.* (2020)
Physical Treatments			
Thermal shock combined with antibiotics	*P. aeruginosa*	Reduction in biofilm formation	Ricker and Nuxoll (2017)
Ultrasound with acid electrolyzed water	*Salmonella* spp. e *S. aureus*	Reduction in biofilm formation	Shao *et al.* (2020) Mizan *et al.* (2020)
Ultrasound with sodium hypochlorite	*L.monocytogenes*	Reduction in biofilm formation	
Ultraviolet radiation	*Alicyclobacillus acidocaldarius*	Reduction in biofilm formation	Prado *et al.* (2019)
Chemical Treatments			
Peracetic acid	*Staphylococcus aureus*	Removal of adherent cells from the biofilm	Lee *et al.* (2016)
Sodium hypochlorite	*Enterococcus faecalis*	Reduction in biofilm formation	Bukhary and Balto (2017)
Sodium hypochlorite	*S. aureus, S. epidermidis, Enterobacter cloacae, K. pneumoniae, P. aeruginosa*, and *E. coli*	Reduction in biofilm formation	Al-Hasani and Al-Gburi (2019)
Ammonium quaternary	*Streptococcus mutans* and *Lactobacillus acidophilus*	Reduction in bacterial growth and biofilm formation	Daood *et al.* (2020)
Quorum Sensing Inhibitors			
Essential oil (*Lippia origanoides*)	*E. coli* O157:H7, *E. coli* O33, *Staphylococcus epidermidis* ATCC 12228 *Chromobacterium violaceum* CVO26	Reduction in biofilm formation in all tested microorganisms. Reduction in violacein formation	Cáceres *et al.* (2020)

(*Continued*)

TABLE 14.2 *(Continued)*

Methods	Biofilm-Forming Microorganism	Activity	Reference
Essential oil *(Citrus aurantium)*	*S.mutans*	Reduced growth and biofilm formation	Benzaid *et al.* (2021)
Essential oil *(Thymus daenensis)*	*Streptococcus pneumoniae*	Reduction in biofilm formation Reduced expression of QS regulatory genes: *LuxS* and *pfs*	Sharifi *et al.* (2018)
Zinc oxide nanoparticles	*P. aeruginosa* and *C. violaceum*	Reduction in biofilm formation Reduced expression of QS regulatory genes: *las* and *pqs*	Khan *et al.* (2020)
Gold nanoparticles	*S. aureus*	Antimicrobial activity and inhibition of biofilm formation	Salam *et al.* (2020)
Gold nanoparticles	*P. aeruginosa*	Reduction in biofilm formation	Ali *et al.* (2020)
Silver nanoparticles	*P. aeruginosa*	Reduction in rhamnolipid production and biofilm formation.	LewisOscar *et al.* (2021)
Silver nanoparticles	*P. aeruginosa*	Reduction in biofilm formation, various virulence factors (protease, elastase, pyocyanin, rhamnolipids, and alginate)	Singh and Saini (2017)
Phagotherapy			
Bacteriophages BΦ-R656 and BΦ-R1836	*P. aeruginosa*	Reduction in biofilm formation	Jeon and Yong (2019)
Bacteriophages vB_EcoM_10C2, vB_EcoM_10C3, vB_EcoM_118B, 112 vB_EcoM_11B2, vB_EcoM_12A1, vB_EcoM_366B, vB_EcoM_366V e vB_EcoM_3A1	*E. coli* O177	Reduction in biofilm formation	Montso *et al.* (2021)

et al., 2020). In general, postbiotics can act on biofilms in two ways: by inhibiting their formation and/or by destroying the formed biofilm. It is believed that one of the strategic points for the performance of postbiotics is the alteration of quorum sensing and the disruption of the exopolysaccharide matrix, as it would result in instability and then affect the surface adhesion and expression of genes important for biofilm synthesis. Furthermore, changes in structural permeability and formation of small pores would increase the fragility of the biofilm EPS matrix (Freire *et al.*, 2018).

Despite being considered a new topic, some studies bring relevant data on the application of postbiotics in combating biofilms in several areas (Moradi *et al.*, 2020). Some of them suggest some mechanisms of action as shown in Figure 14.3.

Overall, the presence of biofilms is a major threat to health systems, food, and the environment. Alternatives capable of combating biofilms in these areas bring benefits to the population as a whole. Until the present moment, some articles have been published on the activity of postbiotics against different types of clinical biofilms. Díaz *et al.* (2020) investigated the use of postbiotic products against *Pseudomonas aeruginosa*, and the authors observed that they were capable of interfering in the quorum sensing and then reducing metabolic activity, virulence factors, and the biofilm. Ishikawa *et al.* (2021) obtained similar results and demonstrated that postbiotics can decrease the expression of virulence genes, reducing biofilm formation by *Aggregatibacter actinomycetemcomitans*, bacteria associated with periodontitis.

Biofilm is also a critical point associated with food quality and safety, since it allows the microorganism to persist in the environment and on processing equipment. The study of Tazehabadi *et al.* (2021) evaluated the biofilm inhibition capacity of three different *Salmonella* strains by postbiotic LABs and obtained a decrease in pathogen persistence. Another important agent in food systems is *Listeria monocytogenes*, a

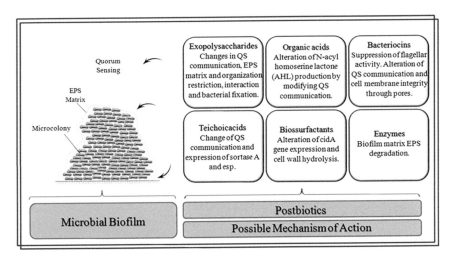

FIGURE 14.3   Critical points for the mechanisms of action of postbiotics in biofilms.

foodborne pathogen known for its resistance to environmental stress, including antimicrobial agents, sanitizers, and disinfectants, making it difficult to eliminate the microorganism. For this reason, the use of postbiotic LABs proved to be a possible alternative method, whereby bacteriocins were shown to have an inhibitory effect on biofilm formation when applied in vitro at specific concentrations (Caballero Gómez et al., 2012; Shao et al., 2020; Moradi et al., 2020; Rad et al., 2021).

The persistence of biofilms in pipes of different types – such as pipelines for petroleum products, crop irrigation, water distribution, sewage collection, etc. – can result in its obstruction and corrosion, causing huge economic losses. In addition, it may carry pathogens along with their products and allow their dissemination. Alternative ways to combat biofilms in pipelines is not yet a widely studied area, but the possibility of combating biofilms with postbiotic substances is known (Heireman et al., 2020; Khawas et al., 2021; Li et al., 2020; Neria-González et al., 2006). *Pseudomonas aeruginosa* and *Listeria monocytogenes* are pathogens capable of causing a wide range of health problems and ones that commonly establish biofilms in drinking water distribution pipes. Their susceptibility to probiotic metabolites is already known, this being a possible solution to the problem (Fu et al., 2021; Hemdan et al., 2021; Nisar et al., 2020).

14.7   FUTURE PERSPECTIVES

Despite the different methods and strategies used to combat biofilms, this chapter brings an important approach to the application of postbiotics in this scenario. The results obtained from the use of postbiotics have aroused great interest in research and in the industrial area due to their commercial appeal. The expectation is that new in vitro and in vivo studies will be carried out to better understand their metabolic pathways and mechanisms of action: in addition to obtaining knowledge about their safety, developing regulations for their use, and establishing the relationship between postbiotics and the fight against microbial biofilms.

REFERENCES

Abatenh E, Gizaw B, Tsegaye Z, Wassie M. (2017) The Role of Microorganisms in Bioremediation- A Review. Open Journal of Environmental Biology, 2(1), 038–046.

Abreu AC, Tavares RR, Borges A, Mergulhão F, Simões M (2013) Current and emergent strategies for dis infection of hospital environments. J Antimicrob Chemother. 68(12): 2718–32.

Aguilar-Toalá JE, Garcia-Varela R, Garcia HS, Mata-Haro V, González-Córdova AF, Vallejo-Cordoba B, Hernández-Mendoza A. (2018) Postbiotics: An evolving term within the functional foods field. Trends in Food Science and Technology, 75: 105–114.

Al-Hasani HMH, Al-Gburi A. (2019) Ethel alcohol and sodium hypochlorite activity against pathogenic biofilm formation. Biochem. Cell. Arch. 19(2): 4451–4456.

Ali SG, Ansari MA, Alzohairy MA, Alomary MN, AlYahya S, Jalal M et al. (2020) Biogenic gold nanoparticles as potent antibacterial and antibiofilm nano-antibiotics against *pseudomonas aeruginosa*. Antibiotics (Basel). 27(9):100.

Al-Shabib NA, Husain FM, Rehman MT, Alyousef AA, Arshad M, Khan A, et al. (2020) Food color 'Azorubine' interferes with quorum sensing regulated functions and obliterates biofilm formed by food associated bacteria: An in vitro and in silico approach. Saudi J BiolSci. 27: 1080–1090.

Alvarez-Sieiro P, Montalbán-López M, Mu D, Kuipers OP. (2016) Bacteriocins of lactic acid bacteria: extending the family. Applied Microbiology and Biotechnology, 100(7), 2939–2951.

Barros CP, Guimarães JT, Esmerino EA, Duarte MCK, Silva MC, Silva R *et al.* (2019) Paraprobiotics, postbiotics and psychobiotics: concepts and potential applications in dairy products. Curr. Opin. Food Sci. 32:1–8.

Barzegari A, Kheyrolahzadeh K, Hosseiniyan Khatibi SM, Sharifi S, Memar MY, Zununi Vahed S. (2020) The Battle of Probiotics and Their Derivatives Against Biofilms. Infect Drug Resist. 13: 659–672.

Benzaid C, Belmadani A, Tichati L, Djeribi R, Rouabhia M. (2021) Effect of Citrus *aurantium* L. Essential Oilon *Streptococcus mutans* Growth, Biofilm Formation and Virulent Genes Expression. Antibiotics. 8,10(1): 54.

Bermudez-Brito M, Plaza-diaz J, Munoz-Quezada S, Gomez-Llorente C, Gil A. (2012) Probiotic Mechanisms of Action. Ann Nutr Metab. 61: 160–174. https://doi.org/10.1159/000342079

Bhargava N, Mor RS, Kumar K, Sharanagat VS (2021) Advances in application of ultrasound in food processing: a review. *Ultrason Sonochem.* 70:105293.

Bhat MI, Kumari A, Kapila S *et al.* (2019) Probiotic lactobacilli mediated changes in global epigenetic signatures of human intestinal epithelial cells during *Escherichia coli* challenge. Ann Microbiol, 69:603–612.

Bintsis T. (2018) Lactic acid bacteria as starter cultures: An update in their metabolism and genetics. AIMS Microbiology, 4(4), 665–684.

Brindhadevi K, Lewisoscar F, Mylonakis E, Shanmugam S, Verma TN, Pugazhendhi A. (2020) Biofilm and Quorum sensing mediated pathogenicity in *Pseudomonas aeruginosa*. Process Biochem. 96: 49–57.

Bukhary S, Balto H. (2017) Antibacterial Efficacy of Octenisept, Alexidine, Chlorhexidine, and Sodium Hypochlorite against *Enterococcus faecalis* Biofilms. J Endod. 43(4): 643–647.

Buron-Moles G, Chailyan A, Dolejs I, Forster J, Mikš MH. (2019) Uncovering carbohydrate metabolism through a genotype-phenotype association study of 56 lactic acid bacteria genomes. Applied Microbiology Biotechnology, 103(7): 3135–3152.

Caballero Gómez N, Abriouel H, Grande MJ, Pérez Pulido R, Gálvez A. (2012) Effect of enterocin AS-48 in combination with biocides on planktonic and sessile *Listeria monocytogenes*. Food Microbiology, 30(1): 51–58.

Cáceres M, Hidalgo W, Stashenko E, Torres R, Ortiz C. (2020) Essential Oils of Aromatic Plants with Antibacterial, Anti-Biofilm and Anti-Quorum Sensing Activities against Pathogenic Bacteria. Antibiotics. 30,9(4): 147.

Carradori S, Di Giacomo N, Lobefalo M, Luisi G, Campestre C, Sisto F. (2020) Biofilm and Quorum Sensing inhibitors: the Road so far. Expert Opin Ther Pat. 30(12): 917–930.

Carrascosa C, Raheem D, Ramos F, Saraiva A, Raposo A. (2021) Microbial Biofilms in the Food Industry-A Comprehensive Review. Int J Environ Res Public Health. 18(4): 2014.

Castellano P, Ibarreche PM, Massani BM, Fontana C, Vignolo G. (2017) Strategies for Pathogen Biocontrol Using Lactic Acid Bacteria and Their Metabolites: A Focus on Meat Ecosystems and Industrial Environments. Microorganisms, 5(3): 38.

Catania V, Lopresti F, Cappello S, Scaffaro R, Quatrini P. (2020) Innovative, eco friendly biosorbent-biodegrading biofilms for bioremediation of oil- contaminated water. N Biotechnol. 25,58: 25–31.

Chang S, Jinchun C, Lin S (2017) Using thermal shock to inhibit biofilm formation in the treated sewage source heat pump systems. *Appl Sci.* 7(4):343.

Chaudhary S, Jyoti A, Shrivastava V, Tomar RS. (2020) Role of Nanoparticles as Antibiofilm Agents: A Comprehensive Review. Curr. Trends Biotechnol. Pharm. 14(1): 97–110.

Chen F, Zhang M, Yang CH (2020) Application of ultrasound technology in processing of ready-to-eatfresh food: a review. *Ultrason Sonochem*. 63:104953.

Cuevas-González PF, Liceaga AM, Aguilar-Toalá JE (2020) Postbiotics and paraprobiotics: from conceptsto applications. *Food Res Int*. 136:109502.

Cui T, Bai F, Sun M, Lv X, Li X, Zhang D, Du H (2020) *Lactobacillus crustorum* ZHG 2–1 as novel quorum-quenching bactéria reducing virulence factors and biofilms formation of *Pseudomonas aeruginosa*. *LWT*. 117:108696.

Daood U, Matinlinna JP, Pichika MR, Mak, K-K, Nagendrababu V, Fawzy A (2020) A quaternary ammonium silane antimicrobial triggers bacterial membrane and biofilm destruction. *Sci Rep*. 10:10970.

Das T, Manoharan A, Whiteley G, Glasbey T, Manos J (2020) Chapter 3 – *Pseudomonas aeruginosa* biofilms and infections: roles of extra cellular molecules. In Yadav MK, Singh BP (Eds.), *New and future developments in microbial biotechnology and bioengineering: microbial biofilms*. Elsevier, 29–46.

De Filippis F, Pasolli E, Ercolini D. (2020) The food-gut axis: Lactic acid bacteria and their link to food, the gut microbiome and human health. FEMS Microbiology Reviews 44(4): 454–489.

Deng Z, Luo XM, Liu J, Wang H (2020) Quorum sensing, biofilm, and intestinal mucosal barrier: involvementthe role of probiotic. *Front Cell Infect Microbiol*. 10:538077.

Deptula P, Chamlagain B, Edelmann M, Sangsuwan P, Nyman TA, Savijoki K, Piironen V, Varmanen P (2017) Food-like growth conditions support production of active vitamin B12 by *Propionibacterium freudenreichii* 2067 without DMBI, the lower ligand base, or cobalt supplementation. *Front Microbiol*. 8:368.

Díaz MA, González SN, Alberto MR, Arena ME. (2020) Human probiotic bacteria attenuate *Pseudomonas aeruginosa* biofilm and virulence by quorum-sensing inhibition. Biofouling, 36(5): 597–609.

Eiteman MA, Ramalingam S. (2015) Microbial production of lactic acid. In Biotechnology Letters, 37(5): 955–972.

Emmanuel EC, Priya SS, George S. (2019) Isolation of biosurfactant from *Lactobacillus* sp.and study of its inhibitory properties against *E. coli* biofilm. J Pure Appl Microbiol.,3: 403–11.

Evivie SE, Huo GC, Igene JO, Bian X. (2017) Some current applications, limitations and future perspectives of lactic acid bacteria as probiotics. Food and Nutrition Research 61(1)

Field D, Ross RP, Hill C. (2018) Developing bacteriocins of lactic acid bactéria intonext generation biopreservatives. Current Opinion in Food Science, 20: 1–6.

Freire NB, Pires LCSR, Oliveira HP, Costa MM. (2018) Antimicrobial and antibiofilm activity of silver nanoparticles against *Aeromonas* spp. Isolated from aquatic organisms. Brazilian Journal of Veterinary Research, 38(2): 244–249.

Fu Y, Peng H, Liu J, Nguyen TH, Hashmi MZ, Shen C (2021) Occurrence and quantification of culturable and viable but non-culturable (VBNC) pathogens in biofilm on different pipes from a metropolitan drinking water distribution system. *Sci Total Environ*. 764:142851.

Gabra FA, Abd-Alla MH, Danial AW, Abdel-Basset R, Abdel-Wahab AM (2019) Production of biofuel from sugarcane molasses by diazotrophic *Bacillus* and recycle of spent bacterial biomass as biofertilizer inoculants for oil crops. *Biocatal Agric Biotechnol*. 19:101112.

Galbe M, Wallberg O. (2019) Pretreatment for biorefineries: A review of common methods for efficient utilisation of lignocellulosic materials. Biotechnology for Biofuels, 12(1)

Ganzle MG (2015) Lactic metabolism revisited: metabolism of lactic acid bacteria in food fermentations and food spoilage. *Current Opinion in Food Science*, 2: 106–117.

Garcia-Gonzalez J, Sommerfeld M. (2016) Biofertilizer and biostimulant properties of the microalga Acutodesmus dimorphus. Journal of Applied Phycology, 28(2): 1051–1061.

Garcia-Lara B, Saucedo-Mora MA, Roldan-Sanchez JA, Perez-Eretza B, Ramasamy M, Lee J, Coria-Jimenez R, Tapia M, Varela-Guerrero VV, Garcia-Contreras R. (2015) Inhibition of quorum-sensing dependent virulence factors and biofilm formation of clinical and environmental *Pseudomonas aeruginosa* strains by Zn O nano particles. Lett Appl Microbiol. 61(3): 299–305.

Gayán E, Serrano MJ, Monfort M, Álvarez I, Condón S. (2012) Combining ultra violet light and mild temperatures for thein activation of *Escherichia coli* in Orange juice. J. Food Eng. 113 (4): 598–605.

Giori GS, LeBlanc JG (2018) Folate production by lactic acid bacteria. In Watson RR, Preedy VR, Zibadi S (Eds.), *Polyphenols: prevention and treatment of human disease.* Acadmeic Press, 15–29.

Gu J, Liu X, Li Y, Han W, Lei L, Yang Y. *et al.* (2012) A method for generation phage cocktail with great therapeutic potential. PLoSOne. 7(3):e31698.

Han W, Mao Y, Wei Y, Shang P, Zhou X. (2020) Bioremediation of Aquaculture Waste water with Algal-Bacterial Biofilm Combined with the Production of Selenium Rich Biofertilizer. Water. 12(7): 2071.

Hasan F, Reza M, Masud HA, Uddin MK, Uddin MS. (2019) Preliminary characterization and inhibitory activity of bacteriocin like substances from *Lactobacillus casei* against multi-drug resistant bacteria. Bangladesh Journal of Microbiology, 36(1): 1–6.

Hatti-Kaul R, Chen L, Dishisha T, Enshasy HE. (2018) Lactic acid bacteria: From starter cultures to producers of chemicals. FEMS Microbiology Letters, 365(20): 213.

Heireman L, Hamerlinck H, Vandendriessche S, Boelens J, Coorevits L, De Brabandere E, De Waegemaeker P, Verhofstede S, Claus K, Chlebowicz-Flissikowska MA, Rossen JWA, Verhasselt B, Leroux-Roels I. (2020) Toilet drain water as a potential source of hospital room-to-room transmission of carbapenemase-producing *Klebsiella pneumoniae*. Journal of Hospital Infection, 106(2): 232–239.

Hemdan BA, El-Taweel GE, Goswami P, Pant D, Sevda S. (2021) The role of biofilm in the development and dissemination of ubiquitous pathogens in drinking water distribution systems: an overview of surveillance, outbreaks, and prevention. World Journal of Microbiology and Biotechnology, 37(2): 36.

Hill C, Guarner F, Reid G, Gibson GR, Merenstein DJ, Pot B, Morelli L, Canani RB, Flint HJ, Salminen S, Calder PC, Sanders ME. (2014) Expert consensus document: The international scientific association for probiotics and prebiotics consensus statement on the scope and appropriate use of the term probiotic. *Nat Rev Gastroenterol Hepatol*, 11(8): 506–514.

Høiby N, Bjarnsholt T, Moser C, Jensen PØ, Kolpen M, Qvist T, Aanaes K, Pressler T, Skov M, Ciofu O. (2017) Diagnosis of biofilm infections in cystic fibrosis patients. APMIS. 125(4): 339–343.

Hu X, Qiao Y, Chen LQ, Du JF, Fu YY, Wu S, Huang L. (2020) Enhancement of solubilization and biodegradation of petroleum by biosurfactant from *Rhodococcus erythropolis* HX-2. Geomicrobiology Journal,37:2,159–169.

Ishikawa KH, Bueno MR, Kawamoto D, Simionato MRL, Mayer MPA. (2021) Lactobacilli postbiotics reduce biofilm formation and alter transcription of virulence genes of *Aggregatibacter actinomycetemcomitans*. Molecular Oral Microbiology, 36(1): 92–102.

Jeon J, Yong D. (2019) Two Novel Bacteriophages Improve Survival in *Galleria mellonella* Infectionand Mouse Acute Pneumonia Models Infected with Extensively Drug-Resistant *Pseudomonas aeruginosa*. Appl Environ Microbiol. 18,85(9):e02900–18.

Jung S, Park OJ, Kim AR *et al.* (2019) Lipoteichoic acids of lactobacilli inhibit *Enterococcus faecalis* biofilm formation and disrupt the preformed biofilm. J Microbiol.57, 310–315.

Kaczorek E, Pacholak A, Zdarta A, Smułek W (2018) The impact of biosurfactants on microbial cell properties leading to hydrocarbon bioavailability increase. *Colloids Interfaces.* 2:35.

Kareem KY, HooiLing F, TeckChwen L, Foong OM, Asmara SA (2014) Inhibitory activity of postbiotic produced by strains of *Lactobacillus plantarum* usingre constituted media supplement edwithinulin. *GutPathog.* 6:23.

Kasahara K, Krautkramer KA, Org E *et al.* (2018) Interactions between *Roseburia intestinalis* and diet modulate atherogenesis in a murine model. Nat Microbiol, 3: 1461–1471.

Khalil ES, Manap MY, Shuhaimi M, Alhelli AM, Shokryazdan P. (2018) Probiotic Properties of Exopolysaccharide-Producing *Lactobacillus* Strains Isolated from Tempoyak. Molecules 23(2): 398.

Khan MF, Husain FM, Zia Q, Ahmad E, Jamal A, Alaidarous M, Banawas S *et al.* (2020) Anti-quorum Sensing and Anti-biofilm Activity of Zinc Oxide Nanospikes. ACS omega, 5(50): 32203–32215.

Khawas N, Tripathi VK, Kumar A (2021) Impact of biofilm on clogging of drip irrigation emitters. *Field Pract Wastewater Use Agri.* 113–129.

Kim KW, Kang SS, Woo SJ, Park OJ, Ahn KB, Song KD, Lee HK, Yun CH, Han SH (2017) Lipoteichoic acid of probiotic *lactobacillus plantarum* attenuates poly I:C-induced IL-8 production in porcine intestinal epithelial cells. *Front Microbiol.* 8:1827.

Kumar N, Kumari V, Ram C, Thakur K, Tomar SK. (2018) Bio-prospectus of cadmium bioadsorption by lactic acid bacteria to mitigate health and environmental impacts. Applied Microbiology and Biotechnology, 102(4): 1599–1615.

Lamont JR, Wilkins O, Bywater-Ekegärd M, Smith DL. (2017) From yogurt to yield: Potential applications of lactic acid bacteria in plant production. In Soil Biology and Biochemistry, 111: 1–9.

LeBlanc JG, Chain F, Martín R *et al.* (2017) Beneficial effectson host energy metabolism of short-chain fatty acids and vitamins produced by commensal and probiotic bacteria. *Microb Cell Fact.* 16:79.

LeBlanc JG, Levit R, Savoy de Giori G, de Moreno de LeBlanc A. (2020) Application of vitamin-producing lactic acid bacteria to treat intestinal inflammatory diseases. Applied Microbiology and Biotechnology, 104(8): 3331–3337.

Lee SHI, Cappato LP, Corassin CH, Cruz AG, Oliveira CAF (2016) Effect of peracetic acidon biofilms formed by *Staphylococcus aureus* and *Listeria monocytogenes* isolated from dairy plants. J Dairy Sci. 99(3): 2384–2390.

Lei X, Zhang H, Hu Z, Liang Y, Guo S, Yang M, Du R, Wang X. (2018) Immunostimulatory activity of exopolysaccharides from probiotic *Lactobacillus casei* WXD030 strain as a novel adjuvant in vitro and in vivo. Food Agric. Immunol., 29: 1086–1105.

LewisOscar F, Nithya C, Vismaya S, Arunkumar M, Pugazhendhi A, Nguyen-Tri P *et al.* (2021) In vitro analysis of green fabricated silver nano particles (AgNPs) against *Pseudomonas aeruginosa* PA14 biofilm formation, their application onurinary catheter. *Prog Org Coat.* 151:106058.

Li Y, Wang Y, Xiao P, Narasimalu S, Dong ZL. (2020) Analysis of biofilm-resistance factors in Singapore drinking water distribution system. IOP Conference Series: Earth and Environmental Science, 558(4): 042004.

Lianou A, Nychas GE, Koutsoumanis KP (2020) Strain variability in biofilm formation: a food safety and quality perspective. *Food Res Int.* 137:109424.

Liu J, Dantoft SH, Würtz A, Jensen PR, Solem C. (2016) A novel cell factory for efficient production of ethanol from dairy waste. Biotechnology for Biofuels, 9(1)

Liu Y, Shi L, Su L, van der Mei HC, Jutte PC, Ren Y *et al.* (2019) Nanotechnology-based antimicrobials and delivery systems for biofilm-infection control. *Chem Soc Rev.* 48:428.

Mącik M, Gryta A, Frąc M. (2020) Biofertilizers in agriculture: An overview on concepts, strategies and effects on soil microorganisms. In Advances in Agronomy, 162: 31–87.

Makarova K, Slesarev A, Wolf Y, Sorokin A, Mirkin B, Koonin E *et al.* (2006) Comparative genomics of the lactic acid bacteria. Proceedings of the National Academy of Sciences of the United States of America, 103(42): 15611–15616.

Malaekeh-Nikouei B, Bazzaz BSF, Mirhadi E, Tajani AS, Khameneh B (2020) The role of nanotechnology in combating biofilm-based antibiotic resistance. *J Drug Deliv Sci Technol.* 60:101880.

Malagón-Rojas JN, Mantziari A, Salminen S, Szajewska H. (2020) Postbiotics for Preventing and Treating Common Infectious Diseases in Children: A Systematic Review. Nutrients., 12(2): 389.

Mantziari A, Salminen S, Szajewska H, Malagón-Rojas JN (2020) Postbiotics against pathogens commonly involved in pediatric infectious diseases. *Microorganisms.* 8:1510.

Masashi M, Satomi I, Masami K, Tomoyuki F, Yukio A, Yuki M, Tatsuya S. (2020) Exopolysaccharides from milk fermented by lactic acid bacteria enhance dietary carotenoid bioavailability in humans in a randomized crossover trial and in rats. The American Journal of Clinical Nutrition, 111: 903–14.

Mazzoli R, Bosco F, Mizrahi I, Bayer EA, Pessione E. (2014) Towards lactic acid bacteria-based biorefineries. In Biotechnology Advances, 32(7): 1216–1236.

Mizan MFR, Cho HR, Ashrafudoulla M, Cho J, Hossain MI, Lee DU *et al.* (2020) The effect of physico-chemical treatment in reducing *Listeria monocytogenes* biofilms on lettuce leaf surfaces. Biofouling. 36(10): 1243–1255.

Mokoena MP. (2017) Lactic Acid Bacteria and Their Bacteriocins: Classification, Biosynthesis and Applications against Uropathogens: A Mini-Review. Molecules, 22(8): 1255.

Montso PK, Mlambo V, Ateba CN (2021) Efficacy of novel phages for control of multi-drug resistant *Escherichia coli* O177 on artificially contaminated beef and their potential to disrupt biofilm formation. *Food Microbiol.* 94:103647.

Mora-Villalobos JA, Montero-Zamora J, Barboza N, Rojas-Garbanzo C, Usaga J, Redondo-Solano M, Schroedter L, Olszewska-Widdrat A, López-Gómez JP. (2020) Multi-Product Lactic Acid Bacteria Fermentations: A Review. Fermentation, 6(1): 23.

Moradi M, Kousheh SA, Almasi H, Alizadeh A, Guimarães JT, Yılmaz N, Lotfi A. (2020) Postbiotics produced by lactic acid bacteria: The next frontier in food safety. Comprehensive Reviews in Food Science and Food Safety, 19(6): 3390–3415.

Nataraj BH, Ali SA, Behare PV, Yadav H. (2020) Postbiotics-parabiotics: The new horizons in microbial biotherapy and functional foods. Microbial Cell Factories, 19(1)

Neria-González I, Wang ET, Ramírez F, Romero JM, Hernández-Rodríguez C. (2006) Characterization of bacterial community associated to biofilms of corroded oil pipelines from the southeast of Mexico. Anaerobe, 12(3): 122–133.

Nisar MA, Ross KE, Brown MH, Bentham R, Whiley H (2020) Water stagnation and flow obstruction reduces the quality of potable water and increases the risk of legionelloses. *Front Environ Sci.* 8:611611.

Oleskin AV, Shenderov BA. (2019) Probiotics and psychobiotics: the role of microbial neuro-chemicals. Probiotics Antimicrob Proteins. 11, 1071–1085.

O'Toole P, Marchesi J, Hill C (2017) Next-generation probiotics: the spectrum from probiotics to live biotherapeutics. *Nat Microbiol.* 2:17057.

Păcularu-Burada B, Georgescu LA, Vasile MA, Rocha JM, Bahrim GE. (2020) Selection of wild lactic acid bacteria strains as promoters of postbiotics in gluten-free sourdoughs. Microorganisms, 8(5)

Paluch E, Rewak-Soroczyńska J, Jędrusik I, Mazurkiewicz E, Jermakow K. (2020) Prevention of biofilm formation by quorum quenching. Appl Microbiol Biotechnol. 104(5): 1871–1881.

Papadimitriou K, Alegría Á, Bron PA, de Angelis M, Gobbetti M, Kleerebezem M *et al.* (2016) Stress Physiology of Lactic Acid Bacteria. Microbiology and Molecular Biology Reviews, 80(3): 837–890.

Pires DP, Melo L, Vilas Boas D, Sillankorva S, Azeredo J. (2017) Phagetherapy as an alternative or complementary strategy to prevent and control biofilm-related infections. Curr Opin Microbiol. 39: 48–56.

Prado DB, Szczerepa MMA, Capeloto AO, Astrath NGC, Santos NCA, Previdelli ITS *et al.* (2019) Effect of ultraviolet (UV-C) radiationon spores and biofilms of *Alicyclobacillus* spp. in industrialized orange juice. *Int J Food Microbiol.* 305:108238.

Principi N, Silvestri E, Esposito S (2019) Advantages and limitations of bacteriophages for the treatment of bacterial infections. *Front Pharmacol.* 8(10):513.

Przekwas J, Wiktorczyk N, Budzyńska A, Wałecka-Zacharska E, Gospodarek-Komkowska E. (2020) Ascorbic Acid Changes Growth of Food-Borne Pathogens in the Early Stage of Biofilm Formation. Microorganisms. 8(4): 553.

Rad AH, Abbasi A, Kafil HS, Ganbarov K (2020) Potential pharmaceutical and food applications of postbiotics: a review. *Curr Pharm Biotechnol.* 21:1576.

Rad AH, Aghebati-Maleki L, Kafil HS, Gilani N, Abbasi A, Khani N. (2021) Postbiotics, as Dynamic Biomolecules, and Their Promising Role in Promoting Food Safety. 11(6): 14529–14544.

Ricker EB, Nuxoll E. (2017) Synergistic effects of heat and antibiotics on *Pseudomonas aeruginosa* biofilms. Biofouling. 33(10): 855–866.

Riley MR, Gerba CP, Elimelech M (2011) Biological approaches for addressing the grand challenge of providing access to clean drinking water. *J Biol Eng.* 5:2.

Rossoni RD, de Barros PP, Mendonça I, Medina RP, Silva D, Fuchs BB *et al.* (2020) The postbiotic activity of *Lactobacillus paracasei* 28.4 against *Candida auris*. *Front Cell Infect Microbiol.* 10:397.

Sabater C, Ruiz L, Delgado S, Ruas-Madiedo P, Margolles A (2020) Valorization of vegetable food waste and by-products through fermentation processes. *Front Microbiol.* 11:581997.

Salam FD, Vinita MN, Puja P, Prakash S, Yuvakkumar R, Kumar P (2020) Anti-bacterial and anti-biofilm efficacies of bioinspired gold nanoparticles. *Mater Lett.* 261:126998.

Sarikhani M, Kermanshahi RK, Ghadam P, Gharavi S. (2018) The role of probiotic *Lactobacillus acidophilus* ATCC 4356 bacteriocin on effect of HB suon planktonic cells and biofilm formation of *Bacillus subtilis*. Int J Biol Macromol. 115: 762–766.

Sarlak Z, Khosravi-Darani K, Rouhi M, Garavand F, Mohammadi R, Sobhiyeh MR (2021) Bioremediation of organophosphorus pesticides in contaminated foodstuffs using probiotics. *Food Control.* 126:108006.

Sauer M, Russmayer H, Grabherr R, Peterbauer CK, Marx H. (2017) The Efficient Clade: Lactic Acid Bacteria for Industrial Chemical Production. Trends in Biotechnology, 35(8): 756–769.

Saxena P, Joshi Y, Rawat K, Bisht R. (2019) Biofilms: Architecture, Resistance, Quorum Sensing and Control Mechanisms. Indian J Microbiol. 59: 3–12.

Settanni L, Corsetti A. (2008) Application of bacteriocins in vegetable food biopreservation. International Journal of Food Microbiology, 121(2): 123–138.

Shao L, Dong Y, Chen X, Xu X, Wang H (2020) Modeling the elimination of mature biofilms formed by *Staphylococcus aureus* and *Salmonella* spp: using combined ultrasound and disinfectants. *Ultrason Sonochem.* 69:105269.

Shao X, Fang K, Medina D, Wan J, Lee JL, Hong SH. (2020) The probiotic, Leuconostoc mesenteroides, inhibits *Listeria monocytogenes* biofilm formation. Journal of Food Safety, 40(2): e12750.

Sharifi A, Ahmadi A, Mohammadzadeh A. (2018) *Streptococcus pneumoniae* quorum sensing and biofilm formation are affected by *Thymus daenensis, Satureja hortensis,* and *Origanum vulgare* essential oils. Acta Microbiol Immunol Hung. 1,65(3): 345–359.

Sharma A, Gupta G, Ahmad T, Kaur B, Hakeem KR. (2020) Tailoring cellular metabolism in lactic acid bacteria through metabolic engineering. In Journal of Microbiological Methods, 170: 105862.

Sharma V, Harjai K, Shukla G. (2018) Effect of bacteriocin and exopolysaccharides isolated from probiotic on *P. aeruginosa* PAO1 biofilm. Folia Microbiologica, 63(2): 181–190.

Shenderov BA. (2013) Metabiotics: novel idea or natural development of probiotic conception. Microbial Ecology in Health & Disease, 24(0)

Shenderov BA, Gabrichevsky GN. (2017) Metabiotics: an overview of progress, opportunities and challenges. J Microb Biochem Technol. 9:11–21.

Singh P, Saini P. (2017) Food and Health Potentials of Exopolysaccharides Derived from Lactobacilli. Microbiology Research Journal International, 22(2): 1–14.

Singh VP. (2018) Recent approaches in food bio-preservation-A review. In Open Veterinary Journal, 8(1): 104–111.

Stewart PS, Bjarnsholt T. (2020) Risk factors for chronic biofilm-related infection associated with implanted medical devices. Clin Microbiol Infect. 26(8): 1034–1038.

Sumrall ET, Shen Y, Keller AP, Rismondo J, Pavlou M, Eugster MR *et al.* (2019) Phage resistance atthe cost of virulence: *Listeria monocytogenes* serovar 4b requires galactosylated teichoic acids for InlB-mediated invasion. PLoS Pathog 15(10): e1008032.

Takamura A, Thuy-Boun PS, Kitamura S, Han Z, Wolan DW (2021) A photoaffinity probe that targets folate-binding proteins. *Bioorg Med Chem Lett.* 40:127903.

Tazehabadi MH, Algburi A, Popov IV, Ermakov AM, Chistyakov VA, Prazdnova EV, Weeks R, Chikindas ML (2021) Probiotic bacilli inhibit *salmonella* biofilm formation without killing planktonic cells. *Front Microbiol.* 12:615328.

Teame T, Wang A, Xie M, Zhang Z, Yang Y, Ding Q, Gao C, Olsen RE, Ran C, Zhou Z (2020) Paraprobiotics and postbiotics of probiotic lactobacilli, their positive effects on the host and action mechanisms: a review. *Front Nutr.* 7.

Thi MTT, Wibowo D, Rehm BHA. (2020) *Pseudomonas aeruginosa* Biofilms. Int J Mol Sci. 21: 1–25.

Tomar SK, Anand S, Sharma P, Sangwan V, Mandal S (2015) Role of probiotic, prebiotics, synbiotics and postbiotics in inhibition of pathogens. In Méndez-Vilas A (Ed.), *The battle against microbial pathogens: basic science, technological advances and educational programs.* Formatex Research Center, 717–732.

Tomusiak-Plebanek A, Heczko P, Skowron B, *et al.* (2018) Lactobacilli with superoxide dismutase-like or catalase activity are more effective in alleviating inflammation in na inflammatory bowel disease mouse model. Drug Des Devel Ther, 12: 3221–3233.

Upadhyaya BP, DeVeaux LC, Christopher LP (2014) Metabolic engineering as a tool for enhanced lactic acid production. In Trends in Biotechnology, 32(12): 637–644.

Varjani SJ, Upasani VN. (2017) Critical review on biosurfactant analysis, purification and characterization using rhamnolipid as a model biosurfactant. Bioresource Technology, 232: 389–97.

Wu C, Huang J, Zhou R. (2017) Genomics of lactic acid bacteria: Current status and potential applications. In Critical Reviews in Microbiology, 43(4): 393–404.

Yordshahi AS, Moradi M, Tajik H, Molaei R. (2020) Design and preparation of antimicrobial meat wrapping nanopaper with bacterial cellulose and postbiotics of lactic acid bacteria. Internacional Journal of Food Microbiology, 321: 108561

Zeng H, Umar S, Rust B, Lazarova D, Bordonaro M. (2019) Secondary Bile Acids and Short Chain Fatty Acids in the Colon: A Focus on Colonic Microbiome, Cell Proliferation, Inflammation, and Cancer. *Int J Mol Sci.* 20(5): 1214.

Zhang Y, Sass A, Van Acker H, Wille J, Verhasselt B, Van Nieuwerburgh F, Kaever V, Crabbé A, Coenye T (2018) Coumarin reduces virulence and biofilm formation in *Pseudomonas aeruginosa* by affecting quorum sensing, type III secretion and C-di-GMP levels. *Front Microbiol.* 9:1952.

Zhao X, Zhao F, Wangb J, Zhongc N. (2017) Biofilm formation and control strategies of food borne pathogens: food safety perspectives. RSC Adv. 7: 36670–36683.

Żółkiewicz J, Marzec A, Ruszczyński M, Feleszko W. (2020) Postbiotics – A Step Beyond Pre- and Probiotics. Nutrients, 12(8): 2189.

15 Impact and Control of Microbial Biofilms in the Oil and Gas Industry

Abioye O.P., Aransiola S.A., Victor-Ekwebelem M.O., Auta S.H., and Ijah U.J.J.

CONTENTS

15.1 INTRODUCTION

Biofilm is an association of microorganisms that are immovably appended to the biotic or abiotic surface, encased within an extracellular polymeric substance (EPS) matrix, which can display new character with respect to metabolic activities, gene expression, growth rate, and protein synthesis (Oxaran *et al.*, 2018; Gedif, 2020). EPSs are composed of polysaccharides, lipids, proteins, and extracellular DNA (Table 15.1), and play an important feature within the pathogenesis of the numerous microbial infections (Ch'ng *et al.*, 2019).

DOI: 10.1201/9781003184942-18

TABLE 15.1

Chemical Composition of Biofilm

S/n	Component	Percentage of Matrix
1	Microbial cells	2.5%
2	DNA and RNA	< 1–2%
3	Polysaccharides	1–2%
4	Proteins	< 1–2% (including enzymes)
5	Water	Up to 97%

Biofilm production can be motivated and influenced by a number of factors, such as surface conditions, chemical and physical growth factors, cellular structures, and any other challenges. The interaction between these and other factors determines its fate (Kostakioti *et al.*, 2013; Mishra *et al.*, 2020); since biofilms are surrounded by high molecular weight EPSs that connect and attach cells, these cells in biofilm can survive harsh growth conditions (Mishra *et al.*, 2020). This takes place due to structural and physiological changes that take place after cells have been attached to conditioned surfaces, with the produced structural polymeric substances acting as a barrier (Yin *et al.*, 2019) preventing its control (Figure 15.1).

It has also been reported that microbial cells within the biofilms are observed to be resistant against ultraviolet (UV) light, metal toxicity, acid exposure, desiccation, pH gradients, etc. (Mishra *et al.*, 2020). Furthermore, biofilm's mode of growth induces microbial resistance to disinfection, which can result in widespread economic and health concerns (Abebe, 2020); for instance, research done on *Listeria monocytogenes* revealed that its biocide resistance and cap-potential to cooperate with other species forming heterogeneous communities allowed this bacterium to survive and struggle within industrial areas (Cabo *et al.*, 2018). Microbial action has been identified as a contributor to rapid corrosion of metals and alloys exposed to soils; seawater, distilled water, and freshwater; crude oil, hydrocarbon fuels, and process chemicals; and sewage. Many industries and elements of infrastructure are affected by microbiologically influenced corrosion (MIC), including oil production, power generation, transportation, and water and wastewater.

Bioremediation is an environmentally friendly, cost effective, sustainable technology that utilizes microbes to decontaminate and degrade a wide variety of pollutants into less harmful products. Relative to free-floating planktonic cells, microbes existing in biofilm mode are advantageous for bioremediation because of greater tolerance to pollutants, environmental stress, and ability to degrade varied harsh pollutants via diverse catabolic pathways. In biofilm mode, microbes are immobilized in a self-synthesized matrix which offers protection from stress, contaminants, and predatory protozoa. Contaminants ranging from heavy metals, petroleum, explosives, and pesticides have been remediated using microbial consortia of biofilms. In the industry, biofilm-based bioremediation is used to decontaminate polluted soil and groundwater.

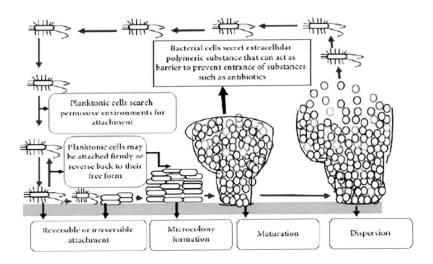

FIGURE 15.1   Biofilm formation and structure (adapted from Kostakioti *et al.* [2013]; Welch and Maunders [2017]; Abebe [2020]).

15.2   MICROBIAL BIOFILM AND ITS FORMATION

Formation of biofilm mainly involves three stages (Figure 15.2). The first stage is the adhesion stage, when cells attach to a surface; in the second stage (sessile growth stage), micro-colonies are formed due to the assemblage of these cells. The adhesion and sessile stages of growth are reversible and the cells can cluster loosely but can detach and return to a planktonic state (Kumar *et al.*, 2017; Zhang *et al.*, 2020). Thereafter, the attached cells secrete EPS, which includes extracellular DNA (eDNA), polysaccharides, and proteins, which develop into a biofilm in the third stage. This stage is irreversible, because the cells are attached within a thick and stable complex bio-molecular layer (Roy *et al.*, 2018). After a biofilm is completely developed, its dispersion or disassembly occurs via both active and mechanical processes; these processes occur in the fourth stage (dispersal stage). The cells within the biofilm secrete not only cell-adhesive matrix components, but also disruptive factors, including proteases, nucleases, phenol-soluble modulins, and regulators (Graf *et al.*, 2019). These disruptive factors can also promote biofilm detachment. During the process of detachment, biofilms can shed individual cells and slough off pieces into the bloodstream and the surrounding tissues, which are associated with many acute and chronic infections (Zhang *et al.*, 2020). Cells with different phenotypes and genotypes co-express individual metabolic pathways, stress responses, and other distinct biological properties within the biofilms. Some of these cells alter extracellular polysaccharide and organelle production, and even cell morphology, after they sense growth within the biofilm community. DNA transfer and genetic recombination between the multiple microbial species within a biofilm occur without direct cell–cell contact through the extracellular matrix, and in this manner, antibiotic resistance genes can be transferred (Zhang *et al.*, 2020).

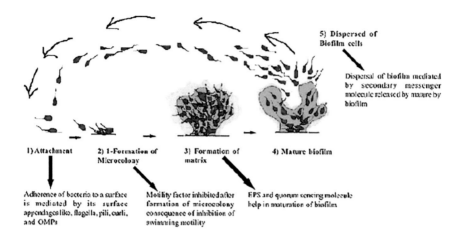

FIGURE 15.2   Microbial biofilm and its formation, showing step-by-step formation.

15.3   FACTORS RESPONSIBLE FOR THE FORMATION OF MICROBIAL BIOFILMS

Biofilm influences the physic-chemical interactions between metal and environment, frequently enhances corrosion, and leads to deterioration of the metal. Biofilm induces many changes in the type and concentration of ions, pH values, and oxidation reduction potential.

Parameters affecting the development of biofilms include:

- Hydrophobicity/properties of the cells
- Surface of the substratum
- Characteristics of the aqueous medium
- Conditioning films forming on the substratum
- Hydrodynamics
- Environmental factors such as temperature of the system or ambient temperature, water flow rate past the surface, nutrient availability effects, and pH of water in the system
- Effectiveness of biofouling remedial measures

15.3.1   HYDROPHOBICITY/PROPERTIES OF THE CELLS

The rate and extent of adherence of microbes depends on the properties of cells like cell surface hydrophobicity, as hydrophobic interactions tend to increase with an increasing nonpolar nature of one or both involved surfaces and adhesion increases with increase in hydrophobicity and presence of fimbriae and flagella, as fimbriae contribute to cell surface hydrophobicity probably by overcoming the initial electrostatic repulsion barrier that exists between the cell and substratum and production of EPS (Maric and Vranes, 2007; Choudhary *et al.*, 2020). A hydrophobic cell will be able to overcome the initial electrostatic repulsion with the solid surface and adhere more readily. The presence of fimbriae, proteinaceous bacterial appendages high in hydrophobic amino acids, can increase cell surface hydrophobicity (Nasib, 2009; Choudhary et al., 2020)

15.3.2 SURFACE OF THE SUBSTRATUM EFFECTS

Rough surfaces and pores inside porous supports tend to enhance biofilm formation. As the surface roughness increases, microbial colonization increases because as the roughness increases, surface area increases and shear forces get diminished; rate of attachment increases because there is a larger surface area to which cells can adhere and multiply (Choudhary et al., 2020).

15.3.3 CHARACTERISTICS OF THE AQUEOUS MEDIUM

Characteristics of the aqueous medium such as temperature, pH, nutrient level, and ionic strength possibly play important roles in attachment of microbes with the substratum. It has been found that the attachment of *Pseudomonas fluorescens* to glass and metal surface is affected by an increase in the concentration of several cations (sodium, calcium, lanthanum, ferric iron), thereby reducing the repulsive forces between the negatively charged bacterial cells and the surfaces (Nasib, 2009; Choudhary et al., 2020).

15.3.4 HYDRODYNAMICS

The hydrodynamic flow layer is the zone of negligible flow which is found immediately adjacent to the substratum/liquid interface. The flow velocity of this zone is negligible, and its thickness is inversely proportional to the linear velocity. Substantial mixing or turbulence is the main characteristic shown by the region outside the boundary layer. The hydrodynamic boundary layer can considerably affect the interaction between cells and substratum. The velocity characteristic of the liquid governs the association of cells with the submerged surfaces. At very low linear velocities, the cells must navigate through the hydrodynamic boundary layer, and cell size and cell motility govern its association with the surface. The boundary layer decreases as the velocity increases, and cells will be exposed to progressively greater turbulence and mixing. Therefore, higher linear velocities form a more rapid association with the surface and provides abundant shear forces on the attaching cells, resulting in detachment of these cells (Choudhary et al., 2020).

15.3.5 NUTRIENT AVAILABILITY

One function of the biofilm is to anchor cells in a friendly, nutrient-rich environment. Phosphorus is a particularly important nutrient. Cells saturated with phosphate have a greater tendency to flocculate and adhere due to their increased hydrophobicity, while those cells depleted in phosphate are more hydrophilic and less likely to adhere (Nasib, 2009).

15.3.6 Presence of Oxygen

Presence of oxygen regulates biofilm formation in many organisms. In the absence of sufficient oxygen supply, biofilm does not form, as bacteria cannot adhere to the substrate surface (Maric and Vranes, 2007).

15.3.7 pH

from various studies, it was observed that different microbes develop biofilm on different pH level; for example, optimal pH for biofilm multiplication of *V. cholerae* is 8.2, below pH 7 – i.e. in an acidic environment, the bacteria lose their ability to form biofilm as they lose mobility. On the other hand, bacteria like *S. epidermidis* and *E. coli* do not need an alkaline environment for multiplying, and hence, they easily form an acidic biofilm pH (Maric and Vranes, 2007).

15.4 EFFECTS OF MICROBIAL BIOFILM ON OIL AND GAS INDUSTRY FACILITIES

Biofilm formation is a universal and fundamental survival mechanism that provides microorganisms with critical advantages over planktonic growth, including greater access to nutrients and other resources, enhanced interaction between organisms, greater environmental stability, and protection from predators, viruses, antibiotics, and other biocidal compounds. Figure 15.3 explains the general mechanism in which biofilm uses to achieve its role (Dang and Lovell, 2016; Vigneron *et al.*, 2018).

Damage to offshore production facilities is one of the effects of microbial biofilm on oil and gas facilities. Offshore infrastructure harbors a complex network of pipes, pipelines, and other engineered structures that are often prone to biofilm formation and pitting corrosion (Duncan *et al.*, 2017).

Microbial biofilm leads to biofouling, slow fluid flow which facilitates the sedimentation of bacteria: the flow regimen of multiphase fluids greatly influences the location of the biofilm formation and associated corrosion risk. Low-velocity flow on production lines allows sediment and bacteria to settle at the bottom of the pipe and at retention points, providing a favorable site for biofilm colonization (Vigneron *et al.*, 2016). This can lead to a vicious circle whereby biofilm formation will further slow fluid flow (biofouling) and promote bacterial sedimentation (Vigneron *et al.*, 2018). Therefore, biofilms associated with high corrosion rates have been observed in heat-affected zones of welds, pipe elbows, separators, or valves present in steel infrastructure which represent areas of locally reduced flow (Duncan *et al.*, 2017). Additionally, the inner surfaces of storage tanks are also susceptible to biofilm formation as a result of stagnant flow conditions (Duncan *et al.*, 2017).

Microbial biofilm can act as electron shuttle: the main role of EPSs in biofilms is to enhance adhesion of the cells and to protect against environmental stresses (Sutherland, 2001). However, when an EPS binds to a metallic surface, it binds iron particles, stimulating interactions between microorganisms and metal ions (Kumar and Mody, 2009). Therefore, EPS bound metal ions can potentially act as electron shuttles and transfer electrons to more distant microorganisms (Kumar and Mody, 2009).

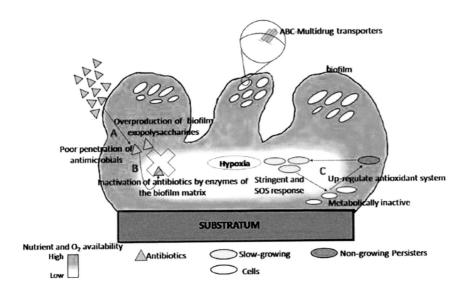

FIGURE 15.3 The general mechanism of biofilm tolerance to various antimicrobials. A – Physical tolerance: biofilm matrix limits the diffusion of antimicrobials. B – Passive tolerance: matrix enzymes inactivate the penetrated antibiotic molecules. C – Physiological tolerance: persisted cells in the deeper layer of biofilm induce adaptive SOS responses and thus become more tolerant (Mishra *et al.*, 2020).

Numerous micro-habitats with different redox potential and chemical gradients are present within biofilms (Schwermer *et al.*, 2008). The presence of these micro-habitats and gradients in key biogeochemical parameters (e.g. pH, electron donors, and acceptors) allows the establishment of microorganisms with complementary but also antagonistic physiologies, and enhances the metabolic versatility of the resident microbial community (Lewandowski, 2000). Indeed, the spatial heterogeneity of biofilms contributes to the formation of optimal, suboptimal, and adverse micro-habitats for a given microorganism within the biofilm three-dimensional structure (Dang and Lovell, 2016).

15.4.1 CORROSIVE EFFECTS ON STEEL SURFACES

Changes in chemical species in a biofilm (e.g. decrease of fresh iron availability after formation of an iron sulfide layer) may also change the microenvironments and thus influence the activities of microorganisms involved in MIC (Lewandowski, 2000; Keresztes *et al.*, 2001; Little and Lee, 2014). Naturally occurring biofilms generally exhibit greater corrosiveness than biofilms of pure cultures, indicating that high corrosion rates are the results of multiple processes and interactions between taxonomically and metabolically different microorganisms (Kip and van Veen, 2015). Each biofilm is unique and its composition and corrosiveness can vary depending on environmental factors (e.g. temperature, salinity, biocide treatment, oil composition, nitrate, and sulfate concentration in fluids). Furthermore, molecular investigation of microbial community composition by multigenic high-throughput amplicon sequencing in corrosive biofilms indicated that most members of a corrosive biofilm can potentially have a direct or indirect corrosive effect on the steel surface (Vigneron *et al.*, 2016).

Metal weight loss is another effect of microbial biofilm on oil and gas industry facilities. Methanogens have been identified as one of the list of microbial lineages associated with biofilms; these methanogens led to methane formation, which is correlated with metal weight loss in incubation experiments (Mand *et al.*, 2015; Okoro *et al.*, 2016). Numerous studies have reported methanogens in corrosive biofilms (Dinh *et al.*, 2004; Davidova *et al.*, 2012; Vigneron *et al.*, 2016).

15.5   BIOFILM AND MICROBIOLOGICALLY INDUCED CORROSION (MIC) IN THE PETROLEUM INDUSTRY

Microbiologically induced corrosion has a potential impact on a wide range of industrial operations. Problems associated with MIC afflict water handling operations and manufacturing processes in oil and gas production, pipelining, refining, petrochemical synthesis, and other industrial sectors. Most of the commercially used metals and alloys – such as stainless steel, nickel and aluminum-based alloys – and materials such as concrete, asphalt, and polymers are readily degraded by microorganisms. Protective coatings, inhibitors, oils, and emulsions are also subject to microbial degradation.

Microbiologically induced corrosion is not in itself a form of corrosion, but rather a process that can influence and even initiate corrosion. It can accelerate most forms of corrosion, including uniform corrosion, pitting corrosion, crevice corrosion, galvanic corrosion, intergranular corrosion, dealloying, and stress corrosion cracking. MIC deteriorates pipes, tanks or vessel surfaces by pitting corrosion. The formation of slime or tuberculation nodules can cause blockages or reduce flow (Rawat *et al.*, 2014). Low flow or stagnant conditions make systems more susceptible to microbial growth. Microbiologically induced corrosion can degrade or cause to fail many different types of system. The ultimate effect is the premature failure of metal components.

Microbiologically induced corrosion–causing organisms are sulfate-reducing bacteria (*Desulphovibrio*, *Desulphotomaculum*, and *Desulphomonas* sp.), iron-reducing bacteria (*Gallionellea ferrugine* and *Ferrobacillus* sp.), acid-producing bacteria (*Pseudomonas*, *Aerobacter*, and *Bacillus*), and sulfur-oxidizing bacteria (*Thiobacillus* sp.). Fungi and algae may also be involved in metal deterioration. In fuel and oil storage tanks, fungal species such as *Aspergillus*, *Penicillium*, and *Fusarium* may grow on fuel components and produce carboxylic acids which corrode iron.

The mechanisms of microbial corrosion of metals may be site-specific and tend to vary with the environment, type of organism, type of metallurgy, and the surface characteristics of the metal. Microorganisms may induce corrosion processes directly or indirectly. Based on pure culture experimentation and molecular investigations, the following two important mechanisms are currently recognized as being involved in MIC.

1. Metal can be indirectly attacked by microorganisms through the production of corrosive metabolites (Figure 15.4). In this process, called chemical microbiologically influenced corrosion (CMIC) or type II corrosion (Xu and Gu, 2011), iron reacts with corrosive compounds such as hydrogen sulfide, generating loose deposits (iron sulfide in the case of reaction with hydrogen sulfide) (Venzlaff *et al.*, 2013). Sulfidogenic microorganisms such as sulfate, sulfite, thiosulfate, and sulfur reducers – but also sulfur oxidizers that produce hydrogen sulfide directly as an end-product of their metabolism – also can be involved in CMIC. Depending on their concentrations and operating conditions, these sulfur derivatives can be more or less potent in causing CMIC (Zhang *et al.*, 2012). In addition to sulfidogenic microorganisms, fermentative bacteria producing hydrogen that will react with elemental sulfur leading to H_2S formation can also be implicated in CMIC. Corrosion via other corrosive metabolic products (e.g. CO_2, nitrite, oxidized sulfur

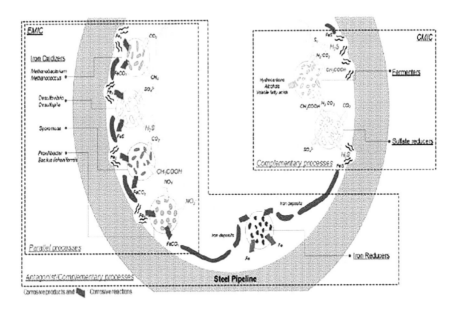

FIGURE 15.4 Conceptual model of microbial communities and processes involved in microbiologically influenced corrosion of the iron and steel infrastructures. Each microbial group is characterized by its potential metabolic functions: fermentation, iron oxidation, iron reduction, and sulfate reduction.

EMIC: electrical microbiologically influenced corrosion; CMIC: chemical microbiologically influenced corrosion. Corrosive products and reactions are labeled in red.

compounds, or volatile fatty acids produced by acid-producing fermentative bacteria) are also implicated in CMIC (Figure 15.4).

2. Metal can be directly attacked by specific microorganisms via direct electron uptake. In this process, referred to as electrical microbiologically influenced corrosion (EMIC) or type I MIC, metabolism of the bacteria is directly fueled by electrons from iron and steel oxidation (Figure 15.4). Mechanical cleaning using various types of maintenance pigs to disrupt and remove biofilms is one of the most effective ways to control MIC in pipelines.

A combination of chemical treatment and pigging may be used when the threat of MIC is significant. Chemical treatment with biocides and surfactant compounds is sometimes applied continuously at a low dose and/or intermittently as a batch treatment. Facility piping dead legs may be flushed on a regular basis to remove water and solids, and then have biocides added to help control biofilms in between flushing treatments.

15.6   ROLES OF BIOFILMS IN REMEDIATION OF OIL-CONTAMINATED SOIL

Bioremediation is an environmentally friendly, cost effective, sustainable technology that utilizes microbes to decontaminate and degrade a wide variety of pollutants into less harmful products. Relative to free-floating planktonic cells, microbes existing in biofilm mode are advantageous for bioremediation because of greater tolerance to pollutants and environmental stress, and ability to degrade varied harsh pollutants via diverse catabolic pathways.

In biofilm mode, microbes are immobilized in a self-synthesized matrix which offers protection from stress, contaminants, and predatory protozoa. Contaminants ranging from heavy metals, petroleum, explosives, and pesticides have been remediated using microbial consortia of biofilms. In the industry, biofilm based bioremediation is used to decontaminate polluted soil and groundwater.

15.6.1   BIOFILMS EMPLOYED TO TREAT OILY SEAWATER

Often, there are oil leaks into seas or streams that are toxic to marine life or may ultimately reach ground water. To avoid toxicity of such oil leaks, contaminated water or seawater should be treated to degrade toxic chemicals.

Removal of oil from seawater has been reported by Al-Awadhi et al. (2003), who used established biofilms rich in hydrocarbon-degrading bacteria. The biofilms were established on gravel particles and glass plates. The microbial consortia in the biofilms included filamentous cyanobacteria, picoplankton, and diatoms. Hydrocarbon-utilizing bacteria Acinetobacter calcoaceticus and nocardioforms were, in part, attached to filaments of cyanobacteria. These studies were performed in batch cultures, and it was demonstrated that the attached biofilms were able to degrade and remove crude oil from contaminated seawater. Radwan et al. (2002) employed a mineral medium containing crude oil as the sole carbon and energy source to iden-tify ten different macroalgae and bacteria capable of using oil. It was concluded that macroalgae found in Arabian Gulf seawater were coated with biofilms rich in oil-utilizing/degrading bacteria. In another report, bacterial biofilms were found associated with corroded oil pipelines where about 60% degradation was reported (Neria-Gonzalez *et al.*, 2006).

15.6.2 Methods of Control of Microbial Biofilm in the Oil and Gas Industry

The most common methods for controlling microbial biofilm in the oil and gas industry include the following.

1. **Chemical method:** The most common chemical method for controlling microorganisms in industrial water systems is the use of biocides. These can either be oxidizing or non-oxidizing chemicals; oxidizing biocides include chlorine, ozone, iodine, hydrogen peroxide, and bromine, which are the types of oxidizing biocides for industrial use – they act by depolymer-izing the EPS matrix, thereby disrupting biofilm integrity. However, the non-oxidizing biocides commonly used include: quaternary ammonium compounds (QACs) and formaldehyde, which are more effective than oxi-dizing ones for controlling algae, fungi, and bacteria because they persist longer and are pH-independent (Romero *et al.*, 2005). For this method to be very efficient and effective, periodical biocide dosing is applied. Recently, a study conducted by Pavlova *et al.* (2020) revealed that the use of antiseptics like chlorhexidine in the form of salts – biacetate, bigluconate, or dihydro-chloride – can effectively inhibit the formation of biofilm. It is known that chlorhexidine acts on cell membranes, increasing their permeability. Acting as a cationic antiseptic having a high degree of affinity for cell walls, it is rapidly adsorbed on the microbial wall.

2. **Biological method:** This method, called bio-competitive exclusion, consists of injecting nitrate, which stimulates the growth of competitive bacteria (nitrate-reducing bacteria, NRB) in water systems to inhibit the develop-ment of SRB (Grigoryan *et al.*, 2008). Using NRB to control MIC may be an attractive option for industry because it is an environmentally friendly technique that can replace biocides. In this technique, nitrate is converted to nitrogen, which is an inert chemical compound.

3. **Natural products:** Natural anti-biofilm agents either act solely or synergis-tically by diverse mechanisms, as illustrated in Figure 15.5.

   i. **Phytochemicals:** There are broadly five classes of natural compounds that have high anti-biofilm properties. These are phenolics, essential oils, terpenoids, lectins, alkaloids, polypeptides, and polyacetylenes (Paluch *et al.*, 2020). Phytochemicals inhibit the quorum sensing mechanism mainly by blocking the quorum sensing inducers like AHL, autoinducers, and autoinducers type 2 (Ciric *et al.*, 2019). Garlic extracts play a vital

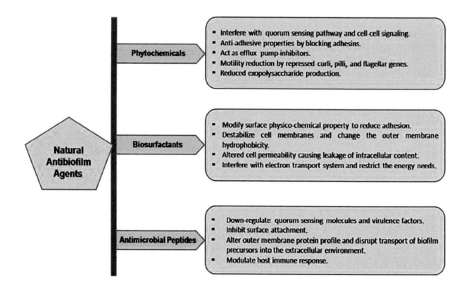

FIGURE 15.5   Workflow of the portrayed natural anti-biofilm agents based on their mode of actions (Mishra *et al.*, 2020).

role in the inhibition of quorum sensing signaling molecules of *Pseudomonas* and *Vibrio* spp. biofilms (Lu *et al.*, 2019). Emodin helps in the proteolysis of transcription factors associated with the quorum sensing and acts as its potent inhibitor. Quorum quenchers, along with antibiotics, are the best alternative anti-biofilm agents, as discussed by many researchers (Paluch *et al.*, 2020). Phytochemicals also play a significant role in inhibiting bacterial adhesions and suppression of genes related to biofilm formation (Adnan *et al.*, 2020). A very recent study on *Adiantum philippense* L. crude extract has shown a promising role in decreasing the content of biofilm exopolysaccharides (Adnan *et al.*, 2020). They observed that *Adiantum philippense* L. crude extract restrains biofilm at the initial stages by targeting adhesin proteins, deforming the pre-formed biofilms, and obstructing EPS production.

ii. **Biosurfactants:** Biosurfactants (BS) hinder biofilm formation by varying the cell adhesion ability through less cell surface hydrophobicity, membrane disruption, and inhibited electron transport chain, thus restricting cellular energy demand (Paraszkiewicz *et al.*, 2019). Biosurfactants of different classes are produced by various microorganisms that exhibit antibacterial, antifungal, and anti-biofilm activities (Paraszkiewicz *et al.*, 2019). Biosurfactants also control the expression of hyphal specific genes, and mainly act by decreasing cellular surface hydrophobicity (Janek *et al.*, 2020).

4. **Use of corrosion inhibitors:** To mitigate biofilms, biocides are dosed together with other chemicals such as corrosion inhibitors for CO_2, H_2S

corrosion, oxygen scavengers, and scale removers in pigging operations (Xu and Gu, 2015). Biocides are also used in the hydraulic fracturing fluids for shale gas and oil production to prevent the biofouling that plugs fissures (Aminto and Olson, 2012).

5. **Biocide enhancers other than surfactants and dispersants:** The use of biocide enhancers other than surfactants and dispersants is another active method to control biofilm. The biocide enhancers themselves may be non-biocidal, but they make the sessile cells more vulnerable to biocides. The lack of toxicity actually means that a biocide enhancer may encounter fewer discharge problems. Chelators such as ethylenediamine-N,N'-disuccinic acid (EDDS) are capable of making bacteria more vulnerable to antimicrobial attacks. However, their effective concentrations are rather high in the mitigation of oilfield biofilms (Wen *et al.*, 2009; Xu and Gu, 2015).

15.7 CONCLUSION

The presence of biofilm in the oil and gas industry induces corrosion of carbon steel and metals, and increases maintenance and operational problems, which leads to reduction in flow, heat transfer rates changes, fouling, and scaling. Biofilm and biofilm producers have been successfully used to remediate and clean up oily seawater, which is an indication that they can be employed in the oil and gas industry for bioremediation of soil.

REFERENCES

Abebe, G. M. (2020) The Role of Bacterial Biofilm in Antibiotic Resistance and Food Contamination. *International Journal of Microbiol*, https://doi.org/10.1155/2020/1705814

Adnan, M., Patel, M., Deshpande, S., Alreshidi, M., Siddiqui, A. J., Reddy, M. N. (2020) Effect of *Adiantum philippense* extract on biofilm formation, adhesion with its antibacterial activities against foodborne pathogens, and characterization of bioactive metabolites: an *in vitro-in silico* approach. *Front Microbiol.* 11:823. doi: 10.3389/fmicb.2020.00823

Al-Awadhi, H., Al-hasan, R. H., Sorkhoh, N. A., Salamah, S. and Radwan, S.S. (2003) 'Establishing oil-degrading biofilms on gravel particles and glass plates', Int Bioremediation & Biodegradation, 51, 181–185.

Aminto, A., and Olson, M. S. (2012) Four-compartment partition model of hazardous components in hydraulic fracturing fluid additives. J Nat Gas Sci Eng. 7: 16–21.

Cabo, M. L. Rodríguez-López, P. Rodríguez-Herrera, J. J. and Vázquez-Sánchez, D. (2018) "Current knowledge on Listeria monocytogenes biofilms in food-related environments: incidence, resistance to biocides, ecology and biocontrol," *Foods*, vol. 7, no. 6, 2018

Ch'ng, J. H., Chong, K. K., Lam, L. N., Wong, J. J., and Kline, K. A. (2019) Biofilm-associated infection by *Enterococci*. *Nat Rev Microbiol.* 17, 82–94. doi: 10.1038/s41579-018-0107-z

Choudhary Princy, Sangeeta Sing, H, Vishnu Agarwal (2020) Microbial Biofilms, Bacterial Biofilms, Sadik Dincer, Melis Sümengen Özdenefe and Afet Arkut, IntechOpen. doi: 10.5772/intechopen.90790. Available from: https://www.intechopen.com/books/bacterial-biofilms/microbial-biofilms

Ciric, A. D., Petrovic, J. D., Glamoclija, J. M., Smiljkovic, M. S., Nikolic, M. M., Stojkovic, D. S., *et al.* (2019) Natural products as biofilm formation antagonists and regulators of quorum sensing functions: a comprehensive review update and future trends. *South Afr J Bot.* 120, 65–80. doi: 10.1016/j.sajb.2018.09.010

Dang, H., and Lovell, C. R. (2016) Microbial surface colonization and biofilm development in marine environments. *Microbiol Mol Biol Rev.* 80(1):91–138. https://doi.org/10.1128/MMBR.00037-15

Davidova, I.A., Duncan, K. E., Perez-Ibarra, B. M, and Suflita, J.M. (2012) Involvement of thermophilic archaea in the biocorrosion of oil pipelines. *Environ Microbiol.* 14(7):1762–1771. https://doi.org/10.1111/ j.1462–2920.2012.02721.x

Dinh, H. T, Kuever, J., Muszmann, M., Hassel, A. W., Stratmann, M., Widdel, F. (2004) Iron corrosion by novel anaerobic microorganisms. *Nature.* 427(6977):829–832. https://doi.org/10.1038/nature02321

Duncan, K. E., Davidova, I. A., Nunn, H. S., Stamps, B. W., Stevenson, B. S., Souquet, P. J., Suflita, J. M. (2017) Design features of offshore oil production platforms influence their susceptibility to biocorrosion. Appl Microbiol Biotechnol. 101(16):6517–6529. https://doi.org/10.1007/s00253-017-8356-8.

Gedif Meseret Abebe (2020) The Role of Bacterial Biofilm in Antibiotic Resistance and Food Contamination. *International Journal of Microbiol*, 2020, 1–10 pages https://doi.org/10.1155/2020/1705814

Graf, A. C., Leonard, A., Schäuble, M., Rieckmann, L. M., Hoyer, J., Maass, S., *et al.* (2019) Virulence factors produced by *staphylococcus aureus* biofilms have a moonlighting function contributing to biofilm integrity. *Mol Cell Proteomics.* 18, 1036–1053. doi: 10.1074/mcp.RA118.001120

Grigoryan, A.A., Cornish, S.L., Buziak, B., Lin, S., Cavallaro, A., Arensdorf, J.J., and Voordouw, G.,(2008) Competitive oxidation of volatile fatty acids by sulfate- and nitratereducing bacteria from an oil field in Argentina, 2008, Appl. Environ. Microbiol. 74, 4324–4335.

Janek, T., Drzymała, K., and Dobrowolski, A. (2020) In vitro efficacy of the lipopeptide biosurfactant surfactin-C15 and its complexes with divalent counterions to inhibit *Candida albicans* biofilm and hyphal formation. *Biofouling.* 36, 210–221. doi: 10.1080/08927014.2020.1752370

Keresztes Z, Felhősi I, and Kálmán E (2001) Role of redox properties of biofilms in corrosion processes. Electrochim Acta. 46(24–25): 3841–3849. https://doi.org/10.1016/S0013-4686(01)00671-5

Kip N, van Veen JA (2015) The dual role of microbes in corrosion. ISME J 9(3):542–551. https://doi.org/10.1038/ismej.2014.169

Kostakioti, M., Hadjifrangiskou, M. and Hultgren, S. J. (2013) "Bacterial biofilms: development, dispersal, and therapeutic strategies in the dawn of the postantibiotic era," *Cold Spring Harbor Perspectives in Medicine*, 3 (40), 103–106

Kumar, AS., Alam, A., Rani, M., Ehtesham, N. Z., and Hasnain, S. E. (2017) Biofilms: survival and defense strategy for pathogens. *Int J Med Microbiol.* 307, 481–489. doi: 10.1016/j.ijmm.2017.09.016

Kumar AS, Mody K (2009) Microbial exopolysaccharides: variety and potential applications. Microb Prod Biopolym Polym Precursors Appl Perspect. 10:229–253

Lewandowski Z (2000) MIC and biofilm heterogeneity. Proc Corros. 400: 1–7 Liang B, Wang L-Y, Mbadinga SM, Liu J-F, Yang S-Z, Gu J-D, Mu B-Z (2015) Anaerolineaceae and Methanosaeta turned to be the dominant microorganisms in alkanes-dependent methanogenic culture after long-term of incubation. AMB Express. 5(1):37. https://doi.org/10.1186/s13568-015-0117-4

Little BJ, Lee JS (2014) Microbiologically influenced corrosion: an update. Int Mater Rev. 59(7):384–393. https://doi.org/10.1179/ 1743280414Y.0000000035

Lu, L., Hu, W., Tian, Z., Yuan, D., Yi, G., Zhou, Y., *et al.* (2019) Developing natural products as potential anti-biofilm agents. *Chin Med.* 14:11. doi: 10.1186/s13020-019-0232-2

Mand J, Park HS, Okoro C, Lomans BP, Smith S, Chiejina L, Voordouw G (2015) Microbial methane production associated with carbon steel corrosion in a Nigerian oil field. *Front Microbiol.* 6:1538.

Maric S, Vranes J. (2007) Characteristics and significance of microbial biofilm formation. Periodicum Biologorum. 2007,109:115–121

Mishra, R., Panda, A, K., Mandal, S. D., Shakeel, M., Bisht S. S. and Khan, J. (2020) Natural Anti-biofilm Agents: Strategies to Control Biofilm-Forming Pathogens. Front. Microbiol.,‌ https://doi.org/10.3389/fmicb.2020.566325

Nasib Q (2009) 18 Beneficial biofilms: wastewater and other industrial applications. In *Biofilms in the food and beverage industries.* Woodhead Publishing.

Neria-Gonzalez, I., Wang, E.T., Ramirez F., Romero, J. M.and Hernandez-rodriguez C. (2006) 'Characterization of bacterial community associated to biofi lms of corroded oil pipelines from the southeast of Mexico', *Anaerobe,* 12, 122–133.

Okoro CC, Samuel O, Lin J (2016) The effects of Tetrakishydroxymethyl phosphonium sulfate (THPS), nitrite and sodium chloride on methanogenesis and corrosion rates by methanogen populations of corroded pipelines. *Corros Sci.* 112.

Oxaran V, Dittmann KK, Lee SHI *et al.* (2018) Behavior of foodborne pathogens Listeria monocytogenes and Staphylococcus aureus in mixed-species biofilms exposed to biocides. *Appl Environ Microbiol.* 84.

Paluch, E., Rewak-Soroczynska, J., Jędrusik, I., Mazurkiewicz, E., and Jermakow, K. (2020) Prevention of biofilm formation by quorum quenching. *Appl Microbiol Biotechnol.* 104, 1871–1881. doi: 10.1007/s00253-020-10349-w

Paraszkiewicz, K., Moryl, M., Płaza, G., Bhagat, D. K., Satpute, S., and Bernat, P. (2019) Surfactants of microbial origin as anti-biofilm agents. *Int J Environ Health Res.* 11, 1–20. doi: 10.1080/09603123.2019.1664729

Pavlova, I. B., Kononenko, A.B., Tolmacheva. G.S. Kardash, G.G., and. Rytsarev, A.Yu. (2020) Formation of biofilms of pathogenic bacteria and the effect of a new disinfectant. BIO Web of Conferences. 17, 00204 (2020) https://doi.org/10.1051/bioconf/20201700204 *FIES 2019*

Radwan, S. S., Al-hasan, R. H., Salamah, S. and Al-dabbous, S. (2002), 'Bioremediation of oily sea water by bacteria immobilized in biofi lms coating macroalgae', *Int Bioremediation & Biodegradation,* 50, 55–59

Rawat JSN, Khandelwal A (2014) *Microbiological causes of corrosion.* Bharat Petroleum Corporate R&D Center.

Romero, J.M., Velazquez, E., Garcia-Villalobos J.L., Amaya, M., Le Borgne, S., (2005) Genetic Monitoring of Bacterial Populations In a Seawater Injection System, Identification of Biocide Resistant Bacteria and Study of Their Corrosive Effect, CORROSION/2005. paper no. 05483, p. 9, Houston, TX.

Roy R, Tiwari M, Donelli G, Tiwari V (2018) Strategies for combating bacterial biofilms: a focus on anti-biofilm agents and their mechanisms of action. *Virulence* 9:522–554. doi: 10.1080/21505594.2017.1313372

Schwermer CU, Lavik G, Abed RMM *et al.* (2008) Impact of nitrate on the structure and function of bacterial biofilm communities in pipelines used for injection of seawater into oil fields. *Appl Environ Microbiol.* 74(9):2841–2851. https://doi.org/10.1128/AEM.02027-07

Sutherland IW (2001) The biofilm matrix – An immobilized but dynamic microbial environment. *Trends in Microbiology.* 9:222–227. https://doi.org/10.1016/S0966-842X(01)02012-1

Venzlaff H, Enning D, Srinivasan J, Mayrhofer KJJ, Hassel AW, Widdel F, Stratmann M (2013) Accelerated cathodic reaction in microbial corrosion of iron due to direct electron uptake by sulfate-reducing bacteria. *Corros Sci.* 66:88–96. https://doi.org/10.1016/j.corsci.2012.09.006

Vigneron A, Alsop EB, Chambers B, Lomans BP, Head IM, Tsesmetzis N (2016) Complementary microorganisms in highly corrosive biofilms from an offshore oil production facility. *Appl Environ Microbiol.* 82(8):2545–2554. https://doi.org/10.1128/AEM.03842-15

Vigneron A, Head IM, Tsesmetzis N (2018) Damage to offshore production facilities by corrosive microbial biofilms. *Appl Microbiol Biotechnol*. 102:2525–2533. https://doi.org/10.1007/s00253-018-8808-9

Welch M, Maunders E (2017) Matrix exopolysaccharides; the sticky side of biofilm formation. *FEMS Microbiol Lett*. 364:120.

Wen J, Zhao K, Gu T, Raad II (2009) A green biocide enhancer for the treatment of sulfate-reducing bacteria (SRB) biofilms on carbon steel surfaces using glutaraldehyde. *Int Biodeter Biodegr*. 63:1102–1106.

Xu D, Gu T (2011) *Bioenergetics explains when and why more severe MIC pitting by SRB can occur*. NACE-2011-11426. NACE International.

Xu D, Gu T (2015) The war against problematic biofilms in the oil and gas industry. *J Microbiol Biotechnol*. 7(5). doi: 10.4172/1948-5948.1000e124

Yin WY, Wang LL, He J (2019) Biofilms: the microbial "protective clothing" in extreme environments. *Int J Mol Sci*. 20(14):3423.

Zhang L, Liang E, Cheng Y, Mahmood T, Ge F, Zhou, K (2020) Is combined medication with natural medicine a promising therapy for bacterial biofilm infection? *Biomed Pharmacother*. 128:110184. doi: 10.1016/j.biopha.2020.110184

Zhang L, Wang X, Wen Z, Liu Z, Li X, Lu M (2012) *Interactive effects of H2S and elemental sulfur on corrosion of steel*. NACE-2012-1575. NACE International. https://www.onepetro.org/ conference-paper/NACE-2012-1575

16 Futurity of Microbial Biofilm Research

Aransiola S.A., Victor-Ekwebelem M.O., Olusegun Julius Oyedele and Naga Raju Maddela

CONTENTS

16.1 INTRODUCTION

Formation of biofilms is one among many significant global challenges to manage infection and healthcare-associated infections due to their inherent tolerance and 'resistance' to antimicrobial treatments. They typically develop on medical device surfaces, and dispersal of single and clustered cells indicates a significant risk of microbial dissemination within the host and increased risk of infection (Percival *et al.*, 2015). Bacteria are ready to colonize and form biofilms on virtually all types of surfaces, including natural and artificial surfaces (Chan *et al.*, 2019). Biofilms are answerable for chronic illness and nosocomial infections, industrial pipe fouling, spoilage of food, contamination of seafood and dairy products, and moreover for ship hull fouling (Schultz *et al.*, 2011; Coughlan *et al.*, 2016). The harmful effects of biofilms on human society are manifold. Therefore, there is a need for development of novel, effected and specific antimicrobial substances which might be utilized to diminish the biofilm-associated pathogenicity in healthcare, food, and other public spaces. Control and mitigation of biofilm has been a challenge in various sectors, and this is discussed in this chapter. The strategies to forestall cell adhesion and biofilm formation as a way of mitigation usually involve the employment of antimicrobial/antiadhesion coatings, release of toxic agents like metal ions at the surface, or smart surfaces (Han *et al.*, 2017). The environmental fate and effects of those biocides during biofouling management and control has posed a challenge in biofilm mitigation.

DOI: 10.1201/9781003184942-19

16.1.1 THE FOOD INDUSTRY

Within the food industry, biofilms can occur on surfaces in contact with or without foods (Muhammad *et al.*, 2020). Biofilms are chargeable for about 60% of foodborne outbreaks (Han *et al.*, 2017). Therefore, the presence of biofilms in food processing environments poses significant risk to food safety and also the food industry (Galie *et al.*, 2018). Biofilms growing in food processing environments may result in spoilage of food, which successively can cause serious public health risks to consumers and heavy economic consequences (Coughlan *et al.*, 2016; Galie *et al.*, 2018). Biofilms also are liable for serious technical challenges of the food industry, wherein they will prevent the flow of warmth across equipment surfaces, increase the fluid frictional resistance at the surfaces, and promote the corrosion rate of the surfaces, resulting in loss of production efficiency (Chmielewski and Frank, 2003; Demirel *et al.*, 2017). As a result, it is imperative to beat these problems. Some challenges have prevented successful mitigation of biofilm, one of which is the formation of multi-species biofilm; most of the organisms that form biofilms in food facilities and industries have the power to determine multi-species biofilms, which are more stable and difficult to manage, thereby posing a challenge in mitigating biofilm (Galie *et al.*, 2018; Khan *et al.*, 2021). Similarly, in industrial, natural environments and the food industry, mitigation of biofilms is a great challenge thanks to their immunity to different anti-microbial agents and disinfectants; for example, one of the most used disinfectants within the food industry is antimicrobial (NaOCl) – but a drawback is that it may be tormented by organic matter because free chlorine might react with natural organic matter and be converted into inorganic chloramines, forming trihalomethanes and thereby reducing its activity against biofilms; hence, making mitigation a challenge (Fernandes *et al.*, 2015). Other reasons which make eradication of biofilm difficult include the following.

1. Slow or restricted penetration of various antimicrobial agents within biofilm.
2. The resistant phenotype, which includes gene transfer and antimicrobial destroying enzymes. In an organic phenomenon, there are unique differences between sessile and planktonic cells which are chargeable for physiological alterations during formation of biofilm.
3. Altered metabolism and cellular environment, in biofilm, some cells face limitation of nutrients and sleep in a starved or slow growing stage (Verderosa *et al.*, 2019; Wolfmeier *et al.*, 2018; Khan *et al.*, 2021).

16.1.2 MEDICINE

In healthcare settings, biofilms are shown to develop on medical device surfaces, dead tissues (e.g. sequestrate of bones), and inside living tissues (e.g. lung tissue, teeth surfaces) (Mohammed *et al.*, 2020). They develop on the surface of biomedical devices like catheters, prosthetic heart valves, pacemakers, breast implants, contact lenses, and body fluid shunts. Methods developed to eradicate and mitigate biofilms, and successful treatment of infections related to biofilms, is troubled due to increased antibiotic resistance in numerous bacterial communities. Also, different classical antibiotics chemotherapy is incapable of eradicating these bacterial cells completely, especially the cells located in central regions of biofilm, which result in a worsening situation globally. Therefore, novel anti-biofilm agents and alternative strategies are required against antibiotic-resistant biofilm communities. Methods including antibiotics, enzymes, plant extracts, sodium salts, metal nanoparticles, acids, and chitosan derivatives have impact on structure of biofilm via numerous mechanisms having different efficiencies (Muhammed *et al.*, 2020).

Another challenge faced in mitigating biofilm within the health sector is the problem of potential injury of patients' tissues by the controlled therapy. As an example, photodynamic therapy has many applications in preventing infections caused by wound biofilm; with relevant photochemical and photosensitizer reactions, it becomes a challenge because if care is not taken, the therapy will affect the patient's surrounding tissues within the body (Roy *et al.*, 2018). Also, the employment of anti-biofilm molecules can interfere with signaling pathways of bacteria in both gram-negative and gram-positive bacteria (Roy *et al.*, 2018).

Contact killing surfaces (antifouling surfaces), which involve the covering of surfaces with coating agents and paints like silver oxide, grapheme, arsenic, mercury oxide, and flowers of zinc nanoparticles, are developed and used effectively to stop the attachment of biofilm bacteria (Kuang *et al.*, 2018), which successively prevents and controls the event of biofilms on biomedicals. Also, polyethylene glycol (PEG) (Zhang *et al.*, 2017) has been utilized in the biomedical industries to resist the adhesion of bacteria, thanks to its hydrophilic surface properties. PEG coatings are ready to repel quite a number of bacterial species like *S. aureus, S. epidermidis, P. aeruginosa, and E. coli*. Unfortunately, contact killing surfaces have the disadvantage that some microorganisms are ready to develop resistance against these surfaces (Hasan *et al.*, 2013), thereby making mitigation a controversy.

16.1.3 Water and Wastewater Treatment

Biofilms are the predominant mode of microbial growth within beverage distribution systems (Coughlan *et al.*, 2016). it is well documented that biofilms represent one of the foremost problems in drinkable distribution systems (Douterelo *et al.*, 2016; Prest *et al.*, 2016). The consumption of contaminated water with pathogenic biofilms has been linked to human infections and waterborne outbreaks (Angles *et al.*, 2007; Prest *et al.*, 2016). The major biofilm-producing bacteria in drinkable water are *P. aeruginosa, Campylobacter jejuni, Legionella pneumophila*, Mycobacteria, *Aeromonas hydrophila*, and *Klebsiella pneuminiae* (Prest *et al.*, 2016; Chan *et al.*, 2019). Since bacterial cells can attach and develop biofilms on the inner surfaces of piping systems, from which cells may be detached into the majority water, they will cause biocorrosion of pipes, with undesirable water quality changes affecting color, taste, turbidity, odor, and reduction of warmth exchange efficiency (Prest *et al.*, 2016). More specifically, the main biofilm-producing bacteria known to market corrosion of metals are sulfate-reducing bacteria, sulfur-oxidizing bacteria, iron oxidizers, iron reducers, and manganese oxidizers (Angles *et al.*, 2007). All in all, biofilms can affect the security of drinkable water supplies and adversely affect water pipelines.

Marine biofouling is the undesirable accumulation of organisms on any natural or human-made objects exposed to seawater. Biofouling has been a significant challenge within the naval industry and for civilian oceangoing ships (Hopkins and Forrest, 2010; Schultz *et al.*, 2011) Generally, accumulation of biofoulers by biofilms on ship hulls can increase the hydrodynamic drag of the ships, which causes challenges for the shipping industry, including speed reduction, increased cleaning times, and greater fuel consumption (Schultz *et al.*, 2011; Demirel *et al.*, 2017). Additionally, biofouling of ship hulls has been considered as a vital vector for the spread of invasive marine species to new habitats. These transported organisms can adversely affect native species through competition and predation. Therefore, biofilms will affect the value of ship usage and also the balance of marine environments. Besides indirect repercussions from increased energy consumption necessary to e.g. overcome increased frictional drag and warmth and mass transfer limitations (Schultz *et al.*, 2011), biofouling organisms reduce water flow and increase biodeposition beneath aquaculture farms (Fitridge *et al.*, 2012), and will be fish pathogens (Floerl *et al.*, 2016).

16.2 BIOFILM MITIGATION APPROACHES: FUTURE DIRECTIONS

According to the latest information, biofilm formation is the major virulence factor in a wide range of chronic infections (Koo *et al.*, 2017). In the context of biofilm control and antibiotics, there is a need for clarity between 'antimicrobial resistance' and 'antimicrobial tolerance' (Gilmore *et al.*, 2018); in fact, these are two different scenarios. Antimicrobial resistance is based on genetics, and there it is irreversible, heritable, genotypic change. In contrast, antimicrobial tolerance is a phenotypic trait which is reversible and helps microorganisms to survive in antibiotic environments. As there is no effective developmental platform for the discovery of antibiotics, control of infectious diseases is becoming more difficult (Lewis, 2013).

In order to control bacterial biofilms, several quorum sensing inhibitors (QSIs) have emerged, and this strategy provides promising hope for the treatment of biofilm-based infections (Subhadra *et al.*, 2018). However, future research is necessary in the area of QSIs. QSIs are effective in the control of biofilms, but effects of QSIs on different stages of biofilm formation are poorly understood, which restricts the applicability of QSIs on humans. Therefore, validation of QSIs' effects on different stages of biofilm formation is obligatory for the full-fledged applications of QSIs in biofilm mitigation.

Quorum quenching (QQ) is one of the emerging strategies for the control of biofouling control (so-called biofilm mitigation), using membrane bioreactors (MBRs) installed in the water and wastewater treatment facilities. However, regarding applications of QQ MBRs, here are yet still many challenges to be addressed. One of the biggest challenges in the QQ MBR is the lifetime of QQ media. Generally, QQ media are composed of PVA and alginate composites, and show significant weight loss over time (Lee *et al.*, 2018). Thus, the lifetime of QQ media is questionable under harsh conditions such as high shear and salinity; therefore, future investigations are necessary to improve the lifetime of QQ media for a sustainable application of QQ MBR. Another important thing regarding QQ research is that as of now, QQ activities have been tested with the biofilms of very few model species of bacteria such as *Escherichia coli, Pseudomonas* sp., etc. Therefore, such insights will not help in understanding the behavior of QQ response in multi-species biofilms or at community level. Very recently, it has been found that inter-species interactions between *Rhodococcus* sp. BH4 (so-called QQ bacterium) and sludge bacteria showed 76% of competitive interactions and 24% of cooperative interactions (Maddela and Meng, 2020), which implies that QQ bacteria do not interact similarly with all biofilm-forming bacteria, and QQ effects are strain-dependent. The increments of biofilm biomass, amount of extracellular polymeric substances, and particle size were 21%, 25–30 times, and 3.5 times, respectively, in a dual-species biofilm containing *Rhodococcus* sp. BH4 and *Serratia* sp. JSB1 (Maddela and Meng, 2020), which greatly warrants the validation of QQ responses against different species of bacteria in various environmental settings. Also, it is necessary to test whether QQ strategy alone is effective in the mitigation of biofilm; accordingly, combinations of biofilm mitigation strategies should be investigated.

At a species level, it is necessary to characterize the extracellular polymeric substances (EPS), because there is a strong correlation between certain functional groups of EPSs and their biofouling potential (Maddela *et al.*, 2018). EPS functional groups such as α-1,4-glycosidic linkage (920 cm^{-1}) and amide II (1,550 cm^{-1}) were positively correlated (r = 0.67 and 0.60, respectively), with the fouling potential of pure species sludge bacteria (Maddela *et al.*, 2018). There is enhanced biofouling potential of bacterial species due to the gelling capacities of certain EPS functional groups like uronic acid (1,020 cm^{-1}) and O-acetyl (1,250 cm^{-1}). Thus, EPSs are the important molecules of bacterial cell aggregation. Fourier-transform infrared spectroscopy (FTIR) analysis of EPS helps in the identification of different environmental isolates and their fouling potentials. As of now, studies have focused on purified EPS and their characteristics, but whole-cell approaches have not been given importance in the area of EPS and fouling relationships.

Another promising area in biofilm mitigation is the use of nanomaterials, or nanoparticles (NPs) (Qayyum and Khan, 2016). Metal nanoparticles (silver NPs, iron NPs, copper NPs, zinc NPs, magnesium NPs), natural compound-based metal NPs, non-metallic inorganic NPs, green synthesized metal NPs, and polymer NPs have proved their anti-biofilm propensities against a broad range of microbial biofilms in both in vivo and in vitro conditions (Qayyum and Khan, 2016). NPs are familiar in biofilm mitigation and drug delivery. NPs can either carry the drugs on their surface or entrap the drug inside and can disrupt the biofilms after their controlled delivery through the EPS layer. Likewise, nanoparticles exert an effective biofilm-disrupting action on the pre-formed biofilms. However, though these NPs have wonderful applications, sustainable application is still doubtful. NPs have large surface areas, high reactivity, and toxicity (chemically synthesized NPs), which do cause significant side effects. Thus, future research must be focused on synthesizing NPs with less toxicity and those which are environmentally friendly. Another important issue is that metal NPs are not specific in their action, and therefore cannot distinguish unwanted (i.e. pathogenic) biofilms from useful biofilms (i.e. symbiotic microbes). In addition, most research results that are available related to NPs versus biofilm mitigation are of in vitro experiments; therefore, in vivo experimental studies are greatly warranted for the full-fledged implications of NPs in the biofilm control.

16.3 CONCLUSION

Bacterial biofilm formation takes place in a series of well-ordered steps, and it is the most common bacterial lifestyle in both natural and human-made environments. The ability of bacteria to colonize surfaces and form biofilms is regarded as a serious issue that has been linked to negative consequences in a variety of fields, including food, water, pharmacy, and healthcare. Various techniques and approaches have been developed in an effort to eliminate harmful biofilms, with the majority of them focusing on bacterial attachment and QS interference, as well as biofilm matrix destruction. Bacterial biofilms, on the other hand, have far-reaching consequences for the environment. Their control in the health and food industries, as well as water treatment, has presented some difficulties, making it difficult to achieve. One of these is the accumulation of biofoulers by biofilms on ship hulls which can increase the hydrodynamic drag of the ships, causing challenges for the shipping industry. Resistance to antibiotics, disinfectants, and antimicrobials has increased the challenge for their mitigation in various industries and settings.

REFERENCES

Angles ML, Chandy JP, Cox PT, Fisher IH, Warnecke MR (2007) Implications of biofilm-associated waterborne *Cryptosporidium* oocysts for the water industry. *Trends Parasitol.* 23:352–356.

Chan S, Pullerits K, Keucken A, Persson KM, Paul CJ, Radstrom P (2019) Bacterial release from pipe biofilm in a full-scale drinking water distribution system. *NPJ Biofilms Microbiomes.* 5:9. doi:10.1038/s41522-019-0082-9

Chmielewski RAN, Frank JF (2003) Biofilm formation and control in food processing facilities. *Compr Rev Food Sci Food Safety.* 2:22–32.

Coughlan LM, Cotter PD, Hill C, Alvarez-Ordonez A (2016) New weapons to fight old enemies: novel strategies for the (Bio)control of bacterial biofilms in the food industry. *Front Microbiol.* 7:1641. doi:10.3389/fmicb.2016.01641

Demirel YK, Uzun D, Zhang Y, Fang HC, Day AH, Turan O (2017) Effect of barnacle fouling on ship resistance and powering. *Biofouling.* 33:819–834. doi:10.1080/08927014.2 017.1373279

Douterelo I, Husband S, Loza V, Boxall J (2016) Dynamics of biofilm regrowth in drinking water distribution systems. *Appl Environ Microbiol.* 82:4155–4168. doi:10.1128/ AEM.00109-16

Fernandes MS, Fujimoto G, Souza LP, Kabuki DY, Silva MJ, Kuaye AY (2015) Dissemination of enterococcus faecalis and enterococcus faecium in a ricotta processing plant and evaluation of pathogenic and antibiotic resistance profiles. *J Food Sci.* 80:M765–M775.

Fitridge I, Dempster T, Guenther J, de Nys R (2012) The impact and control of biofouling in marine aquaculture: a review. *Biofouling.* 28:649–669. doi:10.1080/08927014.2012 .700478

Floerl O, Sunde L, Bloecher N (2016) Potential environmental risks associated with biofouling management in salmon aquaculture. *Aquac Environ Interact.* 8:407–417. doi:10.3354/aei00187

Galie S, García-Gutiérrez C, Miguélez EM, Villar CJ, Lombó F (2018) Biofilms in the food industry: health aspects and control methods. *Front Microbiol.* 9:898.

Gilmore BF, Flynn PB, O'Brien S, Hickok N, Freeman T, Bourke P (2018) Cold plasmas for biofilm control: opportunities and challenges. *Trends Biotechnol.* 36:627–638.

Han Q, Song X, Zhang Z, Fu J, Wang X, Malakar PK *et al.* (2017) Removal of foodborne pathogen biofilms by acidic electrolyzed water. *Front Microbiol.* 8:988. doi:10.3389/fmicb.2017.00988

Hasan J, Crawford RJ, Ivanova EP (2013) Antibacterial surfaces: the quest for a new generation of biomaterials. *Trends Biotechnol.* 31:295–304.

Hopkins GA, Forrest BM (2010) A preliminary assessment of biofouling and non-indigenous marine species associated with commercial slow-moving vessels arriving in New Zealand. *Biofouling.* 26:613–621.

Khan J, Tarar SM, Gu I, Nawaz U, Arshad M (2021) Challenges of antibiotic resistance biofilms and potential combating strategies: a review. *Biotech.* 11:169. https://doi.org/10.1007/s13205-021-02707-w

Koo H, Allan RN, Howlin RP, Stoodley P, Hall-Stoodley L (2017) Targeting microbial biofilms: current and prospective therapeutic strategies. *Nat Rev Microbiol.* 15:740–755.

Kuang X, Chen V, Xu X (2018) Novel approaches to the control of oral microbial biofilms. *Biomed Res Int.* doi: 10.1155/2018/6498932

Lee K, Yu H, Zhang X, Choo KH (2018) Quorum sensing and quenching in membrane bioreactors: opportunities and challenges for biofouling control. *Bioresource Technol.* 270:656–668.

Lewis K (2013) Platforms for antibiotic discovery. *Nat Rev Drug Discov.* 12:371–387.

Maddela NR, Meng F (2020). Discrepant roles of a quorum quenching bacterium (Rhodococcus sp. BH4) in growing dual-species biofilms. *Sci Total Environ.* 713:136402.

Maddela NR, Zhou Z, Yu Z, Zhao S, Meng F (2018) Functional determinants of extracellular polymeric substances in membrane biofouling: experimental evidence from pure-cultured sludge bacteria. *Appl Environ Microb.* 84.

Muhammad MH, Idris AL, Fan X, Guo Y, Yu Y, Jin X, Qiu J, Guan X, Huang T (2020) Beyond risk: bacterial biofilms and their regulating approaches. *Front Microbiol.* 11:928. doi:10.3389/fmicb.2020.00928

Percival SL, Suleman L, Vuotto C (2015) Healthcare–associated infections, medical devices and biofilms: risk, tolerance and control. *J Med Microbiol.* 64(Pt 4):323–334.

Prest EI, Hammes F, Van Loosdrecht MC, Vrouwenvelder JS (2016) Biological stability of drinking water: controlling factors, methods, and challenges. *Front Microbiol.* 7:45. doi:10.3389/fmicb.2016.00045

Qayyum S, Khan AU (2016) Nanoparticles vs. biofilms: a battle against another paradigm of antibiotic resistance. *MedChemComm.* 7:1479–1498.

Roy R, Tiwari M, Donelli G, Tiwari V (2018) Strategies for combating bacterial biofilms: a focus on anti-biofilm agents and their mechanisms of action. *Virulence.* 9:522–554. doi:10.1080/21505594.2017.1313372

Schultz MP, Bendick JA, Holm ER, Hertel WM (2011) Economic impact of biofouling on a naval surface ship. *Biofouling.* 27:87–98.

Subhadra B, Kim DH, Woo K, Surendran S, Choi CH (2018) Control of biofilm formation in healthcare: recent advances exploiting quorum-sensing interference strategies and multidrug efflux pump inhibitors. *Materials (Basel, Switzerland).* 11:1676.

Verderosa AD, Totsika M, Fairfull-Smith KE (2019) Bacterial bioflm eradication agents: a current review. *Front Chem.* 7:824.

Wolfmeier H, Pletzer D, Sarah C, Mansour REW (2018) New perspectives in bioflm eradication. *ACS Infect Diseas.* 4(2):93–106.

Zhang X, Brodus D, Hollimon V, Hu, H (2017) A brief review of recent developments in the designs that prevent bio-fouling on silicon and silicon-based materials. *Chem Cent J.* 11:18.

Index

Page numerals in **bold** indicate a table
Page numerals in *italics* indicate a figure